Rubber Recycling
Challenges and Developments

Green Chemistry Series

Editor-in-chief:
James H. Clark, *Department of Chemistry, University of York, UK*

Series editors:
George A. Kraus, *Iowa State University, USA*
Andrzej Stankiewicz, *Delft University of Technology, The Netherlands*
Peter Siedl, *Federal University of Rio de Janeiro, Brazil*

Titles in the series:
1: The Future of Glycerol: New Uses of a Versatile Raw Material
2: Alternative Solvents for Green Chemistry
3: Eco-Friendly Synthesis of Fine Chemicals
4: Sustainable Solutions for Modern Economies
5: Chemical Reactions and Processes under Flow Conditions
6: Radical Reactions in Aqueous Media
7: Aqueous Microwave Chemistry
8: The Future of Glycerol: 2nd Edition
9: Transportation Biofuels: Novel Pathways for the Production of Ethanol, Biogas and Biodiesel
10: Alternatives to Conventional Food Processing
11: Green Trends in Insect Control
12: A Handbook of Applied Biopolymer Technology: Synthesis, Degradation and Applications
13: Challenges in Green Analytical Chemistry
14: Advanced Oil Crop Biorefineries
15: Enantioselective Homogeneous Supported Catalysis
16: Natural Polymers Volume 1: Composites
17: Natural Polymers Volume 2: Nanocomposites
18: Integrated Forest Biorefineries
19: Sustainable Preparation of Metal Nanoparticles: Methods and Applications
20: Alternative Solvents for Green Chemistry: 2nd Edition
21: Natural Product Extraction: Principles and Applications
22: Element Recovery and Sustainability
23: Green Materials for Sustainable Water Remediation and Treatment
24: The Economic Utilisation of Food Co-Products
25: Biomass for Sustainable Applications: Pollution Remediation and Energy
26: From C–H to C–C Bonds: Cross-Dehydrogenative-Coupling
27: Renewable Resources for Biorefineries
28: Transition Metal Catalysis in Aerobic Alcohol Oxidation
29: Green Materials from Plant Oils
30: Polyhydroxyalkanoates (PHAs) Based Blends, Composites and Nanocomposites
31: Ball Milling Towards Green Synthesis: Applications, Projects, Challenges
32: Porous Carbon Materials from Sustainable Precursors

33: Heterogeneous Catalysis for Today's Challenges: Synthesis, Characterization and Applications
34: Chemical Biotechnology and Bioengineering
35: Microwave-Assisted Polymerization
36: Ionic Liquids in the Biorefinery Concept: Challenges and Perspectives
37: Starch-based Blends, Composites and Nanocomposites
38: Sustainable Catalysis: With Non-endangered Metals, Part 1
39: Sustainable Catalysis: With Non-endangered Metals, Part 2
40: Sustainable Catalysis: Without Metals or Other Endangered Elements, Part 1
41: Sustainable Catalysis: Without Metals or Other Endangered Elements, Part 2
42: Green Photo-active Nanomaterials
43: Commercializing Biobased Products: Opportunities, Challenges, Benefits, and Risks
44: Biomass Sugars for Non-Fuel Applications
45: White Biotechnology for Sustainable Chemistry
46: Green and Sustainable Medicinal Chemistry: Methods, Tools and Strategies for the 21st Century Pharmaceutical Industry
47: Alternative Energy Sources for Green Chemistry
48: High Pressure Technologies in Biomass Conversion
49: Sustainable Solvents: Perspectives from Research, Business and International Policy
50: Fast Pyrolysis of Biomass: Advances in Science and Technology
51: Catalyst-free Organic Synthesis
52: Hazardous Reagent Substitution: A Pharmaceutical Perspective
53: Alternatives to Conventional Food Processing: 2nd Edition
54: Sustainable Synthesis of Pharmaceuticals: Using Transition Metal Complexes as Catalysts
55: Intensification of Biobased Processes
56: Sustainable Catalysis for Biorefineries
57: Supercritical and Other High-pressure Solvent Systems: For Extraction, Reaction and Material Processing
58: Biobased Aerogels: Polysaccharide and Protein-based Materials
59: Rubber Recycling: Challenges and Developments

How to obtain future titles on publication:
A standing order plan is available for this series. A standing order will bring delivery of each new volume immediately on publication.

For further information please contact:
Book Sales Department, Royal Society of Chemistry, Thomas Graham House, Science Park, Milton Road, Cambridge, CB4 0WF, UK
Telephone: +44 (0)1223 420066, Fax: +44 (0)1223 420247
Email: booksales@rsc.org
Visit our website at www.rsc.org/books

Rubber Recycling
Challenges and Developments

Edited by

Jin Kuk Kim
Gyeongsang National University, Republic of South Korea
Email: rubber@gnu.ac.kr

Prosenjit Saha
IIEST, Shibpur, India
Email: senjitiitkgp@gmail.com

Sabu Thomas
Mahatma Gandhi University, Kottayam, India
Email: sabuchathukulam@yahoo.co.uk

Józef T. Haponiuk
Gdansk University of Technology, Poland
Email: jozhapon@pg.gda.pl

and

M. K. Aswathi
Mahatma Gandhi University, Kottayam, India
Email: aswathimk87@gmail.com

ROYAL SOCIETY OF CHEMISTRY

Green Chemistry Series No. 59

Print ISBN: 978-1-78801-084-9
PDF ISBN: 978-1-78801-348-2
EPUB ISBN: 978-1-78801-544-8
Print ISSN: 1757-7039
Electronic ISSN: 1757-7047

A catalogue record for this book is available from the British Library

The Royal Society of Chemistry is a charity, registered in England and Wales, Number 207890, and a company incorporated in England by Royal Charter (Registered No. RC000524), registered office: Burlington House, Piccadilly, London W1J 0BA, UK, Telephone: +44 (0) 20 7437 8656

For further information see our web site at www.rsc.org

Printed in the United Kingdom by CPI Group (UK) Ltd, Croydon, CR0 4YY, UK

Preface

The concept of adopting green chemistry and technologies for the sustainable future of mankind has been generating much-needed public concern and interest in recent years. A lack of properly established technologies to control waste generation imparts serious ecological issues that must be addressed through a proper waste management system. Therefore, a focus on sustainability has become common in almost every industry, particularly in the rubber industry.

Undoubtedly, there is an enormous potential for reclamation and reuse of rubber in almost all countries. Tire waste is the most significant source of rubber waste. A large amount of these tires or rubber can be treated with some very interesting and advanced technologies.

This book is divided into 12 chapters that will provide detailed methodologies and data-sources on rubber recycling to fill the gap between the existing technologies and practical requirements for industrial and academic researchers.

This book will present a comprehensive review of the existing technologies for rubber recycling processes and characterizations with a clear indication of the required advanced technologies that must be undertaken in order to improve the recycling process using economic and green techniques.

We sincerely thank all the authors for their valuable research inputs. We would also like to acknowledge the RSC Editorial team for their help with this book. We would like to thank our research students, Jaideep, Anindya, Suprabha, Soumanti, Sukanya, and Minhaj for their help in producing and proofreading the initial drafts of the chapters of this book.

Jin Kuk Kim, Prosenjit Saha, Sabu Thomas, Jozef T. Haponiuk, and M. K. Aswathi

Green Chemistry Series No. 59
Rubber Recycling: Challenges and Developments
Edited by Jin Kuk Kim, Prosenjit Saha, Sabu Thomas, Józef T. Haponiuk and M. K. Aswathi

Published by the Royal Society of Chemistry, www.rsc.org

Contents

Green Chemistry Series No. 59
Rubber Recycling: Challenges and Developments
Edited by Jin Kuk Kim, Prosenjit Saha, Sabu Thomas, Józef T. Haponiuk and M. K. Aswathi

Published by the Royal Society of Chemistry, www.rsc.org

CHAPTER 1

Grinding of Waste Rubber

JAIDEEP ADHIKARI[†a], ANINDYA DAS[†a], TRIDIB SINHA[b], PROSENJIT SAHA*[a] AND JIN KUK KIM*[b]

[a]Indian Institute of Engineering Science and Technology, Shibpur, Dr. M.N. Dastur School of Materials Science and Engineering, Howrah, 711103, India; [b]Gyeongsang National University, Department of Materials Engineering and Convergence Technology, Jinju, 660701, South Korea
*E-mail: prosenjit@matsc.iiests.ac.in, rubber@gnu.ac.kr

1.1 Introduction

The word rubber originates from the South American Indian word, meaning 'weeping wood' and dictionary definition of rubber is, 'a material that when stretched returns quickly to its approximate original shape'.[1] A three dimensional insoluble and infusible chemical network is formed in the manufacturing of rubber products. This network is formed through an irreversible chemical reaction, known as vulcanization chemistry, between the elastomer, sulfur and other chemicals. After cross-linking the material becomes solid, insoluble and infusible. Hence, direct recycling, or rather reclamation of rubber, is a difficult process which ends up costing half that of natural or synthetic rubbers. However, recycled rubber possesses some properties

[†]Both the authors have equal contribution.

Green Chemistry Series No. 59
Rubber Recycling: Challenges and Developments
Edited by Jin Kuk Kim, Prosenjit Saha, Sabu Thomas, Józef T. Haponiuk and M. K. Aswathi

Published by the Royal Society of Chemistry, www.rsc.org

that are better than the virgin rubber.[2] Reprocessing by melting is also not feasible, because of the three-dimensionally crosslinked molecular structure of rubber.[3] Moreover, conservation of petroleum products has led to a drive towards a cleaner environment and greener waste disposal of unwanted rubber products serves as an added advantage for the rubber recycling process. Waste tire rubbers in stock piles can allow good breeding conditions for mosquitoes and cause several fire related issues. Waste tire fires can spread rapidly and they are very difficult to extinguish, in addition, the fumes emitted from waste rubber contain several harmful volatile compounds. Land filling of waste rubber deteriorates soil properties by killing useful bacteria in the soil.[4] Henceforth, recycling becomes inevitable. Nowadays, the amount of used tire recycling by grinding in the EU27, U.S., Japan and China is estimated to be about 3.6 million tons/year.[5] The utilization of powdered rubber in the field of rubber composites has been increasing over the years and now they are no longer considered as a mere filler with an economic advantage, as they have wide applications in several molded, extruded products. Moreover, these composites have a variety of usages such as in playgrounds, in the preparation of artificial turf for several outdoor sports, mulching applications, and in the fields of animal breeding and automobiles.[6] Grounded waste rubber has various uses in the field of composites, in particular for reinforcing cement like products. The sulfur vulcanization process was invented by Goodyear, who also initiated efforts to recycle cured rubber waste through a grinding method. Utilization of waste rubber in a vulcanized state requires a reduction in particle size. Sometimes, reduction in surface area also plays a critical role in the utilization of waste rubber in the vulcanized state. One of the widely used industrial processes for achieving this with scrap tires and rubber waste is grinding of the waste rubber. Force can be exerted on the rubber chips in various ways, these are impact, compression, shear and attrition. Attrition is the rubbing of rubber chips against a solid face. The grinding processes vary according to the requirements of the final product, which are cutting, shredding and impact. Moreover, the process is dependent on several grinding conditions such as ambient, wet, or cryogenic grinding.[4] Reduction in particle size can also be obtained with shredding or cutting applications, however, shredding creates surfaces with smooth edges. On the other hand, particles obtained by the grinding method have larger surface areas therefore the compatibility of rubber granulates with matrices is increased as a better adherence with the matrix implies a strong reinforcing ability. Under stretching the chances of separation of a particle from the matrices is reduced with increased compatibility.[6] The benefits of the grinding method of rubber for material recycling have been described by Karger-Kocsis *et al.* in their review paper. They accumulated data that described how 1 kg of tire accounts for 128 MJ of energy and that regeneration of this 1 kg rubber yields only 30 MJ of energy. Whereas an additional 6.8 MJ of energy is required to produce 1–1.5 kg of ground tire rubber. ASTM 5603 is the measured standard for characterizing particulate rubber, however, the ASTM 5603 does not provide any information about the physical properties of the powdered sample or provide conclusions about the grinding technology involved in the process. ASTM D 5644 is used to determine the particle size measured in mm or mesh, the surface appearance of the particle is determined by microscopy analysis.[5]

1.2 Sources of Waste Rubbers

Early methods for the disposal of waste rubber products were through land filling or sometimes to be reused as fuel sources. Scrap worn out tires from various automobile industries are the major source of reclamation of waste rubber.[7] Apart from tires, rubbers are utilized in the automobile industry in several other aspects such as in conveyer belts, hoses, shock absorbing dampers, and so forth. Rubber based products are utilized in several civil engineering applications, agricultural products, clothing and latex goods. Products discarded during the manufacturing of these goods and utilized wastes are also a source of reclamation for waste rubbers.[8] There are various processes available for recycling these waste rubber products and the details of the processes have been provided in Table 1.1.[9]

The major source of waste rubber comes from the tire industry. Generally, tires contain several steel wires and fibers for reinforcement purposes. After the grinding and shredding operation, the obtained product is a rubber crumb and the byproducts are separated out from the tires. Representative figures of the crumbs and fibers are depicted in Figure 1.1.[10]

1.3 Waste Rubber Grinding Routes

Until the early 1990s high-rate tire shredding was the most popular technique for reclamation of waste rubber. However, it was a very slow process utilizing a popular digester process in the rubber processing industry. Processing of waste tires largely depends on the type and quality of the tires. The processing becomes difficult with dirty tires and stockpiles which have been stored for prolonged periods. Presently, the common grinding methods include three processing stages: primary, secondary and tertiary, to achieve the fine grade of crumbed rubber, for example the formation of rubber powders.

The primary processing stage reduces down the whole tire to a convenient processing size by utilizing any of the following three instruments: the cracker mill, the hammer mill (high impact), or the rotary shearer. The guillotine is another type of instrument which merely cuts the waste rubber. Generally, the rubbers are first shredded and then subjected to a grinding process. The elasticity of rubbers creates more problems in grinding as compared to the shredding process, and cracker mills are engaged to overcome the problems related to grinding. The feed of rubber material is supplied to the nip point of the two corrugated rollers rotating at different speeds. The materials are caught at the nip point and exposed to a shearing force which

Table 1.1 Recycling approaches for waste rubber products. Reproduced from ref. 9 with permission from Elsevier, Copyright 2001.

No.	Approaches	Introduction
1	Recycling energy as fuel	One method is reclaiming the energy through direct burning, this method is simple but it will cause air pollution. The other is to make fuel, that is, waste rubber is blended with other burnable discard for making rubbish solid fuel which is used in a cement revolving tubular kiln instead of coal, and moreover, it can be used in generating electricity. This method can obtain another material, charcoal black. It can be used as activated charcoal black after activating
2	Thermal decomposing	Making use of the products of thermal decomposition of waste rubber, such as: coal gas, oil and charcoal black and so on. However, the cost of facilities and operation is quite high
3	Modification	Tires are renewed, making man-made fish shelters, conservation materials, and lifebuoys. These materials consume a little energy and make the best use of waste rubber, therefore this is a valuable recycling approach
4	Modification asphalt	Roadway materials
5	Regenerated rubber	Making regenerative rubber through desulfurization of waste rubber. Due to the cost of cryo-regenerating, rubber products will decrease, if regenerative rubber is blended with rubber. Currently there are some new regenerating methods, such as, normal temperature regenerating, cryogenic phase transition catalytic desulfurizing, microwave regenerating, irradiation regenerating and extrusion regenerating
6	Powdered rubber	The main recycling approach of waste rubber is to make PR or FPR. In general, cryogenic crushing is the main process of FPR. The temperature in the process is lower than the glass transition temperature (T_g). Of the polymer in rubber. For example, the temperature of the blends of natural rubber (NR) and styrene-butadiene rubber (SBR) is under −90 °C, to −67 °C at the least, usually the freezing agent is liquid nitrogen. Normal temperature continuous crushing and high pressure water impacting are the crushing approaches used for waste rubber

helps to grind the rubber particles. Hammer mills are the next alternative instrument to break down the size of rubber material. Hammer mills can be of horizontal or vertical type according to their design considerations., and utilize impact force to break down the materials. However, high energy consumption limits their usage in the processing industry. Sometimes, high energy water jets are used as an impact grinder for the breaking down of materials. Altering the architecture of knives and hooks varies the chopping efficiency of the rotary shearer. Power consumption depends on the size and type of the unit.

Figure 1.1 General aspects of the tyre crumbs (a) and the fibres (b) used. Reproduced from ref. 10 with permission from Elsevier, Copyright 2011.

The secondary stage converts the coarse waste rubber chips to granular rubber material. The two roll mill is similar to the cracker mill, but it is smaller in size and hence works at a lower energy. Granulators are another type of rugged machine which produces finer particle sizes. These are high speed and high loading machines which work with either close or open rotor assemblies. Water mist sprays or cooling jackets are required for granulators due to the high loading of materials. Another popular method to break down waste rubber material is to convert rubber chips into a brittle, glassy state by exposing rubber chips to a liquid nitrogen atmosphere, the frozen particles are then fed to the milling unit. The whole assembly is termed as a cryo mill. The extrusion technique is also utilized in the rubber reclaiming industry, in which tire chips experience a shearing action while they are transported through a screw. The chips are fed into the hopper and travel through a rotating screw which pushes the material in a forward direction. The screw design alters the applied shearing force on the material. In this shearing process a significant amount of heat is generated which must be removed to ensure that the properties of the end product remain unaffected. In the extrusion process, particles are obtained with high surface areas.

Tertiary grinding is the ultimate breaking down of crumb rubbers to obtained a fine powder, for these 100% conversion in a single unit can be achieved with wet grinding techniques.

Generally, coarse size reduction is achieved by crushing and shredding operations and finer grade size reduction is achieved using the grinding method. Ultrafine particle size reduction is carried out through pulverizing mills.[11]

1.4 Different Grinding Conditions

Grinding can be performed either at room temperature or in the glass transition region through the use of liquid nitrogen in a cryogenic system.

1.4.1 Ambient Grinding

Ambient grinding can be one of two types, using air impact or a water jet. Vulcanized rubbers are passed though the nip gap of a shear mill or two roll mill at room temperature for reduction of the particle size.[12] Ambient grinding does not mean processing at room temperature, the milling temperature may rise up to 130 °C when using this grinding technology.[5] The particle size reduces with the increase in the number of passes through the nip zone. The principle of fragmentation of particles is by acquiring sufficient kinetic energy by which they can collide with each other and with several parts of the instruments and finally disintegrate into smaller particles. In the ambient grinding method this kinetic energy is achieved by accelerating the particles through the air jet. Turbulence in the mill can also split up particles. Overall, the turbulence, upside-downside air current and energy stored by rubber particles generates heat in the system and therefore cooling of the system is necessary.[12] A schematic representation of the overall process has been provided in Figure 1.2.[13]

The standard granulation process uses knives in a granulator. The knife type granulator assembly (Figure 1.3)[14] is a variation of the ambient grinding process. Elastic and thermoplastic wastes can be ground using this process. The key attribute of this process is its high efficiency with a low vibration and noise level.

Another variation on ambient grinding is the flat-die granulation process (Figure 1.4)[14] in which rubber products are first shredded and then pressed through flat dies. The material is passed through the openings of the perforated flat die of the granulator. In this process there is no cutting process involved using the knives.

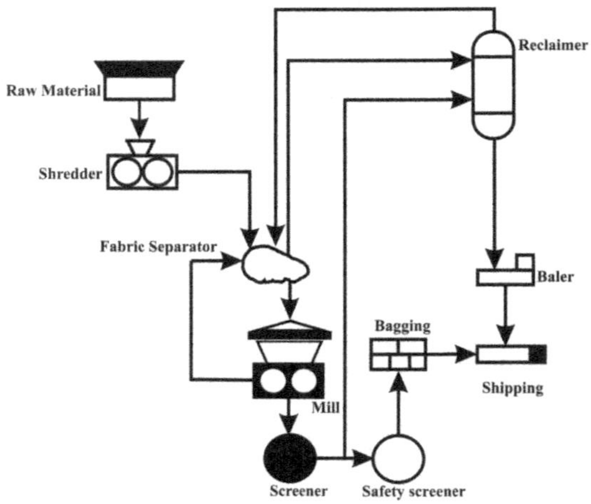

Figure 1.2 Schematic representation of ambient grinding. Reproduced from ref. 13 with permission from Elsevier, Copyright 2013.

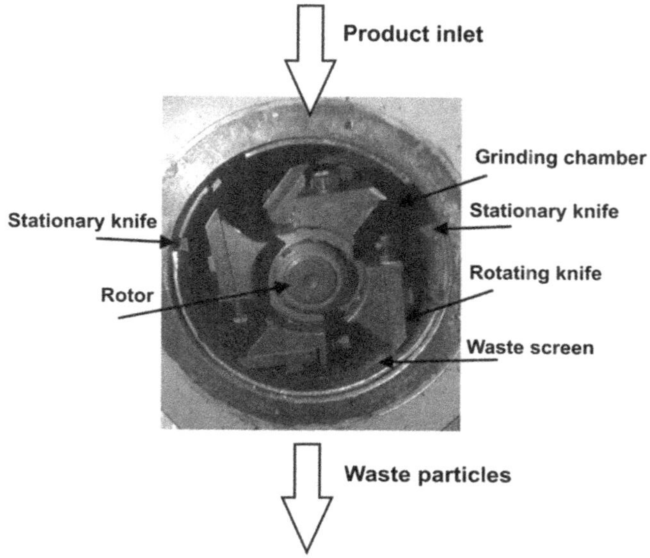

Figure 1.3 Diagram showing a knives granulator (SGP). Reproduced from ref. 14 with permission from Elsevier, Copyright 2017.

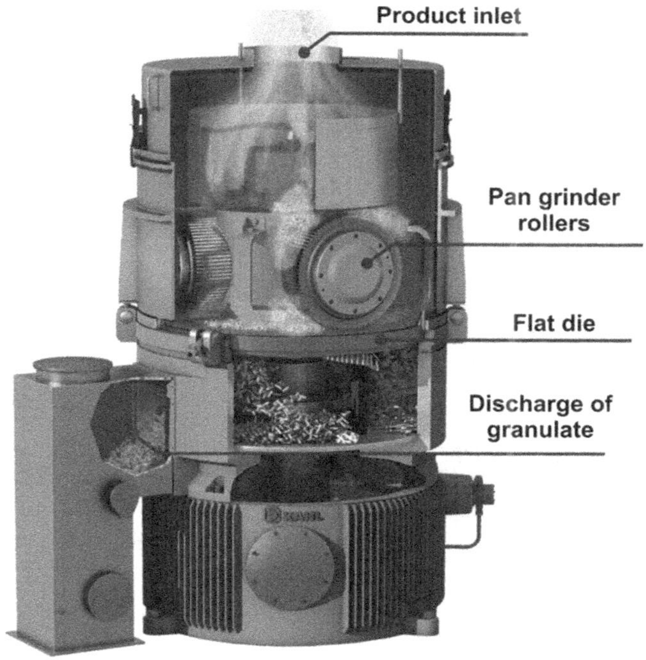

Figure 1.4 Schematic diagram of a flat-die pelleting press (FDGP). Reproduced from ref. 14 with permission from Elsevier, Copyright 2017.

Kinetic energy can be introduced to the rubber particles through the use of a water jet stream. The impact energy exerted by the water jet disintegrates the particles into smaller sizes. The water jets provide the added advantage of cooling down the system during the operation. The major disadvantage of this process is the addition of a processing stage, for example subsequent drying of the particles before finishing. Hence, this process adds to the total energy consumption, making it a costlier process. Moreover, proper separation of the reinforcement material cannot be achieved using this process.[10]

1.4.2 Cryogenic Grinding

The basic principle of cryogenic grinding is the conversion of elastic rubbery chips to a brittle material by exposing them to liquid gases. Generally, liquid nitrogen is used to cool down the rubbery material to below its glass transition region, which is followed by crushing of the frozen brittle material through impact force utilizing a hammer mill. Cooling of the rubber chips can be performed prior to their grinding or during grinding time. The high production rate and low energy utilization makes cryogenic grinding a fascinating process.[12] The process outline is represented in Figure 1.5.[13] However, a pre-grinding and drying operation is required for cryogenically ground material which adds to the overall processing cost. Nitrogen is an

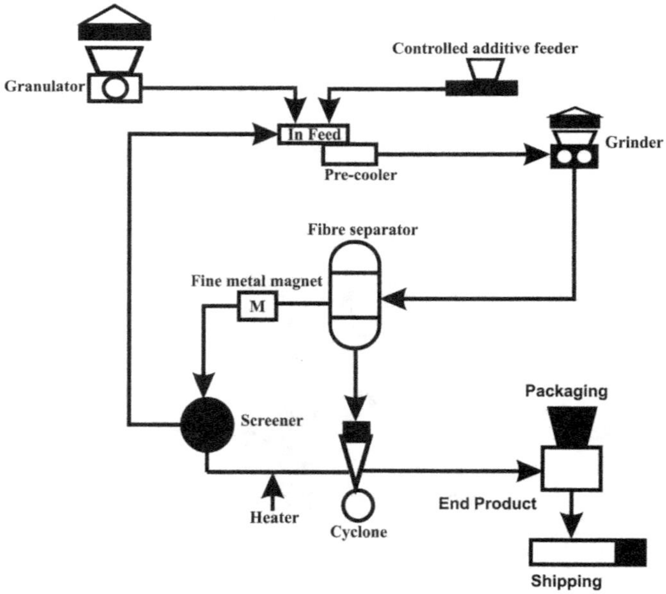

Figure 1.5 Schematic representation of cryogenic grinding. Reproduced from ref. 13 with permission from Elsevier, Copyright 2013.

inert gas which helps to reduce the degree of surface oxidation of the rubber chips.[12]

Cryogenically ground rubber powders possess a very poor physical binding ability with other polymers due to their smoother surface and low surface area. The particle morphology is represented in Figure 1.6.[13] In this respect ambient ground powder is superior to cryogenically ground powder. The physical binding of ambient ground rubber particles with other polymers can be explained by the surface cavities in which polymer chains can easily infiltrate. Generally, the particle size is controlled by the number of grinding cycles or the time spent during the grinding process. However, in cryogenic grinding the immersion time in liquid nitrogen is also a controlling factor for the determination of particle size.[12]

1.4.2.1 Comparative Property Evaluation Between Ambient Grinding and Cryogenic Grinding

Grains of rubber granulates obtained through cryogenic grinding have a smoother surface as compared to the grains of rubber granulates obtained through the ambient grinding method. A smoother surface offers almost no occlusion to the elastomers prepared from this rubber compound, on the contrary ambiently ground rubber granulates offers obstacle to the flowability of the prepared elastomers with these granulates. Ambiently ground granulates are spongy in nature, therefore, when measuring the rheological properties such as the Mooney viscosity, ambiently ground granulates offer a higher viscosity over cryogenically ground granulates. Incorporation

Figure 1.6 Cryogenic ground rubber crumbs. SEM analysis at 200 µm and 400x magnification. Reproduced from ref. 13 with permission from Elsevier, Copyright 2013.

of ambiently ground rubber granulates into fresh rubber decreases the mechanical strength and reduces the ultimate elongation. However, resistance to wear is improved with the addition of ambiently ground rubber granulates to the fresh rubber. Reduction in mechanical strength can be explained by the reduction in the cross link density of the overall compound. Sulfur migrates from the elastomer matrix into the granulate and the cross link density becomes reduced.[6] Properties of the composites can be tailored with variation of the particle size. Composites prepared with rubber ground rubber granulates also experiences some altering properties due to varying particle size and the type of grinding. Ambiently ground granulates offer better interaction with the matrix due to their irregular shape and fuzzy surface. Thus, the mechanical strength of the composites reinforced with ambiently ground rubber granulates have a higher mechanical strength compared to the cryogenically ground granulates. The smooth surfaces of the granulates obtained from cryogenic grinding show limited physical binding with the matrix material and the mechanical properties of the composites deteriorate. Ambiently ground rubber granulates also help to achieve better impact properties through their porous and rougher surfaces. Overall, the process ability of ambient grinding is better than cryogenic grinding.[12]

In ambient grinding, large particles are obtained when the rubber is ground in a toothed-wheel mill and hence the produced powder is restricted to low-strength applications. Cryogenically ground rubber is often roughened by ambient milling to enhance the surface area.[15] The morphologies of the powdered samples obtained through different milling methodologies and different milling instruments are represented in Figure 1.7.[5]

1.4.3 Solution Grinding

The basic principle of grinding remains the same when compared with the other processes, however, in this method rubber chips are first swelled in a solvent, or typically in fatty acids and are then pressed against a finer mesh or subjected to the conventional grinding process.

1.4.4 Grinding by Ozone Cracking

In the ozone cracking process, the waste tire particles are first exposed to a higher concentration of ozone. The material degrades during the exposure to ozone and subsequently mechanical grinding is required for further processing and to obtain a fine powder. However, the obtained fine powder shows a low surface activity due to the oxidation of ozone.[12] Similarly the converse process is also applicable, in which cryogenic grinding is coupled with a subsequent ozone cracking process. Monomers can be recovered in this environmental friendly process for hydrocarbon feed stocks. The initial break down of the vulcanized or cross-linked network through ozone cracking promotes subsequent thermal depolymerization techniques.[16]

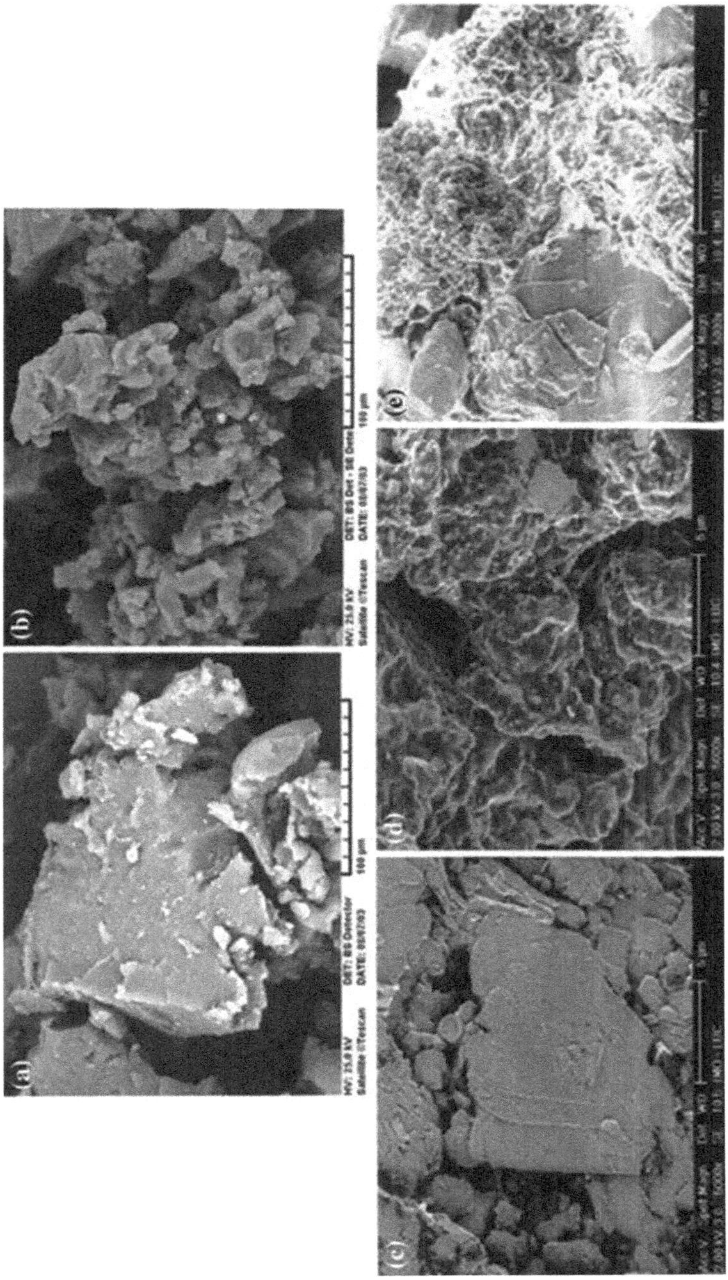

Figure 1.7 SEM pictures of GTR produced in different ways: ambient (mechanically ground) (a); ambient (ground by high pressure water) (b); cryogenically ground in pin mill (c); ambient ground in rotary mill (d); cryogenically ground in rotary mill (e). Reproduced from ref. 5 with permission from Springer Nature, © Springer Science + Business Media, LLC 2012.

The reactivity of the ozone is controlled by the temperature. At low temperatures (near 0 °C) even though the ozone concentration is high; the chemical attack on the double bonds of the rubber become hindered by the low reaction speed. The solubility and diffusion rate of the ozone reduces at low temperatures. Similarly, at higher temperatures (beyond 60–65 °C) ozone becomes unstable and degrades in air. Therefore, the reactivity of ozone cracking reduces. A practical approach is to mimic the actual working conditions of a tire, in which a tire is subjected to sunshine and the vicinity of the air remains cool. Under this condition ozone diffuses quickly through the surface of the rubber and degrades the material. Therefore, a hot rubber surface with cool air in the surrounding environment is best for ozone reactivity.[17]

Ozone attack on rubber granulate is also dependent on the tensile stress values. Over the past few decades several studies have investigated the stress cracking behavior of rubbers. Cracks grow above a certain threshold of strain values, the threshold value depends on the type of polymer. Generally, cracks grow perpendicular to the direction of the tensile stress, crack growth also depends on the ozone concentration. In dry ozone conditions, rapid crack growth is observed initially on the surface of rubber and subsequently the crack grows at a uniform rate. Ozonization in moist environment occurs rapidly by hydrolysis of the rubbery material.[18]

1.4.5 Elastic Deformation Grinding

The concept of elastic deformation grinding (EDG) originates from the principle of a high shear Banbury mixer. A group of Russian researchers observed that under high pressure and shear force, low-density polyethylene (LDPE) forms a fine powder, when it was subjected to cooling instead of heating. In this process elastic energy was forced into the polymer solution and under shear deformation this elastic energy continues to flow through the polymer until the stored energy of the system is released. The energy stored in the elastomers acts as the energy required for fracturing of the material. Thus, a new surface is formed utilizing this energy. The process was initially designed for the grinding of thermoplastics but later it was also used in the rubber recycling industry. Deformability of the elastomers can be varied by altering the process parameters such as temperature, pressure, feed rate and so forth. The consumption of energy in the EDG process is very minimal. Almost two to three times less energy is consumed in the EDG process as compared to the conventional grinding method. Traditionally, high elasticity rubbers are ground in the conventional cryo-grinding method. The processing cost increases as the coolant consumption is also of concern, in addition to its energy utilization. Therefore, the EDG method is advantageous for rubbers such as isoprene rubber, butyl rubber, and siloxane rubbers. The grinding process is carried out either in a single or twin screw extruder. The principle of the fracture mechanism of EDG can be explained by Griffith's theory of fracture. The fracture mechanism deals with initiation of a single crack and subsequently propagation of the crack, which leads to the overall fracture of the material. Every material consists of several pre-existing flaws which act as stress concentration sites. Microcracks initiate in the vicinity of these stress concentration sites and propagate through the flaws. Eventually it turns into a crack and at some point results in the fracture of the overall material.[4] Griffith proposed the forming criterion for the propagation of crack, that is: "A crack will propagate when the decrease in elastic strain energy is at least equal to the energy required to create the new crack surface".[19]

In the case of polymers, a chain scission occurs under the application of stress and free radicals are formed at the rupture end. These free radicals are created due to the application of mechanical load and henceforth they are termed as mechanoradicals. Molecular chain scission depends on the bonding energies associated with the atoms. Chain scission occurs when the applied stress exceeds the critical value required for breaking the covalent bonds. Physical and chemical structures of the polymer become modified and a change in melt rheology is observed. Grinding methodology has an advantage in modifying the surface properties in addition to a reduction in particle size contrary to the conventional milling processes in which only size reduction can be achieved. A change in the physical property is achieved through several sequential stages. Fracture occurs through the weak bonds and the concentration of mechanoradicals decreases exponentially with distance from the fracture surface. The critical value of stress required for chain scission is directly dependent on the temperature, chain length and critical strain rate. In the EDG process the effect of shear can cause a change in the molecular weight distribution (MWD). The rate of mechanochemical reactions is generally higher at low temperature regions.[4,20]

The EDG process is often included in several other process names such as extrusion grinding and solid-state shear extrusion (SSSE). The beauty of this SSSE process is the elimination of the cryogenic coolant. In most cases a twin screw extruder is used in SSSE and a modified version of the twin screw extruder was developed by researchers at the Center of Excellence in Polymer Science and Engineering at the Illinois Institute of Technology, USA. An effective, yet cheaper SSSE process was developed using a single screw extruder by Arastoopour in his patented work, and Bilgili *et al.* also used the single screw version of SSSE in their work and identified the effect of different dependent and independent variables on the pulverization of rubber granulates using SSSE.[15,21]

A schematic diagram of the single screw driven SSSE process has been provided in Figure 1.8.[15] The single screw extruder uses a screw feeder to feed the rubber granulates. The temperature in the extruder zone is adjusted by an electrical heating system and by an air or water cooling system. Generally, the extruder zone is divided into several functional zones with varying amounts of compression ratios. Rotating square-pitched screws are used in the extruder for pushing the material forward into the next zone and differentiating the extruder in several zones. The minimum compression ratio

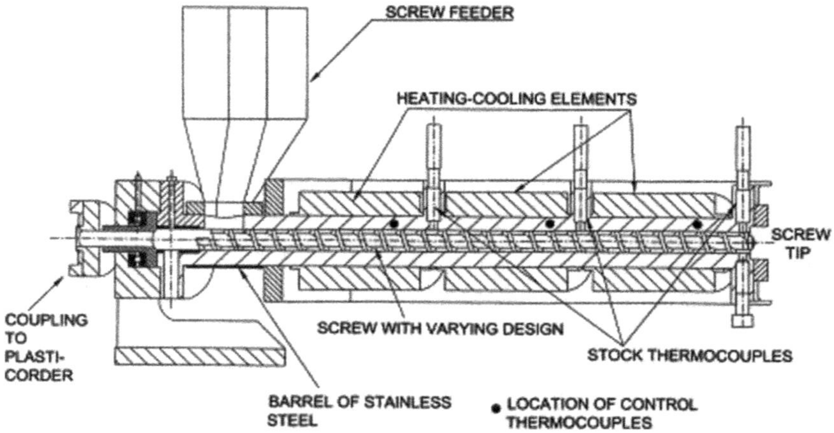

Figure 1.8 Schematic diagram of a single screw extruder used in the pulverization of rubber. Reproduced from ref. 15 with permission from Elsevier, Copyright 2001.

is set to the zone near to the inlet and granulates are dragged along by the rotating screw with minimum compression. The maximum compression is exerted on the rubber granulates in the last zone, near to the outlet. For a specific design of the rotating screw, the compression ratio is dependent on the channel depth. Apart from the compression ratio, the particle size is controlled by several other variables such as temperature, feed rate, rotation rate, and so forth. Development of the shear stresses on rubber granulates are caused by the relative motion of the screw with respect to the barrel wall. Tensile stresses develop on the material as rubber granulates experience high compressive shear strain in the extruder. A significant amount of strain energy is stored in the process, which leads to the formation of new surfaces through the crack opening mechanism (Mode I). The minimum size of the particle is controlled by the design of the extruder and by varying the processing conditions. Granulates can be fragmented several times in a single processing cycle or in a multi-processing cycle, in which the extruded product is fed into the granulator again. The reduction in particle size can be continued until their size becomes so small, that a high compressive strain and consequent high stresses can no longer be applied. Simultaneously, a higher fraction of finer particle enhances the possibility of agglomeration. Agglomeration in the produced powder can also be observed with a greater degree of compaction, and a higher temperature of the pulverization zone. The extent of fragmentation can be enhanced by the storage of a larger amount of strain energy. An enhancement in the stored energy can be achieved by the conversion of rubber to a more elastic or solid-like material. This can be achieved with a screw with a higher compression ratio and a sufficient cooling system in the pulverizing zone. Consumption of more mechanical energy in the SSSE process is necessary to obtain a rubber powder with a smaller particle size.[15]

The concept of EDG of polymeric material using the extrusion process was initiated by Enikolopian who extended the idea proposed by Bridgman for pulverizing metals.[22,23] Bridgman proposed that metals can be pulverized by applying a high pressure and shear force simultaneously. Bridgman established an apparatus with two disks that could function for both hydrostatic pressure and shearing action. The Bridgman anvil is also used to study the pulverization mechanism of the rubbers and a prototype was designed at the Center of Excellence in Polymer Science and Engineering at the Illinois Institute of Technology, USA. In the Bridgman anvil, the desired shearing effect is achieved by coupling two hardened steel surfaces in which one of the surfaces is held stationary while the other surface is free to rotate. Hydrostatic pressure is applied on top of the upper anvil in order to exert pressure on the material by squeezing the two surfaces. The lower anvil has no temperature controller unit and is allowed to rotate only. The desired shearing effect can be achieved by rotating this lower anvil. A schematic diagram of the major components of the apparatus is given in Figure 1.9.[24]

The potential pulverization ability of a rubber disk in a Bridgman anvil based SSSE method can be estimated by mapping the strain energy distribution of the material using the computer aided programming software ANSYS. The process is optimized by using different compression and shear forces. An increment in the normal loading on the rubber sample reduces the residence time in the Bridgman anvil, as the stored elastic energy is increased by an increase in hydrostatic pressure. Similarly, an increment in the rotation

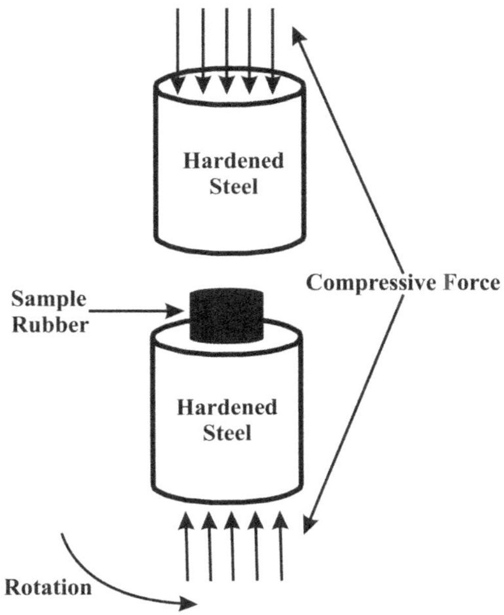

Figure 1.9 Schematic diagram of the Bridgman anvil. Reproduced from ref. 24 with permission from Elsevier, Copyright 1999.

rate reduces the residence time of rubber particles in the anvil. Shearing is related to the angular displacement of the lower anvil. Hence, energy is stored at a faster rate in the material at a higher rotation rate. Subsequently, dissipation of this stored energy occurs at a faster rate through creation of a new surface (*i.e.*, pulverization). Normal force can be minimized by increasing the rotation rates or increasing the residence time. However, a minimal normal force is required for the initiation of the pulverization process. This minimal normal force is required to generate the necessary friction for storing the strain energy.[24]

However, heat generation remains a serious concern in controlling the processing parameters of SSSE. The generated heat can cause partial degradation of the material. Fine particles also tend to agglomerate under exposure to this heat. Heat is generated through the conversion of dissipated elastic deformation energy into heat energy. Hence, a modified version of the single screw extruder based SSSE system was designed at Illinois Institute of Technology; USA by Shahidi *et al.* primarily to provide an isothermal condition at the pulverization zone and to provide an efficient heat removal system. A schematic of the modified SSSE process is provided in Figure 1.10 in which the major working system has been divided into two separate portions; namely, the pulverization section and the extrusion section.[25]

In this model an additional screw is enclosed in the cylindrical housing to increase the efficiency of the extrusion and pulverization processes. In this elongated cylindrical housing multicomponent screws are attached which rotate independently of each other. Each multicomponent screw has an extrusion portion and a pulverization portion. A higher compression ratio and increment in the shearing forces at the pulverization zone is achieved by reducing the clearance between the cylindrical housing and the screw. Simultaneously, the overall production efficiency is enhanced.[25]

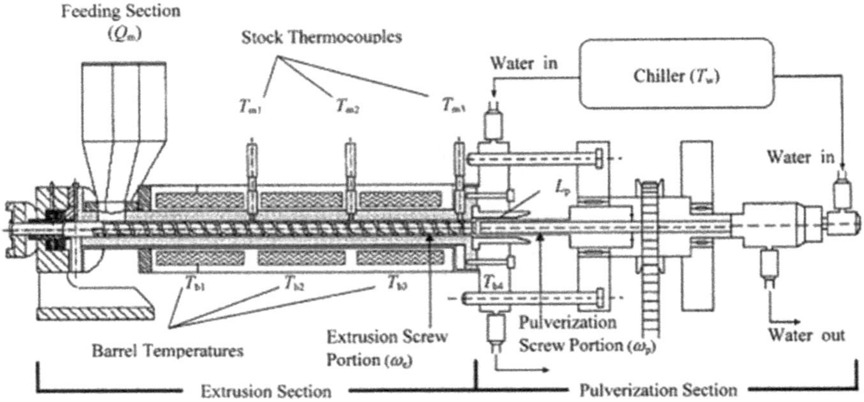

Figure 1.10 A longitudinal schematic of the modified (new design) solid state shear extrusion process. Reproduced from ref. 25 with permission from John Wiley and Sons, © 2006 Wiley Periodicals, Inc. 2006.

1.5 Devulcanization Methods of Rubber

Size reduction of waste rubber is carried out in two steps. In the grinding process, the breakdown of particles is the primary objective so that the obtained crumb rubber can be fed into the subsequent processing stages. The successive stage is breaking of the chemical bonds (primarily sulfur bonds) in a process called devulcanization. Incorporation of grounded rubber particles into polymer blends is often not suitable due to the presence of a sulfur crosslink network. Compatibility becomes affected by sulfur crosslinks which leads to a weak interface and deterioration in the properties of the final product, therefore, the necessity of devulcanization arises. Devulcanization can be achieved by utilizing several means such as chemical, ultrasonic, microwave, biological, and other methods.

1.5.1 Chemical

The initial devulcanization method for the reclamation of waste rubber is based on the utilization of several chemicals. Other means of devulcanization for the purpose of breaking rubber chains involve external energy by means of a two roll mill, or by utilizing microwave or ultrasonic forces. However, these processes cannot distinguish between breaking the crosslinks and the main carbon–carbon bonds that form the backbone of the rubber. Therefore, the properties of reclaimed rubber deteriorate with devulcanization. Organic diallyl disulfide,[26] diphenyl disulfide,[27] inorganic phenylhydrazine–ferrous chloride[28,29] and so forth can be used for production of devulcanized rubber. Organic solvents such as toluene and benzene are generally used for swelling of the rubbers as a first step towards devulcanization. The use of hazardous solvents is a serious draw back in terms of environmental issues. Moreover, solvent recovery after the reaction is a serious challenge which imposes additional costs to the end product.[30] A green medium, in the form of supercritical CO_2 can be used as a reaction medium for some of the chemicals, such as diphenyl disulfide. Supercritical CO_2 is able to swell rubbers properly and can easily distribute the cleaving chemicals. It is non toxic in nature and can be removed easily after devulcanization. Moreover, supercritical CO_2 is cheaper and non flammable, which makes it suitable for industrial processing.[31] Devulcanization can be achieved by controlled oxidation of the carbon framework of rubbers, forming COOH and NO_2 groups. Rios *et al.* achieved this oxidation with nitric acid.[32] Nitrobenzene,[33] benzoyl peroxide,[34] aliphatic amines[35] and so forth are some of the other few chemical agents that have been studied by researchers over the years as a devulcanizing agent for rubbers.

1.5.2 Ultrasonic Devulcanization

Ultrasonic vibration creates a localized energy density that is sufficient to break the cross links of S–S and C–S bonds. Hence, ultrasonic vibration induces cavitations around impurities or voids in the cured rubber. The bond energy of C–C is on the higher side compared to S–S and C–S bonds. Therefore, occurrence of cleavage in the principal backbone of rubber is minimal. Degradation and mechanical property loss is minimal in the ultrasonic devulcanization method. It is a continuous process suitable for industrial applications, as the ultrasonic device is attached in the extrusion path itself. Moreover, no chemicals are used in this process making the process a cleaner and environmentally friendly process. The arrangement and positioning of the ultrasonic device can be varied in the extrusion path. Key variable parameters in the ultrasonic devulcanization method are the devulcanization temperature and amplitude of the ultrasonic waves. Particle size depends on the above mentioned parameters, along with the die pressure in the twin screw extruder.[36]

1.5.3 Microwave Devulcanization Method

In microwave cracking, electromagnetic energy is used to break the crosslinks of the sulfur–sulfur (S–S) and carbon–sulfur (C–S) in the rubber with an aim to restoring the conformation ability of the rubber. In microwave cracking a large amount of material can be processed rapidly in continuous processing. Batch processing is also possible with microwave cracking. The power source and time of processing can be varied to achieve the desired amount of cracking. Molecular interaction with the electromagnetic field is necessary in order to break the crosslinks with microwave energy. In this method rubbers are exposed to a specific amount of microwave energy at a specific frequency in order to cleave the carbon–carbon bonds. Generally, microwave cracking is feasible for rubbers with polar groups. However, for non polar rubbers such as styrene-butadiene rubber (SBR), microwave cracking can be achieved with the use of conducting fillers in the rubber compositions. Carbon black is the most popular reinforcing filler for non polar rubbers in this regard. Depolymerization of rubber wastes can be avoided in the microwave method. Therefore, the physical properties of the final product remain almost equivalent to the original vulcanizate. In molecular interactions the electromagnetic energy is converted into heat energy. Temperature in the range of 260–350 °C can be achieved using microwave treatment and a 300 MHz to 300 GHz frequency is generally used in the microwave method. Higher efficiency of microwave synthesis can be achieved with a high temperature and high pressure sintering process. This process applies the heat energy very quickly and uniformly to the waste rubber.[37,38]

1.5.4 Biological Devulcanization Technique

The negative impact of the common devulcanization process involves the emission of CO_2 and other hazardous byproducts such as SO_x which overall affects the global warming of the environment. In this scenario, the biological process is an environmental friendly process with the utilization of the least energy. The microbial devulcanization processes uses sulfur-oxidizing and sulfur-reducing bacteria.[39] However, vulcanized rubbers can resist bacterial attack. Several studies have been conducted on the devulcanization technique using different types of microorganisms. The sulfur bonds of the vulcanized elastomers are cleaved with the use of microorganisms. A temperature controlled bioreactor is needed, in which finely ground rubber powders are mixed with media containing the appropriate bacterium. The cleaving mechanism through the bacterium varies with the species of bacterium used in the process. Starting with approximately ten days to a few hundred days, the contact time may vary by a wide range. After processing, the newly formed devulcanized rubber materials are rinsed off properly in order to remove the microorganisms.[40]

Actinomadura sp., *Actinomycessp., Gordonia* sp., *Dactylosporangium* sp., *Micromonospora* sp., *Streptomyces* sp., *Thermomonospora* sp., *Xanthomonas* sp. and so forth are some of the rubber degrading bacteria used. Bacteria from the CNM (Corynebacterium, Nocardia, Mycobacterium) group are some of the most effective rubber-degrading strains. Similarly, species of *Penicillium, Aspergillus, Cladosporium* and so forth. are some of the rubber degrading fungi. In microbial attack the whole carbon skeleton is exposed to the microbes and there remains a possibility of degradation of the whole structure.[40]

The concept of biodesulfurization of the sulfur–carbon bonds of dibenzothiophene (DBT) without disrupting the main carbon structure is applied in case of biodesulfurization of tire rubber. Rubber products contain several curing and anti-ageing agents which reduce the efficacy of bacterial treatment. Therefore, additives need to be removed from the rubber before applying any bacterial treatments. Cryo-grinding and successive removal of toxic additives by extraction with ethanol is a common approach used before the microbial devulcanization method.[41]

1.5.5 Other Devulcanization Techniques

Devulcanization through physical deformation is achieved by repeated deformation of rubber particles, temperature and pressure are the main process variables. Apart from varying the process parameters, the design parameters can also be altered to enhance the efficacy of mechanical devulcanization. The efficacy of the process can be enhanced with the utilization of steam for devulcanization. Sometimes chemicals such as caustic soda are used in combination with steam to achieve a pronounced devulcanization effect.

1.6 Relationship Between Energy and Particle Size for Grinding Routes

Shredding and breaking down of waste tire materials produces newer surface areas and the energy consumed during this breaking down process is directly proportional to the newly formed surface areas. For a given quantity of materials with a given reduction ratio (initial particle size to the final particle size) the energy consumption requirement is always constant. Apart from this the required energy for shredding and chopping is also dependent on the generated heat and pressure during the grinding process. Sometimes, heating can induce a positive effect on the overall grinding efficiency. However, excess heat generation can lead to the depolymerization of rubber particles and production fumes, which give the potential risk of fire hazards, in addition too much heat deteriorates the mechanical properties of the ground materials. Surfactants are introduced in the spray method to reduce the particle size, which is similar to liquid cutting additives for metal forming technology.[11]

In grinding two processes occurs simultaneously: (i) formation of smaller particles with fracturing; and (ii) aggregation of particles. A model has been proposed by Prut *et al.* to control the parameters regarding the aggregation of particles. The proposed model assumes a spherical geometry of the ground particle.[42]

If the number of particles per unit volume is "n_0" and r is the respective diameter of each of the obtained particles, then "n_0" will be given as:

$$n_0 = \frac{1}{4/3\pi r^3} \tag{1.1}$$

Similarly, for a volume "V" with a total number of particles "n" it will be:

$$n = \frac{V}{4/3\pi r^3} \tag{1.2}$$

When the particles move in a dispersion medium independent to each other, the change in the number of particles with grinding time can be estimated using an empirical formula:

$$\frac{\mathrm{d}n}{\mathrm{d}t} = k_1 n - k_2 n \tag{1.3}$$

The term $k_1 n$ predicts the grinding rate, and the term k_1 is dependent on the ratio of material deformation to the particle fracture energy. The term $k_2 n$ is associated with particle aggregation rate. With an increase in the grinding time particle formation approaches the aggregation rate and the radius of the formed particles depends on shear deformation, fracture energy and surface energy.[42]

1.7 Classification of Powdered Rubber

The specified particle size of different classes of rubber and their possible applications is provided in Table 1.2.[9]

Table 1.2 Classification, manufacture and use of powdered rubber. Reproduced from ref. 9 with permission from Elsevier, Copyright 2001.

Classification	Particle size μm	Particle size Mesh	Facilities	Application
Coarse PR	1400–500	12–30	Grinding mill, rolling mill, rotary crushing mill	Ballast mats, the raw material of degrading regenerated rubber
Tiny PR	500–300	30–47	Rolling mill, rotary crushing mill	The raw material of oiling regenerated rubber
Fine PR	300–75	47–200	Cryogenic mill, freezing crushing mill	Products of molding and extrusion, rubber mats for crossties, soft pipes for irrigating vases, and modified asphalt for paving
Ultrafine PR	Under 75	Above 200	Rotary colloid mill	Used in renewing of tires (under 20 μm, 30 phr)

1.8 Conclusion

The grinding of rubber products is the preferred waste recycling route due to its economic and environmental advantages. Reduction in particle size can be achieved only by grinding of waste rubber products. However, compatibility with the matrix can be enhanced by breaking the C–S crosslinks. Therefore, grinding of rubber is followed by a devulcanization process. Grinding of rubber can be achieved in ambient conditions, as well as at a sub zero temperature known as ambient grinding. Ambient grinding does not necessarily confine the temperature limit to a room temperature range. The processing temperature can reach up to 130 °C in the ambient grinding process. Ambient grinding imposes a rougher surface on the processed powder and hence physical binding of the rubber particles with other matrix material is observed when using this grinding methodology. Kinetic energy to break the particles can be achieved by applying a water stream jet. Elastic deformation grinding is yet another fascinating technique for size reduction of the particles. Similarly, the devulcanization method is also carried out in several processes. Each of them has their own limitations and advantages. Therefore, selection of the process is purely dependent on the end application and the cost of the product.

References

1. A. Ciesielski, *An Introduction to Rubber Technology*, Rapra Technology Limited, Shawbury, Shrewsbury, Shropshire, SY4 4NR, UK, 1st edn, 1999, vol. 1, p. 7.
2. A. I. Isayev, in *The Science and Technology of Rubber*, ed. B. Erman, J. E. Mark and C. M. Roland, Academic Press Elsevier, 225 Wyman Street, Waltham,

USA, The Boulevard, Langford Lane, Kidlington, Oxford, UK, 4th edn, 2013, vol. 15, p. 697.

3. T. Chaubey and H. Arastoopour, *J. Appl. Polym. Sci.*, 2011, **119**(2), 1075.
4. K. Khait, Tire Rubber Recycling by Mechanochemical Processing, in *Rubber Recycling*, ed. S. K. De, A. I. Isayev and K. Khait, CRC Press, Taylor & Francis Group, 6000 Broken Sound Parkway NW, 2005, p. 4.
5. J. Karger-Kocsis, L. Mészáros and T. Bárány, *J. Mater. Sci.*, 2013, **48**(1), 1.
6. M. Sienkiewicz, H. Janik, K. Borzędowska-Labuda and J. KucińskaLipka, *J. Cleaner Prod.*, 2017, **147**, 560.
7. B. Adhikari, D. De and S. Maiti, *Prog. Polym. Sci.*, 2000, **25**(7), 909.
8. R. Ahmed, A. V. D. Klundert and I. Lardinois, *Rubber Waste: Options for Small-scale Resource Recovery – Urban Solid Waste Series 3*, Backhuys Publishers, Nieuwehaven 201 2801 CW Gouda, the Netherlands, 1996, ch. 1, p. 11.
9. Y. Fang, M. Zhan and Y. Wang, *Mater. Des.*, 2001, **22**(2), 123.
10. A. M. Fernández, C. Barriocanal and R. Alvarez, *J. Hazard. Mater.*, 2012, **203–204**, 236.
11. M. W. Rouse, Manufacturing Practices for the Development of Crumb Rubber Materials from Whole Tires, in *Rubber Recycling*, ed. S. K. De, A. I. Isayev and K. Khait, CRC Press, Taylor& Francis Group, 6000 Broken Sound Parkway NW, 2005, p. 1.
12. W. Dierkes, Untreated and Treated Rubber Powder, in *Rubber Recycling*, ed. S. K. De, A. I. Isayev and K. Khait, CRC Press, Taylor& Francis Group, 6000 Broken Sound Parkway NW, 2005, p. 3.
13. D. L. Presti, *Constr. Build. Mater.*, 2013, **49**, 863.
14. M. Sienkiewicz, K. Borzędowska-Labuda, S. Zalewski and H. Janik, *Constr. Build. Mater.*, 2017, **154**, 144.
15. E. Bilgili, H. Arastoopour and B. Bernstein, *Powder Technol.*, 2001, **115**(3), 265.
16. G. A. Platz, *US Pat.* 5369215 A, 1994.
17. F. Cataldo, *Polym. Degrad. Stab.*, 2001, **72**(2), 287.
18. F. Cataldo, *Polym. Degrad. Stab.*, 2001, **73**(3), 511.
19. K. I. Parashivamurthy, *Material Science and Metallurgy*, Pearson, India, 2012, ch. 6, p. 91.
20. O. A. Serenko, U. A. Muradova, I. B. Meshkov, E. S. Obolonkova, S. N. Zelenetskii and A. M. Muzafarov, *Int. Polymer Sci. Tech.*, 2009, 24.
21. E. Bilgili, H. Arastoopour and B. Bernstein, *Powder Technol.*, 2001, **115**(3), 277.
22. N. S. Enikolopian, *Pure Appl. Chem.*, 1985, **57**(11), 1707.
23. M. Eskandari, H. Arastoopour and J. D. Schieber, *AIChE Annual Meeting, Conference Proceedings*, 2005, p. 4850.
24. D. Schocke, H. Arastoopour and B. Bernstein, *Powder Technol.*, 1999, **102**(3), 207.
25. N. Shahidi, H. Arastoopour and G. Ivanov, *J. Appl. Polym. Sci.*, 2006, **102**(1), 119.
26. G. K. Jana and C. K. Das, *Macromol. Res.*, 2005, **13**(1), 30.

27. G. K. Jana, R. N. Mahaling, T. Rath, A. Kozlowska, M. Kozlowski and C. K. Das, *Polimery*, 2007, **2**, 131.
28. N. Kawabata, B. Okuyama and S. Yamashita, *J. Appl. Polym. Sci.*, 1981, **26**(4), 1417.
29. S. Yamashita, N. Kawabata, S. Sagan and K. Hayashi, *J. Appl. Polym. Sci.*, 1977, **21**(8), 2201.
30. M. Sabzekar, M. P. Chenar, S. M. Mortazavi, M. Kariminejad, S. Asadi and G. Zohuri, *Polym. Degrad. Stab.*, 2015, **118**, 88.
31. I. Mangili, E. Collina, M. Anzano, D. Pitea and M. Lasagni, *Polym. Degrad. Stab.*, 2014, **102**, 15.
32. R. R. V. A. Rios, M. Gontijo, V. P. Ferraz, R. M. Lago and M. H. Araujo, *J. Braz. Chem. Soc.*, 2006, **17**(3), 603.
33. K. Masaki, S. I. Ohkawara, T. Hirano, M. Seno and T. Sato, *J. Appl. Polym. Sci.*, 2004, **91**(5), 3342.
34. S. Rooj, G. C. Basak, P. K. Maji and A. K. Bhowmick, *J. Polym. Environ.*, 2011, **19**(2), 382.
35. V. V. Rajan, W. K. Dierkes, J. W. M. Noordermeer and R. Joseph, *Rubber Chem. Technol.*, 2005, **78**(4), 572.
36. A. I. Isayev, T. Liang and T. M. Lewis, *Rubber Chem. Technol.*, 2014, **87**(1), 86.
37. D. Hirayama and C. Saron, *Ind. Eng. Chem. Res.*, 2012, **51**(10), 3975.
38. A. Zanchet, L. N. Carli, M. Giovanela, R. N. Brandalise and J. S. Crespo, *Mater. Des.*, 2012, **39**, 437.
39. S. Sato, Y. Honda, M. Kuwahara, H. Kishimoto, N. Yagi, K. Muraoka and T. Watanabe, *Biomacromolecules*, 2004, **5**(2), 511.
40. M. Yikmis and A. Steinbüchel, *Appl. Environ. Microbiol.*, 2012, **78**(13), 4543.
41. V. Tatangelo, I. Mangili, P. Caracino, M. Anzano, Z. Najmi, G. Bestetti, E. Collina, A. Franzetti and M. Lasagni, *Appl. Microbiol. Biotechnol.*, 2016, **100**(20), 8931.
42. E. Prut, D. Solomatin, O. Kuznetsova, L. Tkachenko and D. Khalilov, *J. Elastomers Plast.*, 2015, **47**(1), 52.

CHAPTER 2

Surface Treatment of Rubber Waste

X. COLOM*[a], M. MARÍN-GENESCÀ[b], K. FORMELA[c] AND J. CAÑAVATE[a]

[a]Department of Chemical Engineering, Universitat Politècnica de Catalunya Barcelona Tech, Terrassa, Spain; [b]Department of Mechanical Engineering, Universitat Rovira i Virgili, Terragona, Spain; [c]Department of Polymer Technology, Faculty of Chemistry, Gdansk University of Technology, Gdansk, Poland
*E-mail: xavier.colom@upc.edu

2.1 Introduction

In recent years there has been a constantly increasing awareness of finding new ways to recycle materials waste. Polymeric materials do not decompose easily and their disposal implies serious environmental problems. During the production of rubber compounds, natural and synthetic rubbers are vulcanized, becoming crosslinked by sulphur. The rubber formed by this irreversible reaction, combines the different components and acquires the required strength and elasticity over a wide range of temperatures. However, the irreversibility of this process, creates major issues for the recycling of rubber compounds (*i.e.* tires) at the end of their lifespan. Due to this

Green Chemistry Series No. 59
Rubber Recycling: Challenges and Developments
Edited by Jin Kuk Kim, Prosenjit Saha, Sabu Thomas, Józef T. Haponiuk and M. K. Aswathi

Published by the Royal Society of Chemistry, www.rsc.org

crosslinked structure and the presence of stabilizers and other additives, natural degradation of rubber compounds is even longer than other polymeric materials.

Large amounts of rubber products such as hoses, sealants or specialty tires are used in cars, trucks, and airplanes. After some time, the tires become worn. They can be recycled, reconditioned for their reuse, or just classified as end of life (EOL) rubber compounds.

In 2013, about 3.6 million tons of used tires were generated in Europe. An estimated 3.1 million tons of end-of-life tires (ELTs) were recovered and recycled, which represents a treatment rate of 86%.[1] The EU countries with the highest amount of post-consumer tires in 2013 were: Germany with 670 kt, the UK with 445 kt, Italy with 438 kt, France with 411 kt, Spain with 295 kt and Belgium with 70 kt. Dumping tires in landfill has not allowed in the European Community since 2003, the landfill Directive (2003/31/EC) forbade the disposal of tires from 2006. In USA the total amount of ELTs generated was 3.8 million tons, and of this 95.9% were recovered and recycled.[2]

Rubber waste (RW) dumped in landfills leads to mosquito breeding and migration of unhealthy additives to the environment. Moreover, this process is expensive because of the high transport and maintenance costs and does not add to the value of the residua. 1 Kg of tire has a heat value of 32.6 KJ, whereas carbon has only 18.6–27.9 KJ.[3] The possibility of recovering this energy has produced a significant increase in rubber waste recycling practices during the past five years. Some other properties that are considered when approaching the possibility of adding value to waste tires are: (a) vibrations absorption, (b) drain capacity, grounded or in granules, (c) low weight, (d) toughness and high resistance to being cut, (e) weathering endurance, and (f) flexibility.[4]

According to a report produced by the European Tire and Rubber Manufacturers Association (ETRMA)[1] in 2015, typical ways of making use of rubber waste include:

- **Retreading**: This way is preferred, because the rubber is reused in its original function. Only 15% of crude oil is needed to produce a retreaded tire without any loss in quality. The price is reduced by up to 45%. While car tires are retreated once, van tires can be retreated twice and airplanes tires up to seven times. The process consists of adding an already vulcanized thread strip to a prepared carcass. The binding process takes place between 100–120 °C in an autoclave.[5] In Europe, 259 000 tons of tires were retreaded in 2011.[1]
- **Applications with high added value**: this is usually based on the recovery of the materials included in the tire. The rubber waste can be used in its original state, ground or in granules. In the last two cases rubber, metal and fibers included in the tire should be separated.

  As a filler or ground in artificial grass fields: artificial grass requires less maintenance and water consumption than natural grass and its use is popular in many countries. RW can be used as a filler in synthetic fiber carpets or as a base ground on the field below. Average sizes for this application are usually between 0.5 and 2.5 mm, although that may depend on several factors.

Security floors and playpens: UNE-EN1177 regulation specifies that playpens and some areas of senior people homes have to be built with materials that present absorption of impact and weather resistance. Compounds based on rubber waste are perfectly suited to these requirements. The average size used for these applications is usually between 2 and 8 mm.

Ornamental uses: Green areas or roundabouts can be decorated using colored and thick rubber waste compounds (size between 12 and 25 mm). Some of the advantages are: easy application, security (elasticity), low maintenance and null degradability.

Bitumen modification for motorways construction: Noise reduction, better car adhesion to the road, high crack resistance, long life and low maintenance are some of the advantages of including rubber waste into the bitumen blend.

Civil applications: some of these uses can include the whole tire, without separation of components. In this case, the EOL tires must be carefully selected and follow the ASTM 6270-98 guidelines regarding particle distribution and steel content. Some examples include: drain drenches, embankment fillers, drain layers for landfills, and so forth. This practice has been common in the United States for a long time and has become an increasingly popular technique in Europe in the last few years.

- **Energy recovery**: as mentioned above, because of their heat value, EOL tires present an alternative to fuel, providing energy comparable to coal but including lower percentages of sulphur. They can be used as a dedicated fuel in power stations or cement kilns. An environmentally friendly technique free of hazardous emissions was developed to generate valuable fuels or chemicals from waste tires. The process involves breakdown of the rubber materials by selective oxidation decoupling of C–C, C–S and S–S bonds by using water as a solvent or near its supercritical conditions.[6] During 2013, 1361 ktons of tires were used in Europe as combustible energy sources.

Table 2.1 shows the composition of the tires for a typical car and truck. Styrene butadiene rubber (SBR) is the most common synthetic rubber used in the production of passenger car tires (about 20%) combined with other elastomers such as natural rubber (NR), isoprene rubber (IR) and butadiene rubber (BR).

During the last 10 years, all European countries have struggled to find other ways to reuse EOL tires and RW. As reviewed previously, most applications mainly consist of a mechanical transformation of the tire including separation of the components. There is a growing interest in developing

Table 2.1 Typical composition of passenger and truck tires in Europe.[4]

Composition	Passenger car tire	Truck tire
Natural rubber	22%	30%
Synthetic rubber	23%	15%
Carbon black	28%	20%
Other additives (*e.g.* curing system, fibers, processing aids, *etc.*)	14%	10%
Steel	13%	25%
Estimated average weight of new tire	8.5 kg	65 kg

products with added value. Following this trend, there is a worldwide scientific interest in mixing EOL tires with other polymeric materials in order to obtain thermosets or thermoplastics. The main issue is overcoming the lack of adhesion between the components of the mixture. In order to improve the compatibility between RW particles and any polymer matrix, two basic strategies are applied. One is based on the devulcanization of the RW, selectively breaking the sulphur bridges that form the crosslinked structure, followed by re-vulcanization of the compound after the mixing process. The second approach is to increase the bonding between the particles and matrix, allowing surface modification of the RW particles in order to increase the interfacial interaction.

Different treatments are applied to the rubber waste and they can be classified into three categories: physical, biological and chemical. The physical and biological usually have an effect on the whole rubber mass, while the chemical methods only affect the surface. When the treatment affects the complete mass of the rubber waste it includes devulcanization and the product its considered to be reclaimed rubber. This process generally leads to a product with RW fully integrated into the compound. In some cases this process leads to products that are similar to the original (*i.e.* tires). On the other hand the superficial treatments are aimed at modifying the surface of the RW in order to increase the compatibility. Generally the integration of the RW in the produced compound is not total, and different phases are present in the blend.

In *physical devulcanization*, the RW is reclaimed with the help of external energy. The three-dimensional network breaks down and the macromolecular rubber chain is transformed into small molecular weight fragments. The main physical reclaiming processes are: mechanical, thermo-mechanical, cryo-mechanical, microwave and ultrasonic.

The *mechanical reclaiming process* of rubber waste takes place in an open two-roll mixing mill, in which the milling is carried out at high temperatures. During the process, a drastic molecular weight breakdown takes place due to the mechanical shearing at high temperatures. One example is the Maxwell patent where waste rubber is treated with reclaiming agents, such as organic disulfides or mercaptans.[7] The rubber runs between a smooth stator and a cylindrical rotor arranged to provide an axial shear zone in which rubber is frictionally propelled by the rotor action.

The *thermo-mechanical process* consists of the degradation of the vulcanizated network by swelling of the scrap rubber into a suitable solvent (aromatic and paraffinic hydrocarbon) and then milling it to form a fine powder (around 20 µm).

The *cryomechanical reclaiming process* was developed in the mid 1960s. This method only reduces the size of the rubber waste particles, with the devulcanization achieved being very low. The process consists of placing small pieces of vulcanized rubber into liquid nitrogen and then transferring them into a ball mill in order to grind them in the presence of liquid nitrogen. The product obtained is a fine powder from 150–600 µm. The main problem of this method is the elevated cost.

A controlled dose of *microwave energy* at a specific range of frequencies and energy levels can be used to cleave the S–S bonds.[8,9] Frequencies between 915 and 2450 MHz produce energy which is sufficient to break all crosslinking bonds, but insufficient for breaking of the main polymer chain. Therefore, unlike other systems, this produces a material that is not degraded. By using this technique, RW can be reclaimed without depolymerization, obtaining a material capable of being recompounded and revulcanizated and showing physical properties essentially equivalent to the original. At the same time, the microwave method is economically viable and ecological.[10]

The last method included in the mechanical reclaiming category uses *ultrasound*. This devulcanization process requires high energy to break the carbon–sulfur and sulfur–sulfur bonds. A relevant reference on this method was published by Isayev *et al.*[11]

Most of the recent developments in reclaiming of rubbers have relied on *biotechnological processes*. The biological attack of natural rubber latex is quite easy, but as the natural rubber polymer is transformed into an industrial material by vulcanization and inclusion of numerous ingredients, biological degradation is much more difficult. The microbial devulcanization of rubber particles was investigated by Loffer *et al.* in an attempt to increase the possibility of producing high quality rubber products containing a larger percentage of reused rubber waste.[12] Rubber waste particles (mainly SBR for EOL tires) with 1.6% sulphur content were treated with different species of *Thiobacillus*, *T. ferrooxidans*, *T. thiooxidans* and *T. thioparus* in shake flasks in a laboratory reactor. They found that the sulphur oxidation was particle size dependent. The best results were obtained with *T. thioparus* with a particle size of 100–200 µm. At the end of the experiment, 4.7% of the total sulphur of the rubber powder was oxidized to sulphate within 40 days.

Chemical methods can be considered as reclaiming methods or surface treatments depending on their range and intensity. The reclaiming agents are generally disulphides or mercaptans exclusively selected to work at high temperature. Since 1910 many chemical reclaiming agents have been developed for natural and synthetic rubbers, such as diphenyl disulphide, dibenzyl disulphide, diamyl disulfide,[13,14] bis(alcoxy aryl) disulfides,[15] butyl mercaptan and thiophenols,[16] xylene thiols[17] and phenols sulphides and disulphides.

Surface functionalization of RW particles is carried out mainly by oxidation processes, graft polymerization, gas exposure or radiation. All these different treatments aim to modify the RW surface by generating polar groups derived from oxidation, such as peroxy, hydroxyl or carbonyl groups. The presence of these groups on the surface of RW improves the interactions with polar polymers. Furthermore, these groups can also interact with reactive polymer groups such as maleic anhydride to improve the compatibility between the RW and the polymeric matrix.

Oxidation processes include chemical etching by using nitric or sulphuric acids, hydrogen peroxide, ozone or $KMnO_4$.[18–21] On the other hand, grafting onto a RW particle surface can be carried out with solvent containing monomers, which graft through the action of a free radical initiator. Different monomers have been used in this technique, including styrene, acrylamide or methacrylate.

This chapter concerns the surface modification of rubber waste and is dedicated to the oxidation, chlorination, and use of coupling agents and graphitization with dynamic vulcanization of different rubber compounds.

2.2 Experimental

2.2.1 Materials

The materials used in this study were high density polyethylene (HDPE) and ground tire rubber (GTR), the properties of these materials are described below. Other specific materials used in the different purposed methods will be described in the corresponding section. The HDPE used as the matrix was supplied by REPSOL-YPF (ALCUDIA® 4810-B), with a density of 960 Kg m^{-3}. The HDPE was characterized by a melt flow index (MFI190/2.16) of 1.0 g/10 min, a Young's modulus of 927.90 MPa, a tensile strength of 17.17 MPa, a elongation at break of 390.80% and a toughness of 38.4 J. GTR was kindly provided by Gestión Medioambiental de Neumáticos S.L. (GMN) in Maials (Lleida), Spain. The average particle size was between 400 and 600 μm. The mixing process for the composites was carried out in a two roll mill heated at 150–155 °C. The HDPE matrix was melted for a minute and then the GTR particles were added and mixed for another 5 minutes.

Composite sheets ($150 \times 150 \times 2$ mm^3) were prepared by hot press molding at 100 kN and 170 °C for 10 minutes. After that step, the sample sheets were cooled for 5 minutes under pressure using water. After that, the materials were mechanically shaped as test specimens according to ASTM-D-412-98 specifications.

2.2.2 Equipment

2.2.2.1 *Mechanical Testing*

Tensile tests were carried out following ASTM-D-412-98 specifications in order to determine the evolution of the Young's modulus, toughness, tensile strength, and the percentage of breaking deformation. A standard Instron 3300 series tension testing system was used. The testing speed was 20 mm min^{-1}, the temperature was 23 °C, and the relative humidity was 50%. Five replicate samples were analyzed for each test and the average and standard deviation percentages were calculated.

2.2.2.2 *Attenuated Total Reflectance FTIR Spectroscopy*

The Fourier-transform infrared spectroscopy (FTIR) spectra were acquired using a Nicolet Avatar 320 spectrometer equipped with a CsI optical collimated and a horizontal attenuated total reflectance (HATR) accessory. The internal reflecting element (IRE) was a ZnSe crystal set at an incidence angle of 45°. ATR-FTIR spectra were collected in the range of 4000–650 cm^{-1}, with 40 scans and a resolution of 4 cm^{-1}.

2.2.2.3 *Scanning Electron Microscopy Analysis*

Scanning electron microscopy (SEM) was used to qualitatively study the morphology of the fractured surface at room temperature. The samples were examined with a JEOL JSM 6480 SEM. Dry samples were coated with a thin layer of gold before observation under the microscope, in order to increase their conductivity.

2.3 Surface Oxidation of the Rubber Waste Particles

As explained in the introduction, surface treatments are widely applied to improve compatibility between two materials with low interfacial adhesion. Tire rubber is a compound made out of many raw materials and every tire has been exposed to different weather conditions and handling. Therefore, after tire grinding, the chemical nature of the tires can be very different. The applied surface treatment must completely change the whole surface area in one specific way. In this section, we report studies on the chemical and morphological modification of the waste tire rubber by different chemical acids and how this modification influences the adhesion between GTR and HDPE and subsequently the tensile properties of the final blend.

Treatment with sulphuric acid can be used in rubbers as an alternative to other surface treatments (*i.e.* ozone or ionization) to improve adhesion with other polymers. Most polymers that contain tertiary C–H bonds are suitable for sulfonation.[22] During sulfonation with sulphuric acid, the H of a C–H bond is removed and replaced by a SO_3 molecule, which is then hydrogenated, to form sulfonic acid at the site of the attachment. The created sulfonic acid can be neutralized with ammonium in order to form a more stable species. Sometimes this stabilization is a key aspect to obtain an adequate adhesion performance. The neutralization using ammonia extracts the H from the sulfonic acid moieties produced during the immersion of the rubber in the sulphuric acid and leaves a stabilized ion pair $(-C-SO_3^{-}NH_4^{+})$.

Some studies have reported application of surface treatments to rubber and their results have been useful in order to understand the modifications that take place on the rubber waste. Cepeda *et al.* studied the influence of the styrene content of a SBR treated with sulfuric acid.[23] It was observed that the lower the styrene content, the more significant modifications were produced on the surface, and consequently the test specimens showed the best adhesion in the T-peel test. In ATR-FTIR analysis, severe modifications to the rubber surface were observed. A strong absorption was seen in the 1200–1000 cm^{-1} range that was indicative of O=S=O stretching of the sulfonic acid, two bands at 1040 and 1167 cm^{-1} corresponding to the formation of sulfonic acid moieties were observed, and finally, a new band at 1350 cm^{-1}, due to O=S=O linkages, that did not correspond to sulfonic acid and may correspond to a sulfonate $(-C-SO_2-OR-)$, appeared. When observed with SEM, the treated sample with a lower styrene content showed big and deep cracks meaning a strong degradation after the treatment.

One interesting example of a similar treatment applied to tire rubber is related to attempts to use tire rubber as a carbonaceous adsorbent. In tire rubber, carbon black is one of the most abundant components. Due to the action of the acid, a tailored tire rubber surface, in terms of pore size distribution in the mesopore and macropore range, is achieved. Possible applications for these adsorbents in the chemical industry are wide, for example, in the catalysis processes. The desired pore size tends to be big, because these materials can accommodate a variety of host molecules, for example, substances of pharmacological interest, which are much larger than those that can absorb the traditional microporous zeolites.[24]

Two different studies on these subjects are worth mentioning. The first was performed by Manchón *et al.*[25] Tire rubber was subjected to various heat (pyrolysis) and chemical treatments (H_2SO_4, HNO_3 and mixtures of H_2SO_4/HNO_3) and the resultant products were characterized texturally. They found that tire rubber is a nonporous material in all porosity ranges; this fact influences the observed incompatibility of tires with the rest of the polymeric materials. They also concluded that the development of the porosity caused by the heat treatment occurred in the mesopore range while the chemical treatments produced changes in the macropore range. The effect of HNO_3 specifically was determined to create large pores in the material. Other scientists, such as Ariyadejwanich *et al.* also worked on preparing highly mesoporous activated carbons from waste tires.[26]

The use of waste tires as a cheap filler in NR mixes was studied by Yehia *et al.*[27] In order to achieve suitable adhesion, they applied oxidant surface treatments. The treatments were applied in solutions of 30% H_2O_2 and different concentrations of HNO_3 (10, 20, 30 and 60%). They observed that the tensile strength of the treated (20% of HNO_3 and 30% of H_2O_2) waste tires, was improved when compared to the blend containing non-treated samples. They attributed this improvement to the creation of new functional groups due to the chemical oxidation of the surface and their possible posterior reactions. Up to 30 phr of surface modified waste tires could be safely used as a filler, together with high abrasion carbon black in NR formulations. The obtained treated material presented reasonable physical mechanical properties at this level of loading.

A group of investigators led by López Cuesta carried out a study using HDPE and GTR.[21,28,29] They tested different oxidation treatments: a wet process with potassium permanganate ($KMnO_4$) and a dry process using γ-irradiation. They also added maleic anhydride grafted polyethylene (MAgPE) into the blend. The idea of using $KMnO_4$ as an oxidant had been used previously by Liu *et al.*[30] $KMnO_4$ was intended to break the unsaturated C=C bonds from ethylene propylene diene monomer (EPDM) rubber and to create hydroxyl groups. In the case of the work by Lopez Cuesta *et al.*, it was used to break the residual C=C bonds from the GTR. Afterwards, the modified GTR powder was incorporated into a PP matrix containing a small amount of grafted maleic anhydride. The authors found a significant improvement on elongation at break while tensile strength remained constant. As mentioned previously, they also submitted GTR to a radiation treatment, because of its proven oxidation effect in other polymeric materials. Numerous studies show that the radiation energy can induce a macromolecular chain scission[31] and some of the free radicals created, can recombine together to crosslink[32] or react with oxygen from the air creating polar groups.[33] Although the oxidation of the GTR was proven by a specific wettability test and FTIR spectroscopy, no significant improvement in properties was observed when compared with the HDPE/MAgPE/GTR blend without any previous oxidation treatment of the GTR particles.

As already stated, sulphuric and nitric acids have been used in the production of carbonaceous adsorbents from tires. The acids modify the rubber surface from tire wastes resulting in development of a great porosity. These results led to consideration of the possibility of using that modification in order to improve the mechanical adhesion of GTR/HDPE blends. According to the previous results, the chemical reaction produced by the acid on surfaces created an appropriate morphology that could improve the interlocking between the matrix and the particles. From another point of view, acid treatments seemed to be a financially worthwhile way to achieve a suitable material, because their application does not require any specific equipment and is technically simple.

In this section we report our results for the application of acids to the improvement of adhesion between HDPE and GTR. H_2SO_4 and HNO_3, are used to produce chemical etching on GTR. In addition, to elucidate the effect of particle size on tensile properties of non-treated and treated GTR, particles of different size categories (<200, 200–400 and 400–600 µm) were included in the test specimens and studied. The effect of surface modifications on the GTR particles was monitored using FTIR spectroscopy and SEM.

2.3.1 Results and Discussion

2.3.1.1 Tensile Properties

Tables 2.2–2.4 shows the results of the tensile strength, Young modulus, elongation at break and toughness determined by the tensile test. These properties were analyzed as a function of the GTR particles content, acidic treatment and particle size.

Tensile strength was found to decrease when increasing the tire content. Acidic treatments were found to increase the tensile strength in most cases when particle size was below 500 µm. The effect of the acidic treatment is related to the etching of the GTR surface, the elimination of moieties and additives and the achievement of a microporous surface is more suitable for mechanical adhesion.[34] The acid treated samples show similar behavior, obtaining better results when using a 50% sulphuric-nitric solution than for the nitric or sulphuric acid treatments, for the lowest particle size and highest GTR content.

As expected, particle size plays an important role in the tensile strength.[35] A minor decrease of the tensile strength at high content was observed for the small particle size. The effect of particle size is clearly observed when comparing the three categories. Bigger particles provide a higher probability

Table 2.2 Effect of particle size and acidic treatment on GTR surface (Particle size <200 µm).

GTR (%)	Surface treatment	Young's modulus (MPa)	Tensile strength (MPa)	Elongation at break (%)	Toughness (J)
0%	Not treated	1362.28	24.06	513.34	59.34
0%	Sulphuric 50%	1362.28	24.06	513.34	59.34
0%	Nitric 50%	1362.28	24.06	513.34	59.34
0%	Sulphuric/nitric	1362.28	24.06	513.34	59.34
10%	Not treated	1186.80	21.43	35.55	0.53
10%	Sulphuric 100%	1423.57	19.34	17.49	1.75
10%	Nitric 100%	1403.10	20.78	36.69	0.82
10%	Sulphuric/nitric	1501.07	22.09	19.58	0.24
20%	Not treated	879.22	17.67	23.00	2.68
20%	Sulphuric 100%	1346.10	18.76	13.28	2.98
20%	Nitric 100%	1364.29	20.14	17.04	2.78
20%	Sulphuric/nitric	1401.18	20.34	9.23	0.89
40%	Not treated	803.14	14.43	14.09	1.13
40%	Sulphuric 100%	985.24	16.83	13.56	1.25
40%	Nitric 100%	1028.35	19.23	12.98	1.59
40%	Sulphuric/nitric	1409.15	21.56	13.08	1.54

Table 2.3 Effect of particle size and acidic treatment on GTR surface (Particle size 200–500 μm).

GTR (%)	Surface treatment	Young modulus (MPa)	Tensile strength (MPa)	Elongation at break (%)	Toughness (J)
0%	Not treated	1362.28	24.06	513.34	59.34
0%	Sulphuric 50%	1362.28	24.06	513.34	59.34
0%	Nitric 50%	1362.28	24.06	513.34	59.34
0%	Sulphuric/nitric	1362.28	24.06	513.34	59.34
10%	Not treated	1156.89	19.43	12.55	1.53
10%	Sulphuric 100%	1213.12	22.89	15.49	1.86
10%	Nitric 100%	1203.19	21.77	12.19	1.48
10%	Sulphuric/nitric	1261.23	20.02	11.38	1.62
20%	Not treated	939.12	15.07	9.00	0.87
20%	Sulphuric 100%	996.10	18.01	12.28	1.48
20%	Nitric 100%	974.89	15.36	9.04	0.98
20%	Sulphuric/nitric	997.38	15.89	7.23	0.71
40%	Not treated	540.19	10.76	14.26	0.93
40%	Sulphuric 100%	745.20	11.73	13.01	1.05
40%	Nitric 100%	658.39	11.23	9.98	0.66
40%	Sulphuric/nitric	749.55	12.56	7.48	0.51

Table 2.4 Effect of particle size and acidic treatment on GTR surface (particle size > 500 μm).

GTR (%)	Surface treatment	Young's modulus (MPa)	Tensile strength (MPa)	Elongation at break (%)	Toughness (J)
0%	Not treated	1362.28	24.06	513.34	59.34
0%	Sulphuric 50%	1362.28	24.06	513.34	59.34
0%	Nitric 50%	1362.28	24.06	513.34	59.34
0%	Sulphuric/nitric	1362.28	24.06	513.34	59.34
10%	Not treated	866.89	14.43	10.10	0.97
10%	Sulphuric 100%	757.42	13.89	8.23	1.15
10%	Nitric 100%	767.19	15.56	11.32	1.24
10%	Sulphuric/nitric	878.22	13.12	8.57	0.72
20%	Not treated	666.12	12.35	10.92	0.80
20%	Sulphuric 100%	590.17	11.45	9.85	0.78
20%	Nitric 100%	655.39	12.66	8.93	0.65
20%	Sulphuric/nitric	677.31	11.23	7.23	0.48
40%	Not treated	446.19	9.76	11.26	0.68
40%	Sulphuric 100%	440.89	9.03	10.91	0.57
40%	Nitric 100%	398.59	9.43	8.45	0.47
40%	Sulphuric/nitric	455.95	7.86	7.87	0.42

of failure cracks, whereas smaller particles may develop small microcracks with dimensions below the critical length. Quite obvious is the case of particles above 500 μm, showing a significant drop in the mechanical properties values for all studied cases. Also, the possibility of achieving a bigger particle size by conglomeration of particles during the process of blending should be considered. The agglomeration of particles would decrease the mechanical performance. However, at an industrial scale, the classification of particles by size may not be economically viable, as the increase of the properties is not worth the loss of tire material that this implies.

In addition, the effect of acids on tensile properties seems to be related to the particle size. It was observed that acid treatments improve the sample tensile strength, mainly when the smaller particle size type was treated, whereas in particles with a size higher than 500 μm, the effect of the acid treatment does not significantly improve the studied property. For instance, the tensile strength was found to decrease by about 25% (untreated) for compositions of 20% GTR when the particle size is below 200 μm, meanwhile the same particle size treated with sulfo-nitric acid (20% of the GTR) showed a decrease of only 13%.

A significant decrease of tensile strength was observed for the samples with a GTR amount higher than 10% w/w with a particle size higher than 200 μm. In the case of a particle size between 200 and 500 μm, for a GTR content of 5%, a 25% decrease in tensile strength was observed. Taking into consideration acid treatments as a surface etching of the GTR particles, the acid treatment effect is higher when increasing the surface area. In the case of small particles, the increase of surface area allows a more extensive etching of the acid. At the same time, as the etching produced should be, in terms of depth, equivalent in all cases, the roughness relative to the particle size should be more important, improving the GTR/HDPE mechanical adhesion.

As shown Table 2.2, samples with a low GTR content and particle size below 200 μm, resulted in an increase (from 1 to 5%) of the stiffness (Young's modulus) compared with neat HDPE. The data shows that the sulphuric-nitric treatment slightly reduces the stiffness *versus* neat HDPE at high GTR contents (20 and 40%). In contrast to other materials based on the addition of elastomers, the addition of a high concentration of GTR acts by decreasing the deformation ability of the ductile and plastic HDPE matrix. Even without treatment, it was observed that the Young's modulus of the samples was higher than the neat HDPE (without GTR) for samples with a low GTR content (5%) and particle size comprised between 200 and 500 μm. When the content of the GTR particles increases, the poor interfacial adhesion causes a decrease of the stiffness for similar reasons to the exposed samples when considering the tensile strength. Also, the decrease of the Young's modulus value is more drastic when using bigger particle sizes.

The effect of the acid treatment on the Young's modulus is clearly visible in the results observed in Tables 2.2–2.4. The etching of the acid acts by promoting mechanical adhesion, but also rigidizes the rubber by extracting the substances that may act as a plasticizer. The acid treated GTR particles have a rigid nature as has also been observed in the studies by other authors.[23,25] The GTR obtained with previous acid etching are usually stiffer than those that have not been treated. The sulfo-nitric mixture seems to provide the best results, but in general all acids have an improving effect. As exposed previously in the discussion of the tensile strength of the samples, the effect of acid treatment on the Young's modulus was higher at smaller particle sizes.

The addition of GTR in all prepared samples produces a drastic fall in the toughness and elongation at break (Tables 2.2–2.4). Toughness and elongation at break for GTR samples show lower values than the neat HDPE matrix, with considerations regarding the particle size and type of treatment being irrelevant in this case. The decrease of elongation at break is related to the imperfect interfacial adhesion between components. The incidence of the poor adhesion between phases on this property is especially important. On the other hand, the compatibilizing effect achieved by the acid treatments is counteracted by the increase of the stiffness of the rubber. As discussed previously, the extraction of additives, oligomers, or plasticizers of the GTR particles by the acid produces a rigid material. The reduction of the ability of deformation of the rubber influences the decrease of elongation and subsequently the decrement of the toughness.

2.3.1.2 *Scanning Electron Microscopy*

Images of the fracture surface of the GTR samples are shown in Figure 2.1. The pictures are taken at different levels of magnification, which allows the comparison of the particle sizes. The Figure 2.1a depicts a sample containing sulphuric treated GTR particles sizes between 200–500 µm. In the centre appears a big particle showing some cracks and pores big enough to be observed at this level of magnification. The particle is unlinked to the matrix, presenting deep voids around its contour. The tire particle seems to be resting on the HDPE, without being properly attached to it. On the other hand, the matrix has been strained and deformed plastically.

Figure 2.1b is taken from a sample including particles with sizes between 200–500 µm treated with HNO_3 and shows similar features. The interaction between both components of the blends is not good and there are many cavities around the GTR particle. Figure 2.1c, with a particle size lower than 200 µm, shows a different behavior, as the magnification is 10 times

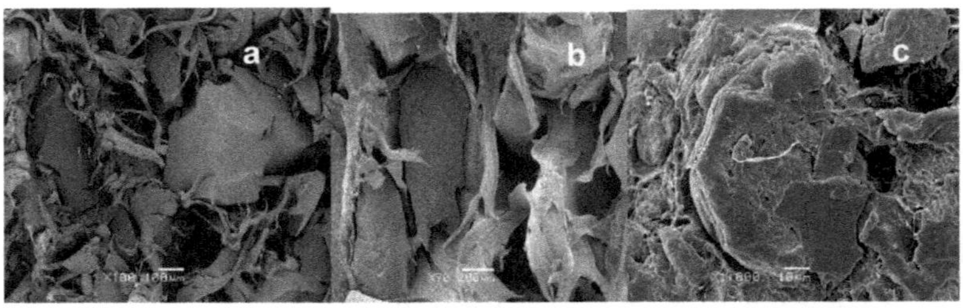

Figure 2.1 SEM images of different fracture surfaces of GTR/HDPE. (a) Particle size between 200–500 µm treated with H_2SO_4; (b) particle size between 200–500 µm treated with HNO_3 and; (c) particle size lower than 200 µm treated with H_2SO_4.

higher, the particle appears much smaller. The particle seems much more integrated into the matrix, and there is an area on the right with a clean cut that indicates that the particle itself has been broken instead of detached, which proves the good performance obtained by the interfacial contact, when the particles have been treated with H_2SO_4. The contour of the particle does not show voids, instead some fragments of the HDPE sprout from the GTR showing points of good attachment between both components. The high magnification allows the appreciation of the roughness achieved by the pre-treatment.

2.3.1.3 Spectroscopy Study of Surface Treatment (FTIR-ATR)

Chemical changes produced by acid treatments on the GTR particles were studied by means of FTIR-ATR spectroscopy. Figure 2.2 shows the FTIR-ATR spectral area of 450–1800 cm^{-1} for the samples treated with H_2SO_4 and HNO_3 and compares these values to those for the untreated GTR.

To obtain this spectrum, the band at 1026 cm^{-1} assigned to carbon black[36] was chosen as a reference. The spectral analysis of untreated GTR shows a weak band at 1739 cm^{-1} that is associated to the thermal oxidation that occurs as a result of the exposure of the surface to oxygen (especially during

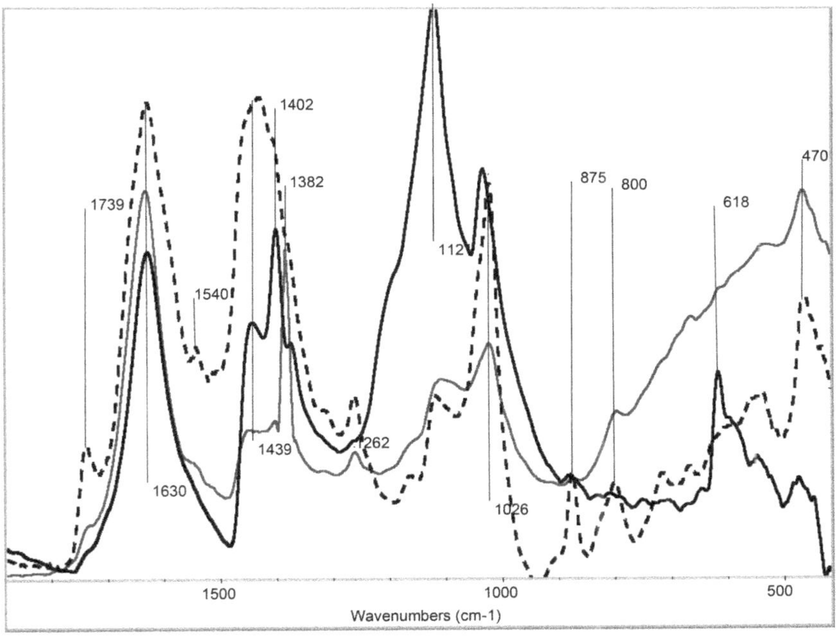

Figure 2.2 Spectral area of GTR particles treated with HNO_3, untreated and treated with H_2SO_4. Dash line spectra relates to not treated particles, dark line spectra relates to particles treated with sulfuric acid and gray line spectra relates to particles treated with nitric acid.

the grinding handling), and which induces the formation of an oxidation skin that includes carbonyl groups.

The strong band at 1630 cm^{-1} is associated to the C=C of polyisoprene, the weak band at 1540 cm^{-1} to zinc sterarate (an anti-adherent compound), the bands at 1439 cm^{-1} with the scissoring vibrations of $=CH_2$ (on butadiene), the band at 875 cm^{-1} with the trans isopropenyl unit ($-C(CH_3)=CH-$) and the band at 470 cm^{-1} with S–S.[37,38]

The analysis shows that treating the tire rubber with sulphuric acid produces several chemical and degradative modifications on the tire surface, mainly the formation of sulfonic acid, a decrease in double bonds (1630 cm^{-1}) due to the degradation process of polybutadiene and other unsaturated components of the tire, and a decrease of minor components. These phenomena may be related to the disappearance of the weak bands at 1739 and 1540 cm^{-1}, the decreasing of bands at 1439 and 875 cm^{-1} and the appearance of new bands at 1402, 1121 and 618 cm^{-1}, which is indicative of the O=S=O stretching of the absorptions of sulfonic acid. These results are in agreement with previous studies.[34,38]

The treatment of GTR with nitric acid shows similar results to the sulphuric acid treatment, except for the sulfur-oxygen derivatives. As shown in Figure 2.2, the spectra of samples exposed to nitric acid show a decrease of the same bands as those exposed to sulphuric acid (1739, 1630, 1540, 1439 and 875 cm^{-1}), plus an increase of the band at 1382 cm^{-1} assigned to N–N=O.[37] The bands assigned to O=S=O do not appear in this case.

The observed chemical modifications agree with those presented in previous studies, sulphuric acid acts as a strong dehydrating substance that can take up hydrogen and oxygen from organic matter and cause carbonization.[39,40] HNO_3 is also a powerful oxidizing agent when used in concentrated solutions. They modify the surface of the material introducing sulphur and nitrogen surface groups as $-SO_3$ and $-NO_2$. There is also an increase of groups such as O=S=O and $C-SO_2$-OR, a decrease in the C–H and an increase of C=C.

2.3.2 Treatment of Gtr Using Oxidation Acids

The main conclusions of this section can be summarized as follows.

Acidic treatments act by improving the compatibility of the components of the blends resulting in better tensile properties. Acid treatments act over the GTR particles causing etching of the surface and improving mechanical adhesion. For small particle sizes, treatment with a sulfo-nitric mixture gives the best results in terms of tensile properties, but H_2SO_4 alone also has a very positive effect. Smaller particle size produces samples with higher tensile strength and stiffness and results obtained using particle sizes below 200 μm, and between 200 and 500 μm are both acceptable in terms of the balance of properties/content of GTR, whereas particles with a size bigger than 500 μm produce a drastic reduction of tensile properties. The effect of the acidic pre-treatment is more intense when using smaller particle sizes.

The results of the analysis of the FTIR-ATR spectral bands were: (i) every treatment studied produces a chemical modification on the GTR surface, inducing the formation of specific groups; (ii) several degradative effects, mainly the decrease in double bonds and the elimination of minor components and moieties, do not depend on the acid used in the treatment.

The SEM images show differences related to the particle size, in which small particles are more attached to the matrix because of the high relative roughness and the relative magnitude of pores and cracks. The etched surface achieved by acids may not be enough to effectively link bigger particles to the matrix.

The results achieved for the proposed samples allow them to be considered as a real alternative to some value added injection molding products. It is worth mentioning that using acid treatment also makes the sample stiffer. This feature could be interesting when designing a determinate component.

Last but not least, there is a question that lingers around this type of treatment. An acid treatment is not an expensive procedure and no special machinery is needed but, is an acid treatment procedure environmentally friendly? These procedures should be very well conducted in order to constitute a suitable option.

2.4 Coupling Agent and Chlorination Treatment on Rubber Waste Particle Surface

Properties and performance of materials are strongly influenced by the nature of their components and their compatibility at an interfacial level. As described in the previous sections, when interfacial adhesion is not enough, some treatments are applied in order to achieve better results. In section 2.3 oxidant acid treatments producing a porous surface were analyzed.[28] These treatments, designed initially to develop other characteristics of the tires surface, may be useful to improve the mechanical interlock with the matrix because of the high porosity obtained. Other surface treatments, such as chlorination applied to SBR rubber, have also been reported as a way to increase the adhesion. In this case, the treatment favors the migration of wax contained in the rubber to the surface and subsequent elimination.[41] The paraffin wax is also responsible for the lack of adhesion of the rubber. Other approaches to the compatibilization in similar materials include grafting of polymers and use of coupling agents.[43] In the case of HDPE-RW samples, a treatment based on the modification of the rubber with acrylamide has been reported.[44] The use of silanes could also improve the matrix–reinforcement interface properties by the development of chemical interactions.

The aim of this section is to study samples of HDPE/GTR that have been surface modified with acid and silane in order to improve the compatibility between both components. The comparative behavior of the GRT surface treated with Trichloroisocyanuric acid (TCI) and using γ-methacriloxypropyltrimethoxysilane (silane A-174) as a coupling agent has been analyzed. The HDPE/GTR samples were monitored by determining its tensile properties. Chemical and physical modifications related to the material behavior were studied by means of FTIR spectroscopy and SEM.

2.4.1 Surface Treatment of GTR by TCI and Silane A-174

Chlorination treatment was performed by: (a) wiping of the rubber with a tissue paper immersed in ethyl acetate, (b) 15 min later the wiped rubber was treated with TCI ethyl acetate solution (2.5 wt%) and (c) 1 h later, an ethanol/water solution (25 wt%) was applied to the chlorinated tire surface.

Treatment using silane as a coupling agent was carried out as follows: every 5 g of oven-dried reused tire and the corresponding proportion of Silane A-174 were added to 50 ml of diethylic ether. The resulting mixture was heated for 1 h at 60 °C until total evaporation of the solvent was achieved.

2.4.2 Results and Discussion

2.4.2.1 Tensile Properties

Tensile analysis of HDPE/GTR modified with TCI and silane (Table 2.5) shows that the addition of GTR to the HDPE in small quantities (10%) improves the stiffness and provides a more brittle behavior, the samples with GTR modified with 2% of silane obtained better results than neat HDPE at concentrations lower than 40%, which is attributed to the improvement of the interactions between both components. The use of a coupling agent gives better results of tensile strength than any other treatment at compositions below 40%. On the other hand, stiffness is not as high as that observed in acidic treatments

Table 2.5 Effect of TCI, silane and untreated GTR surface samples on the mechanical properties of GTR/HDPE samples.

GTR (%)	Surface treatment	Young's modulus (MPa)	Tensile strength (MPa)	Elongation at break (%)	Toughness (J)
0%	Not treated	163.27	7.86	680	22.1
0%	TCI	163.77	8.86	680	22.1
0%	Silane 2%	163.27	7.86	680	22.1
10%	Not treated	172.25	7.44	37.62	1.60
10%	TCI	186.60	4.3	11.55	0.53
10%	Silane 2%	232.47	11.08	34.19	1.75
20%	Not treated	103.10	6.55	33.69	0.82
20%	TCI	101.07	2.59	10.51	0.24
20%	Silane 2%	109.2	9.7	28.00	1.18
40%	Not treated	56.10	3.76	33.92	0.90
40%	TCI	64.39	2.3	12.97	0.22
40%	Silane 2%	49.25	4.29	31.81	0.94

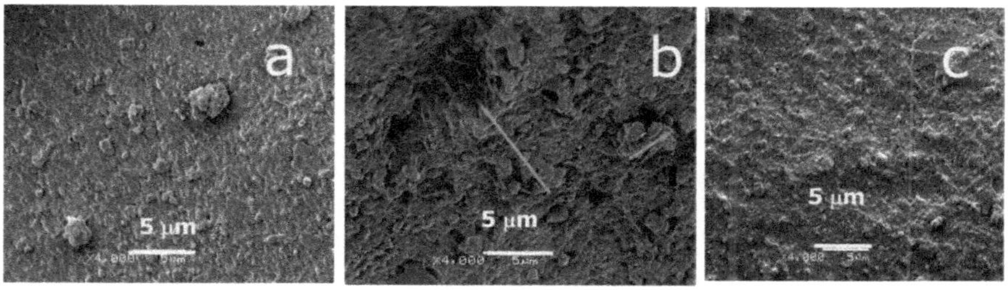

Figure 2.3 SEM pictures of GTR treated with (a) TCI, (b) silane and (c) untreated.

(*i.e.* sulfuric or nitric acid treated rubber) because the silane does not affect the rigidity of the particles of rubber. This is also related to the better results obtained when measuring toughness and elongation with silane compared with any other treatment. In contrast TCI only provides better results than neat HDPE in terms of the Young's modulus values.

2.4.2.2 Scanning Electron Microscopy

As observed in Figure 2.3, the particle surface treatment with TCI does not seem to affect the rubber surface. In contrast, the surfaces particles modified with silane show better interfacial adhesion between the rubber particles and the matrix due to a physical modification of the rubber particles. The observed enhancement of the tensile properties when using a coupling agent is due to the chemical interactions that also produce an increase in compatibility. The mechanical results obtained are justified in the light of these considerations.

2.4.2.3 Spectroscopy Study of Surface Treatment (FTIR-ATR)

The FTIR spectral analysis of GTR treated with silane shows several differences and changes in the bands that suggest interaction between silane and GTR. The dipoles related to the methyl and methylene groups of the polymer and the double bonds (1383–1457 and 875 cm^{-1}) are affected by the presence of silane. In Figure 2.4 the spectra of GTR treated with silane shows an increase in the silane related bands, at 1260, 1095 and 815 cm^{-1} that imply participation of this group in the interactions with GTR. A good interaction of GTR with silane also indicates an effective compatibilization. The same coupling agent used here has already been studied regarding HDPE-silane interactions with positive results.[45] Figure 2.4 also shows the spectra of GTR treated with TCI. The chlorination treatment causes chemical modification. Chlorination of the hydrocarbon chains (increase of the 534 cm^{-1} band and a weak increase of the 1420 cm^{-1} band) is observed and also formation of oxidized species such as C=O and O=S=O (increase of 1720 and 1400 cm^{-1} bands). These results agree with other previous works.[42] The increase of

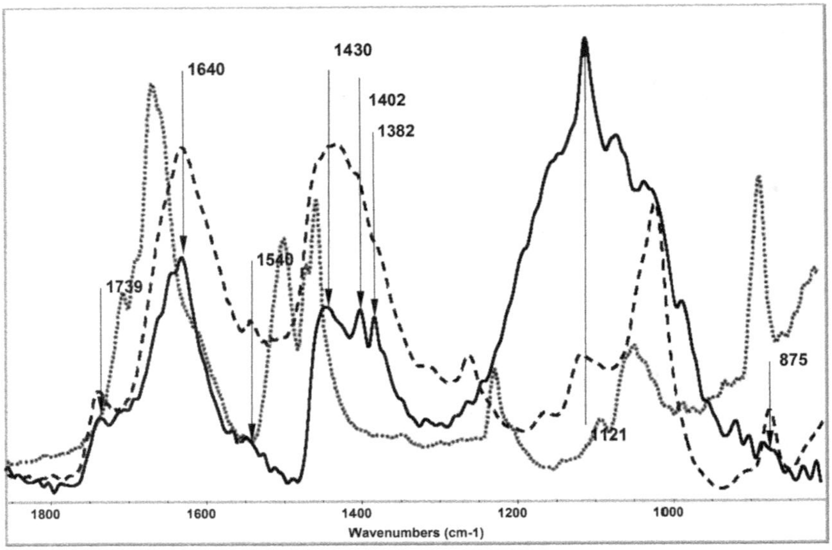

Figure 2.4 Spectral area of GTR particles treated with silane, not modified and treated with TCI. Dashed line spectra relates to not treated particles, dark line spectra relates to particles treated with silane and dotted line spectra relates to particles treated with TCI.

functional groups on the surface of the polymer increases the surface energy and have a negative influence on the adhesion with HDPE. The effect of the removal of paraffins or other weak layers present in rubber is not enough to achieve a better interlocking of the GTR with the HDPE matrix.

2.4.3 Treatment of GTR Using TCI and Silane

The main conclusions of the study can be summarized as follows, the addition of micronized reused tire rubber to the HDPE in small quantities (5%) improves the stiffness and provides a more brittle behavior. Treatment with silane improves the ability to develop interactions between both components obtaining better results than neat HDPE at concentrations lower than 30%. In terms of the tensile strength, the use of a coupling agent gives higher values than any other treatment at compositions below 30%. On the other hand, stiffness is not so high as that observed for TCI rubber treatment because silane does not affect the stiffness of the particles of rubber. This is related to the better results obtained when compared to any other treatment provided by silane when measuring toughness and elongation. TCI only provides better results than neat HDPE for the Young's modulus values.

SEM micrographs show differences depending on the treatment followed. Treatments with TCI do not seem to affect the rubber surface. The fracture surfaces indicate interfacial adhesion of the rubber particles to the matrix mainly due to a physical modification of the surface of the particles. The increase of the tensile properties observed when using a coupling agent is due to the chemical interactions and an increase in compatibility. The particles appear to be coupled to the matrix. Instead, TCI treatment leads to particles not properly linked to the HDPE. The results obtained for mechanical properties are justified in the light of these considerations.

By studying the FTIR spectral bands, the modifications followed different ways according to the treatment. TCI acid seems to produce degradative effects, while silane acts in a different way, developing interactions with the rubber in which methyl, methylene and silane are involved, interpenetrating with the matrix and improving the ability to establish a link between the material components.

2.5 Effect of Surface Modification of Rubber Waste Grafted with EPDM

In this section we study the modification of GTR by grafting. The process is based on EPDM dynamically crosslinked with peroxides. The choice of peroxides is important in the final sample properties. Trigonox 311 (T-311), and Trioxepane, two peroxides developed by Akzo Nobel Polymer Chemicals have been selected.

A common issue when preparing samples with these compositions is the difficulty to include a high content of rubber. For recycling purposes, a high content of rubber is desirable and threfore optimizing the blending process is necessary. Shen *et al.* used a co-rotating twin screw extruder, working at specific settings in order to produce PP/EPDM with a high content of elastomer.[46] Using this method they achieved samples with a higher tensile strength, elongation at break and better processability.

The mechanism of crosslinking polymers by a peroxide can be divided into three steps as shown in Figure 2.5.[47] The first step, which determines the rate of the overall reaction, is the homolytic decomposition of the peroxide and the generation of free radicals. The second step consists of a hydrogen abstraction by the alkoxy radical derived from the peroxide and the third and last step is the combination of two polymers to form the C–C crosslinks. However, additional reactions may be important and must been

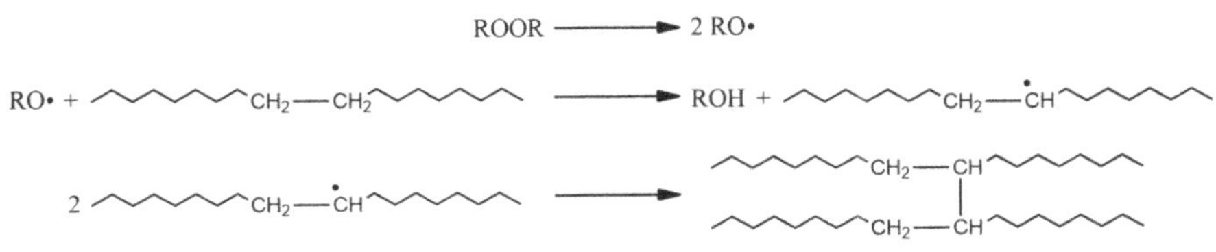

Figure 2.5 Mechanism of crosslinking polymers using peroxide.

taken into account. Yamazaki and Seguchi[48] reported studies on the chemical crosslinking mechanism of polyethylene using peroxides at elevated temperatures and have identified alkyl radicals and allyl radicals, the latter apparently being formed *via* disproportionation of two polymer radicals. The utilization of peroxides can also induce the degradation of the thermoplastic, because peroxides are not selective for unsaturated elastomers. Therefore, undesirable side reactions such as disproportionation or β-chain scission can also take place during the crosslinking process as shown in Figure 2.6.

The chemical structures and commercial names of the peroxides used in present study are:

- Dicumyl Peroxide (DCP) 98% of purity from Sigma-Aldrich (Figure 2.7).
- Trigonox® 311 3,3,5,7,7-pentamethyl-1,2,4-95% pure active oxygen content of 9.18% Trioxepane from Akzo Nobel Polymer Chemicals (T-311) (Figure 2.8).

The dialkyl peroxides, such as DCP, are the most widely used, as they are cost-efficient for most crosslinking applications. Until recently, crosslinking of polyethylene always took place using these conventional crosslinking

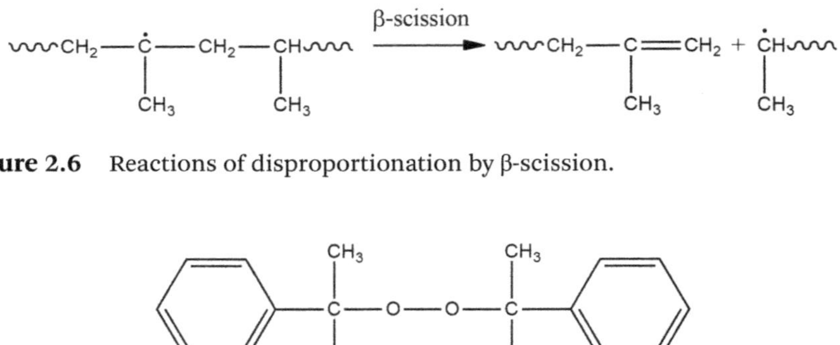

Figure 2.6 Reactions of disproportionation by β-scission.

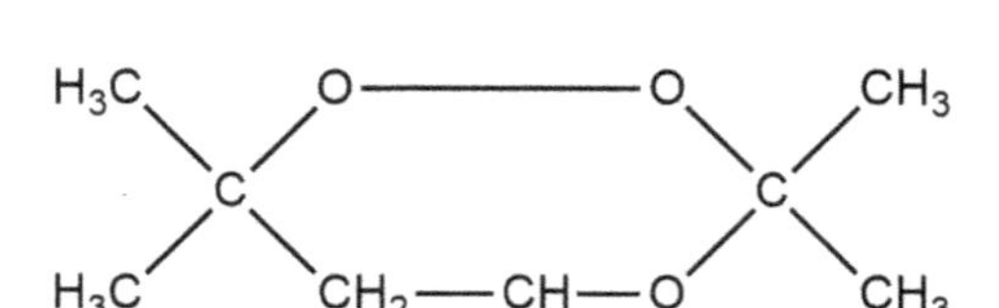

Figure 2.7 Dicumyl Peroxide (DCP) 98% purity from Sigma-Aldrich.

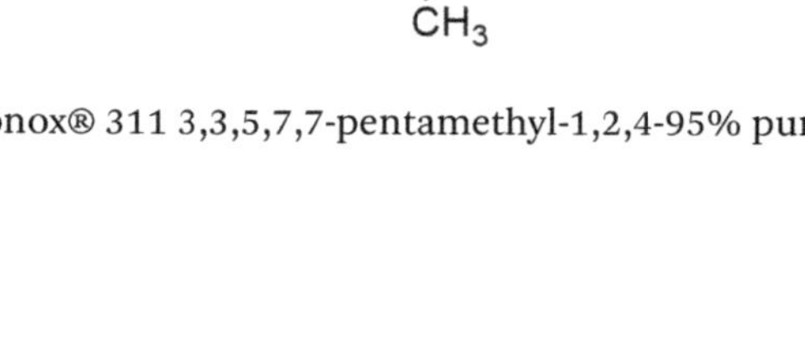

Figure 2.8 Trigonox® 311 3,3,5,7,7-pentamethyl-1,2,4-95% pure (T-311).

peroxides. To prevent premature decomposition of these peroxides in the extruder, processing temperatures were limited to 160–170 °C.

Instead, T-311 has a decomposition temperature significantly higher compared to any other commercially available crosslinking peroxide, resulting in a safe processing temperature of 20–30 °C above those of the conventional organic peroxides. The use of this peroxide at higher temperatures results in a less viscous, more homogeneous melt which not only improves the extruder output, but also has a positive effect on the final properties of the cured sample.

The decomposition rate of peroxides depends on the bond strength of the weakest bond, this is the O–O bond. The chemical nature of the hydrocarbon group attached and the polarity of the substrate in which the peroxide is dispersed will affect the O–O bond strength. The chemical nature of the hydrocarbon group will influence the structural effects such as the steric, inductive and resonance stabilization of the free radicals generated by the peroxide decomposition.

The chemical nature of the peroxides resulting from free radicals is very important for determining the final product properties. The most useful peroxides are those in which the primary free radicals are sufficiently reactive to abstract hydrogen atoms from the polymer chain to form the corresponding macro-radicals. The relative reactivity or stability of the free radicals generated is again related to the hydrogen bond dissociation energy of the parent compound.

The decomposition products of DCP generate cumyloxy radicals, which further undergo β-cleavage to form acetophenone and methyl radicals. Both the cumyloxy and methyl radicals have enough energy to abstract hydrogen atoms from most sites available in a polymer. Moreover, the methyl radical is not as sterically hindered as the initial cumyloxy radical and may be more efficiently involved in reactions where steric hindrance is a limitation.[49] DCP is known as being a very efficient peroxide due to the reactivity and amount of generated radicals. The decomposition products of DCP are methane, acetophenone, 2-phenylpropanol-2, α-methylstyrene and water.

The major decomposition products for T-311 are methane, acetone, isopropyl acetate, 3-hydroxy-1,3-dimethylbuthyl acetate and 3-oxy-1-methylbuthyl acetate. The decomposition reaction can be seen in Figure 2.9.

2.5.1 Results and Discussion

Four different experimental sets were designed in order to perform an accurate study of the mechanical properties. The first set was carried out on composites 40/30/30 (HDPE/EPDM/GTR) with peroxide contents from 0.1–3% using T-311 (Table 2.6). The influence of the T-311 content on the mechanical properties was studied. In order to assess the previous results, a second set of samples containing only DCP was tested. Different temperatures of curing, related to peroxide activity, were examined too (Table 2.7). Moreover, the influence of the HDPE/rubber composition of the composites was tested at a

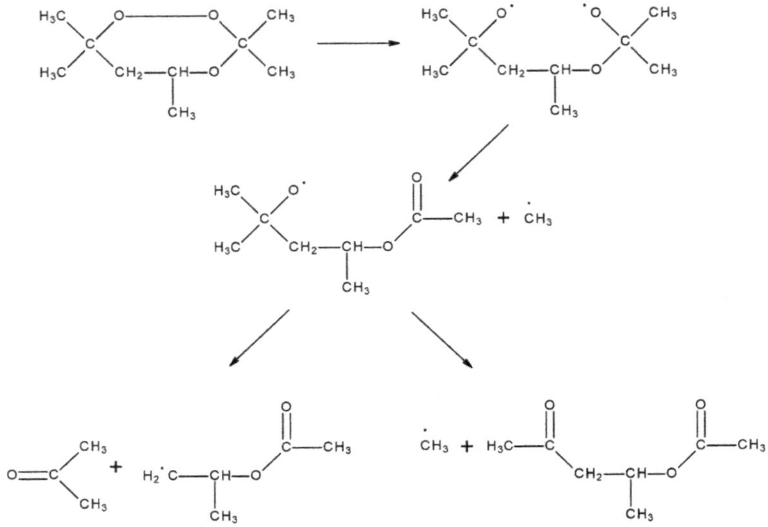

Figure 2.9 Decomposition products of Trigonox 311: methane, acetone, isopropyl acetate, 3-hydroxy-1,3-dimethylbuthyl acetate and 3-oxy-1-methylbuthyl acetate.[33]

Table 2.6 Mechanical properties of the blends containing 40% HDPE, 30% EPDM, 30% GTR and different dosages of Trigonox 311.

Trigonox 311 (%)	Young's modulus (MPa)	Std. dev (%)	Tensile strenght (MPa)	Std. dev (%)	Elongation at break (%)	Std. dev (%)
0	98.99	6.00	5.12	0.25	153.03	21.63
0.1	116.91	10.80	5.1	0.23	145.98	34.81
0.2	126.71	9.70	5.32	0.42	170.14	40.03
0.5	120.99	4.60	5.6	0.22	190.19	32.72
1	126.71	4.45	5.85	0.12	268.36	29.15
2	128.26	7.61	6.22	0.15	260.93	29.21
3	130.83	6.43	6.55	0.95	228.53	48.43

Table 2.7 Mechanical properties of the blends containing 40% HDPE, 30% EPDM, 30% GTR with 0.1% of DCP and the mixtures with 0.1% DCP + 0.5% Tri gonox 311, at 180, 190 and 200 °C.

Trigonox 311 (%)	DCP%	Temperature (°C)	Young's modulus (MPa)	Std. dev (%)	Tensile Strenght (MPa)	Std. Dev (%)	Elongation at break (%)	Std. dev (%)
0.5	0.1	180	90.34	8.49	6.74	0.42	281.41	28.91
0.5	0.1	190	97.51	5.32	7.06	0.67	297.33	30.62
0.5	0.1	200	95.98	10.61	5.79	0.91	186.24	28.74

Table 2.8 Mechanical properties of formulations showing variation of the rubber/plastic blends at fixed peroxide dose and a constant EPDM/GTR ratio of 1 : 1. Numbers in parenthesis corresponds to the standard deviation.

HDPE%	EPDM%	GTR (%)	Trigonox 311 (%)	DCP (%)	Young's modulus (MPa)	Tensile strenght (MPa)	Elongation at break (%)
10	45	45	0.75	0.15	9.78 (0.21)	2.45 (0.21)	360.00 (35.02)
20	40	40	0.67	0.13	23.75 (1.43)	4.00 (0.42)	337.00 (44.81)
40	30	30	0.5	0.1	97.51 (5.32)	7.06 (0.67)	297.33 (30.62)
60	20	20	0.33	0.07	263.15 (18.81)	8.25 (0.22)	107.70 (24.21)
80	10	10	0.17	0.03	637.47 (14.61)	14.59 (0.51)	42.26 (10.83)
100	0	0	0	0	927.90 (27.91)	17.17 (0.52)	39.08 (5.01)

Table 2.9 Mechanical properties of the 40 : 60 plastic : rubber blends at different rubber (EPDM : GTR) ratios. Numbers in parenthesis corresponds to the standard deviation.

HDPE (%)	EPDM (%)	GTR (%)	Trigonox 311 (%)	DCP (%)	Young's modulus (MPa)	Tensile strength (MPa)	Elongation at break (%)
40	60	0	0.5	0.1	41.04 (2.32)	14.22 (0.53)	751.21 (18.84)
40	50	10	0.5	0.1	65.59 (6.95)	10.92 (0.53)	592.65 (19.14)
40	40	20	0.5	0.1	89.19 (11.84)	7.26 (1.14)	417.68 (32.68)
40	20	40	0.5	0.1	132.93 (10.19)	6.14 (0.16)	121.38 (22.27)
40	0	60	0.5	0.1	172.42 (10.32)	6.33 (0.45)	25.78 (4.46)

constant EPDM/GTR ratio and peroxide content (Table 2.8). Finally, the GTR was substituted at different concentrations by EPDM in a fixed plastic/rubber and peroxide content composition (Table 2.9). These last sets of samples were tested in order to determine the amount of EPDM that can be replaced by GTR whilst maintaining a good balance of mechanical properties of the (HDPE/EPDM/GTR).

2.5.1.1 *Tensile Properties*

Table 2.6 shows the influence of the peroxide content on the mechanical properties of samples in which the HDPE/Rubber and EPDM/GTR ratios were kept constant at 40/60 and 50/50, respectively. Previous studies have shown that GTR surface characteristics and particle size provide poor adhesion in composites of HDPE/GTR.[50] When EPDM is incorporated into the composite, elongation at break and toughness increase immediately, however, the modulus and tensile strength decrease due to the incorporation of a rubber-like phase. As peroxide content increases, all properties show higher values due to crosslinking, but at higher contents of T-311 (2 and 3%) elongation at break starts decreasing. This could be related to the reaction of the peroxides with the thermoplastic matrix, because as mentioned above, they are not selective for unsaturated elastomers.[51] Similar results were reported by Radhesh Kumar *et al.* and Naskar *et al.* in previous studies.[52,53] In polymer composites, an increase in the elongation at break is directly related to a better compatibility between phases. Therefore, it can be concluded that samples containing 1% of T-311 show the maximum elastomeric thermoplastic properties. In other words, from this experiment, 1% of T-311 produces the best results to crosslink EPDM and seems to be effective in creating a network to encapsulate the GTR particles.

DCP was also tested in combination with T-311, in order to study the synergetic effects and compare the processing and physical properties of the composites. Although 1% of T-311 showed the maximum performance in the first study, the content was reduced to 0.5% when adding DCP as an additional peroxide, in order to avoid an excess of peroxide, this would also be beneficial for cost savings. Table 2.7 shows the mechanical properties of the DCP alone and in combination with T-311, at different processing temperatures. Sample 9, with 0.1% of DCP, was consolidated at 170 °C and did not show difficulties in handling. However, its properties were not as good as those of sample 6. Composites with higher DCP contents led to premature crosslinking in the two-roll mixer (working temperature 150–155 °C) and those pressed at higher temperatures became very sticky and difficult to remove from the press, being impossible to measure. Table 2.7 shows the synergy effect of both peroxides at 180, 190 and 200 °C for different samples. Consolidation temperature seems to have an important role in the final properties. Different conditions for samples containing 0.5% of T-311 and 0.1% of DCP were tested. The combination based on 0.5% of T-311 and 0.1% of DCP respectively seems to give the maximum elastomeric thermoplastic properties. Nicolini *et al.* also found that the blend of two curing agents, DCP/bismaleimide, in PP/EPDM 35/65 w/w gave the best mechanical properties and much better results than DCP alone.[54]

The HDPE/rubber composition also has an important influence in the final material properties. Table 2.8 shows the mechanical properties of composites in which the EPDM/GTR ratio was kept constant at 1 : 1, and the peroxide combination of 0.5% T-311 and 0.1% DCP was used. As the HDPE content increases from 10 to 40%, the Young's modulus, tensile strength and toughness gradually increase too. Elongation at break decreases due to the incorporation of a plastic phase. With even higher contents of HDPE (more than 50%), the Young's modulus and tensile strength increase, while elongation at break and toughness decrease. The higher the HDPE content the less of a toughening effect of the elastomeric phase. Therefore, the mixture with 40% HDPE and 60% rubber phase shows the highest desired properties. Similar results were found by Naskar *et al.*[53]

The last set of samples were used to study the mechanical properties when fresh rubber is replaced by GTR. Up to this point, the best properties have been obtained with the 40/60 composition of the HDPE/rubber phase. Therefore, the HDPE phase and peroxide content were maintained while the EPDM content was varied from 60–0%. Table 2.9 shows the results. When GTR content increased the loss in the mechanical properties due to the incompatibility between GTR and HDPE is clear. From 60–30% of EPDM, the composite has gradually less elongation at break, but when 40% of GTR is included, EPDM seems to lose its encapsulating and compatible effect leading to an important decrease of the elongation at break. It is interesting to note that the composite with 60% GTR is even more brittle than the one containing 10% EPDM. It seems that peroxides only vulcanize EPDM and do not react with the possible existing active sites of GTR.

After this study, an overall vision of composite behavior was acquired. In order to assess the substitution of EPDM for GTR, two samples were selected for comparison. A composite without GTR (40% HDPE + 60% EPDM + 0.5% T-311 + 0.1% DCP) and a composite that has 30% EPDM substituted by GTR (40% HDPE + 30% EPDM + 30% GTR + 0.5% T-311 + 0.1% DCP). The results show (Table 2.9) that even with the considerable amount of 30% GTR, the 297% value in elongation at break fulfils the major criteria for a thermoplastic elastomer, which is that this value must be higher than 100%.[55] However, if the GTR amount is increased up to 40% the elongation at break drastically drops down to 121% which is almost within the limit. Therefore, 30% of GTR in the composite is the optimum amount for the desired balance of a maximum GTR quantity with acceptable mechanical properties.

2.5.1.2 Scanning Electron Microscopy

Figure 2.10a shows the incompatibility of the sample with HDPE and GTR. These two materials, without affinity due to their different chemical compositions, appear as GTR particles that are almost isolated without being wet by the HDPE matrix. Without any kind of compatibility between the two materials, no adhesion is possible, leading to very poor mechanical properties, as has been described previously.

The graphitization of EPDM into the surface of GTR in the HDPE/GTR sample is shown in Figure 2.10a. The addition of an elastomeric component which, at the same time contains a high amount of polyethylene, should help to improve the compatibility between HDPE and GTR. Particles are not as isolated as in the samples without EPDM. However, this can be observed as some GTR particle parts are still not surrounded by either HDPE or EPDM. In other words, EPDM is still not able to encapsulate the GTR particles.

The encapsulation effect is shown in Figure 2.10b and c with the addition of peroxides. In both microphotographs the GTR particles appear to

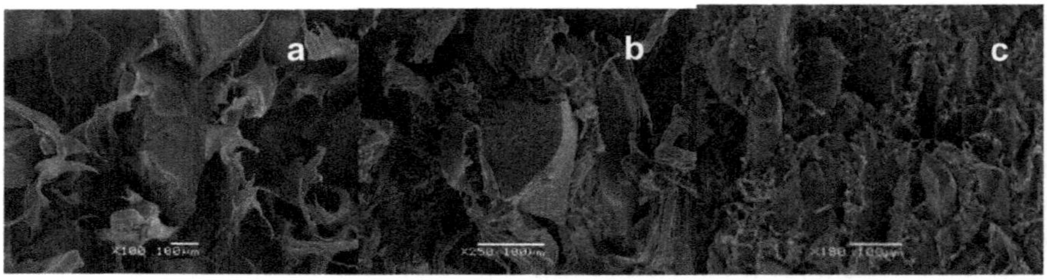

Figure 2.10 SEM images of the samples: (a) 40HDPE/30EPDM/30GTR, (b) 40HDPE/30EPDM/30GTR 3T-311 and (c) 40HDPE/30EPDM/30GTR 0.5T-311 0.1DCP.

be encapsulated by EPDM. In the case of Figure 2.10c, corresponding to the composite with 0.5% T-311 + 0.1% DCP, the encapsulating effect is much more visible. The fact that GTR particles are completely encapsulated means that the EPDM has formed a three-dimensional network within them, and at the same time, EPDM creates a co-continuous phase with HDPE. This effect leads to a much better adhesion and better mechanical properties, as already observed.

2.5.1.3 Atenuatted Total Reflectance-FTIR

ATR-FTIR spectroscopy was used to study the chemistry and reactivity involved between the peroxides and the blends after dynamic vulcanization. As stated above, the main peroxides reaction consists of the abstraction of the hydrogen atoms from the polymer chain and the formation of the corresponding macroradicals. The reactivity of the generated free radicals depends on the hydrogen bond dissociation energy. The T-311 major decomposition products are methane, acetone, isopropyl acetate, 3-hydroxy-1,3-dimethylbuthyl acetate and 3-oxy-1-methylbuthyl acetate, while the major DCP decomposition products are methane, acetophenone, 2-phenylpropanol-2, α-methylstyrene and water, as already commented in previous section.

In order to investigate the chemical changes to samples including T-311, the sample spectra with a constant 40/30/30 ratio of HDPE/EPDM/GTR and an increasing peroxide dose from 0–3% were examined. The studied bands have been referenced by Choudhury *et al.* analyzing similar composites based on milk pouches (LDPE-LLDPE), EPDM and DCP.[56]

Figure 2.11 shows the spectra area of different samples dynamically crosslinked with peroxides, in which the presence of different C=O stretching bands (1592, 1586 cm^{-1}) is associated to carbonylated products such as ketones, ketoesters, aldehydes or other products that can be generated by the decomposition of T-311. The C=O peak intensity is observed to be higher for the composites cured with a higher peroxide amount, indicating an oxidative degradation of the matrix polymer due to the higher peroxide concentration.

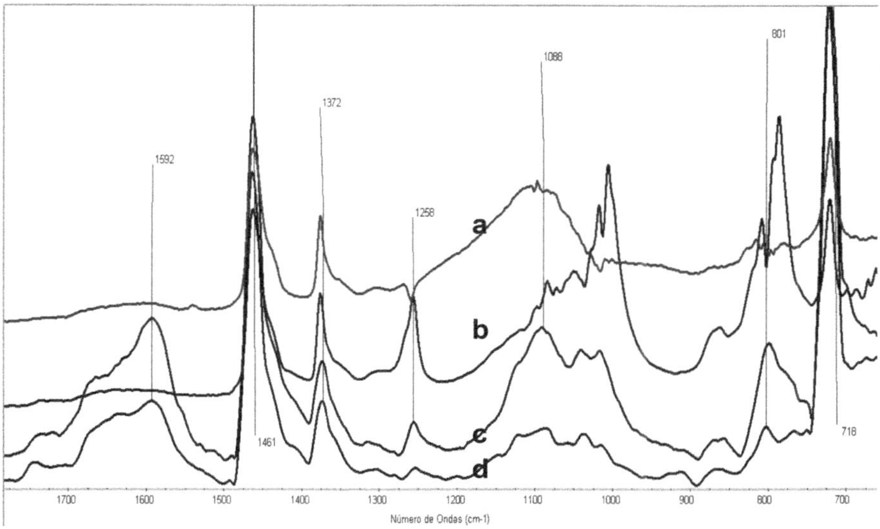

Figure 2.11 Spectra area of different samples dynamically crosslinked with peroxides, Trig1 and Trig3 (indicates 40HDPE/30EPDM/30GTR 1% T-311 and 3% T-311). (a) spectra relates to EPDM50/GTR50/Trigonox3; (b) spectra relates to EPDM100; (c) spectra relates to Trigonox3 and; (d) spectra relates to Trigonox1.

The spectral area between 1400 and 1000 cm^{-1} shows the bands at 1258, 1088 and 1025 cm^{-1} assigned to the asymmetric stretching vibrations of the C–O–C bonds of the esters and ethers linkages during the curing reaction between the peroxide free radicals and the polymeric macroradicals. The C–O–C band absorbance increases up to 1% of T-311, but it decreases at a higher peroxide dose (3%) due to the preferred oxidative degradation reaction of the peroxide with the matrix. The results are in accordance with the mechanical properties, in which the composite with 1% of T-311 showed more interesting mechanical behavior.

Analyzing the spectral area from 900–600 cm^{-1}, it can observed that the band at 797 cm^{-1} corresponds to the double bond –C=C– of the ENB unit of EPDM. Its significant intensity decrease is associated to the disappearance of the double bonds as a consequence of the crosslinking.

2.5.2 Surface Modification of GTR Grafted with EPDM

From this study, the conclusion was that the incorporation of EPDM, as fresh rubber, into the blend helped to encapsulate the GTR particles following a grafting process, creating a co-continuous phase and improving the adhesion between phases. This encapsulating effect should be improved as it was observed to decrease at higher GTR amounts. The composition of 40% HDPE, 30% EPDM, 30% GTR, 0.5% T-311 and 0.1% DCP gave the best balance regarding the high GTR amount (30%) and mechanical properties. Process conditions, such as press temperature, have an important role in the final mechanical properties. The mixture of two different peroxides types gave the best synergistic effect. A high peroxide dosification has a negative impact on the crosslinking due to a possible thermooxidative degradation of the thermoplastic matrix as observed by FTIR-ATR analysis.

The grafting phenomena of EPDM over GTR particles can be observed in SEM images. An optimization of the samples for maximum adhesion between GTR particles and HDPE matrix was studied during the mechanical properties test. Good adhesion is only possible if the GTR particles are encapsulated by fresh rubber, for example EPDM. Good adhesion between phases is directly translated into good mechanical properties in the final composite.

The rubber-like characteristics of the EPDM itself and the high ethylene content can create a good matrix compatibility, with similar surface energies, between the encapsulated GTR particles and the HDPE matrix. The aim is to substitute as much EPDM as possible by GTR without decreasing the properties of the final compound. A detailed analysis regarding the influence of each component in the final mechanical properties has been performed. The ATR-FTIR spectroscopy analysis indicated a high degree of crosslink and good adhesion between the matrix and the rubber phase. Morphology of the blends was assessed by SEM and the results were in concordance with the results.

2.6 Global Conclusions

After the study of the three different methods of surface modification of RW and the comparison of the mechanical properties, the maximum amount of RW particles (*i.e.* GTR) and easiness of production of the final compound blended with thermoplastic (HDPE), the following final conclusions can be extracted:

(i) The surface treatment of WR was studied with GTR of particle size between 400–600 μm. This particle size corresponds to the standard delivery of GTR particles from industrial processes. This is an important influence on the particle size, but a separation would probably mean an economical impediment.

(ii) The effectiveness of the surface treatment method in order to improve the adhesion degree with HDPE can be checked by the study of the mechanical properties of the compounds. Some interesting conclusions are: (a) Acid treatment causes etching on the surface of RW particles, improving the mechanical adhesion. A brittle performance of the compound can be observed at lower GTR particle size (<200 μm). Treatment with H_2SO_4 improves the rubber's ability to interact with HDPE, thus providing the materials with a greater stiffness, which is also a consequence of the rigidity of the rubber after the acidic treatment because of the extraction of additives and degradation. (b) The use of a coupling agent gives higher values in most mechanical properties, specifically toughness and elongation, compared to any other treatment at compositions below 30%. In any case, the stiffness of silane treated samples is not so high as those observed for TCI and acid rubber treatment because the silane does not affect the rigidity of the particles of rubber. TCI only provides better results than neat HDPE for the Young's modulus values. (c) Modification of the GTR by the grafting process based on EPDM dynamically crosslinked with a mixture of peroxides produces good tensile properties. The compound containing 40% HDPE, 30% EPDM, 30% GTR, 0.5% T-311 and 0.1% DCP gave the best mechanical properties.

(iii) SEM micrographs showed differences depending on the treatment followed. Treatment with H2SO4 produces a rough surface that is more suitable for mechanical adhesion, whilst TCI does not seem to affect the rubber surface. The fracture surfaces indicate that there is an interfacial adhesion of the rubber particles to the matrix mainly due to a physical modification of the particles. The increase of the tensile properties observed when using a coupling agent is due to chemical interactions that also produce an increase in compatibility. The particles treated with silane appear to be quite coupled to the matrix. TCI treatment leads to particles not being properly linked to the HDPE. The graphitization of EPDM into the surface of GTR in the HDPE/GTR sample improves the compatibility between the HDPE and GTR due to the effect of encapsulation of EPDM around the RW particles. The results obtained for the mechanical properties are justified in the light of these considerations.

(iv) The ATR-FTIR technique has been used to corroborate mechanical and morphological results. All of these have been justified by this experimental process.

References

1. ETRMA, *End of Life Tire Report*, European Tire and Rubber Manufacturers Association, 2015.
2. Rubber Manufacturers Association, *Tire Service Life for Passenger Car and Light Truck Tires*, 2014.
3. H. F. Mark, N. M. Bikales, C. G. Overberger and G. Menges, *Encyclopaedia of Polymer Science and Engineering*, New York, 1998, pp. 787–804.
4. A. Pehlken and D. H. Müller, *Resour., Conserv. Recycl.*, 2009, **54**, 140.
5. U. Schmidt and D. Reinke, *Wiederverwertung von Altgummi, Status und Tendenzen in der Gummiindustrie. Neue Konzepte für die Autoverwertung, VDI Berichte 934*, VDI Verlag Düsseldorf, 1991, p. 369.
6. S. Lee, F. O. Azzam and B. S. Kocher, US Patent 5,516,952, 1996.
7. B. Maxwell, US Patent 4,146,508, 1979.
8. V. M. Makrov and V. F. Drozdovski, *Reprocessing of Tyres and Rubber Wastes*, Ellis Horwood, New York, 1991.

9. D. S. Novothy, R. L. Marsh, F. C. Masters and D. N. Tally, US Patent 4,104,205, 1978.
10. X. Colom, J. Sans, A. S. B. Norazman and N. Y. I. B. Zailan, *Afinidad*, 2017, **74**, 577.
11. S. P. Yushanov, A. I. Isayev and V. Yu Levin, *J. Polym. Sci.*, 1998, **34**, 2409.
12. M. Loffer, G. Straube and E. Straube, *Biohydrometall Technology Proc. International Symposium*, 1993, vol. 2, p. 673.
13. J. C. Elgin, Can Patent 456,789, 1949.
14. E. F. Sverdrup, US Patent 2,494,593, 1949.
15. L. H. Howland and L. B. Tewksbury, Can Patent 2,469,529, 1949.
16. J. C. Elgin and E. F. Sverdrup, Can Patent 452,085, 1948.
17. P. A. Gibbons and F. H. Cottons, US Patent 2,408,296, 1496.
18. X. Colom, J. Cañavate, F. Carrillo, J. I. Velasco, P. Pagès, R. Mujal and F. Nogués, *Eur. Polym. J.*, 2006, **42**, 2369.
19. X. Colom, J. Cañavate and F. Carrillo, *Composites, Part A*, 2007, **38**, 44.
20. F. Cataldo, O. Ursini and G. Angelini, *Polym. Degrad. Stab.*, 2010, **95**, 803.
21. R. Sonnier, E. Leroy, L. Clerc, A. Bergeret and J. López-Cuesta, *Polym. Test.*, 2007, **26**, 137.
22. B. L. Erickson, H. Asthana and L. T. Drzal, *J. Adhes. Sci. Technol.*, 1997, **11**, 1249.
23. C. M. Cepeda, M. M. Pastor, T. P. Ferrandiz and J. M. Martín, *Int. J. Adhes. Adhes.*, 2001, **21**, 161.
24. J. Pérez, A. L. Doadrio, J. C. Doadrio, P. Horcajada, I. Izquierdo, A. Martínez, B. Muñoz, A. Ramila and M. Valletregí, *Proceedings of the XXIX Biennale Meeting of the Spanish Society of Chemistry*, Madrid, 2003, p. S304.
25. E. Manchón, E. Macías, A. Nadal, C. Fernández and V. Gómez, *Microporous Mesoporous Mater.*, 2004, **67**, 35.
26. P. Ariyadejwanich, W. Tanthapanichakoon, K. Nakagawa, S. R. Mukai and H. Tamon, *Carbon*, 2003, **41**, 157.
27. A. A. Yehia, M. A. Mull, M. N. Ismail, Y. A. Hefny and E. M. Abdel-Bary, *J. Appl. Polym. Sci.*, 2004, **93**, 30.
28. R. Sonnier, E. Leroy, L. Clerc, A. Bergeret and J. M. Lopez-Cuesta, *Polym. Degrad. Stab.*, 2006, **91**(10), 2375.
29. R. Sonnier, E. Leroy, L. Clerc, A. Bergeret, J. M. Lopez-Cuesta and A. S. Bretelle, *Polym. Test.*, 2008, **27**(7), 901.
30. H. S. Liu, J. L. Mead and R. G. Stacer, *Chelsea Center for Recycling and Economic Development*, 2001, vol. 36, p. 10.
31. M. Sen, C. Uzun, O. Kantoglu, S. M. Erdogan, V. Deniz and O. G. Güven, *Nucl. Instrum. Methods Phys. Res., Sect. B*, 2003, **208**, 480.
32. G. Lopitaux, X. Coqueret, F. Boursereau and G. Larnac, *Nucl. Instrum. Methods Phys. Res., Sect. B*, 2003, **208**, 500.
33. I. Banik and A. K. Bhowmickm, *Radiat. Phys. Chem.*, 2000, **58**, 293.
34. J. M. Cepeda, M. M. Pastor, T. P. Ferrrándiz and J. M. Martín, *J. Adhes.*, 2000, **73**, 135.

35. J. Cañavate, F. Carrasco, P. Pagès and X. Colom, *Polym. Polym. Compos.*, 2005, **13**, 669.
36. X. Colom, A. Faliq, K. Formela and J. Cañavate, *Polym. Test.*, 2016, **52**, 200.
37. B. C. Smith, *Infrared Spectral Interpretation: A Systematic Approach*, CRC Press, 2004.
38. J. Coates, Interpretation of infrared spectra. a practical approach, in *Encyclopaedia of Analytical Chemistry*, ed. R. A. Mayers, John Wiley and Sons Ltd, Chichester, 2000.
39. C. M. Cepeda, M. M. Pastor, T. P. Ferrándiz and J. M. Martín, *Polym. Surf. Modif.*, 2000, **2**, 305.
40. O. Figovslq, D. Beilin, N. Blanck, J. Potapo and V. Chernyshe, *Cem. Concr. Compos.*, 1996, **18**, 437.
41. W. K. Dierkes, Rubber recycling, in *Recent Research Developments in Macromolecules*, ed. S. G. Pandalai, Research Signpost, Trivandrum, 2003, vol. 7, p. 265.
42. D. Romero-Sanchez, M. M. Pastor-Blas and J. M. Martin-Martinez, *Int. J. Adhes. Adhes.*, 2001, **21**, 325.
43. C. Nakason, A. Kaesaman and P. Supasanthitikul, *Polym. Test.*, 2004, **23**, 35.
44. J. I. Kim, S. H. Ryu and Y. W. Chang, *J. Appl. Polym. Sci.*, 2000, **77**, 2595.
45. X. Colom, F. Carrasco, P. Pagès and J. Cañavate, *Compos. Sci. Technol.*, 2003, **63**, 161.
46. S. Abdou-Sabet and K. Shen, U.S Patent 4,594,390, 1982.
47. S. Camara, B. C. Gilbert, R. J. Meier, M. Duin and A. Whitwood, *Polymer*, 2006, **47**, 4683.
48. T. Yamazaki and T. Seguchi, *J. Polym. Sci., Part A: Polym. Chem.*, 2000, **18**, 3383.
49. P. R. Dluzneski, *Rubber Chem. Technol.*, 2001, **74**, 451.
50. X. Colom, J. Cañavate, F. Carrillo and J. J. Suñol, *J. Appl. Polym. Sci.*, 2009, **112**, 1882.
51. S. Camara, B. C. Gilbert, R. J. Meier, M. van Duin and A. C. Whitwood, *Polymer*, 2006, **47**, 4683.
52. C. Radhesh Kumar, I. Furrmann and J. Karger-Kocsis, *Polym. Degrad. Stab.*, 2002, **76**, 137.
53. K. Naskar, A. K. Bhowmick and S. K. De, *Polym. Eng. Sci.*, 2001, **41**, 1087.
54. A. Nicolini, T. L. Ávila de Campos and M. A. Maldaner, *J. Appl. Polym. Sci.*, 2008, **109**, 3093.
55. E. Lievana and J. Karger-Kocsis, *Prog. Rubber, Plast. Recycl. Technol.*, 2004, **1**, 1.
56. A. Choudhury and B. Adhikari, *Polym. Int.*, 2007, **56**, 1213.

CHAPTER 3

Thermoplastic Elastomers Filled With GTR

K. FORMELA*[a], J. HAPONIUK[a], S. WANG[b] AND X. COLOM[c]

[a]Gdansk University of Technology, Department of Polymer Technology, G. Narutowicza Str. 11/12, Gdansk, 80-233, Poland; [b]Shanghai Jiao Tong University, Department of Polymer Science and Engineering, 800 Dongchuan RD., Minhang District, 200240, Shanghai, China; [c]Universitat Politècnica de Catalunya Barcelona Tech, Department of Chemical Engineering, Colom 1, 08222, Terrassa, Barcelona, Spain
*E-mail: kformela.ktp@gmail.com, krzysztof.formela@pg.edu.pl

3.1 Introduction

The unique properties of polymers (*e.g.* durability, low density, corrosion and chemical resistance, *etc.*) combined with their easy processing and low-cost manufacturing has resulted in a wide range of industrial applications of plastics. As a consequence, dynamic development of plastic production is observed each year, and has caused problems with the waste management of plastics, which has become a huge environmental, economic and social problem.[1,2] According to a PlasticsEurope report,[3] in Europe around 25.8 million tons of post-production and post-consumer waste plastics are

Green Chemistry Series No. 59
Rubber Recycling: Challenges and Developments
Edited by Jin Kuk Kim, Prosenjit Saha, Sabu Thomas, Józef T. Haponiuk and M. K. Aswathi

Published by the Royal Society of Chemistry, www.rsc.org

generated to multiple solid waste streams. However, only 29.7% of waste plastics in Europe are recovered through material recycling, while 39.5% and 30.8% are used as a fuel substitute (energy recovery) and accumulate in landfill, respectively.

Among various waste plastics, the utilization of cross-linked polymers is a very difficult task requiring a complex solution. The three-dimensional network present in cross-linked polymers significantly improves their performance properties, including their mechanical properties, shape memory properties, thermal stability, chemical resistance, and so forth. On the other hand, these usage features cause significant difficulties during the recycling of cross-linked polymers, for example polyurethane foams, thermosetting resins, cross-linked polyolefins or waste rubber, which is related to their poor fluidity and consequently limited processability.[4–7]

Continuous and dynamic development of the automotive industry leads to an increasing amount of waste rubber, mainly from used tires. The European Tyre & Rubber Manufacturers' Association indicates that more than 3.6 million tons of used car tires are discarded every year in the EU countries.[8] It should be noted that a compact car contains around 60 kg of rubber, the major contributor (around 70%) being the car tires.[9] The main problems are illegally dumped or stockpiled waste tires, which are a serious threat to the natural environment and human health. For example, in 2016 around 9000 people were evacuated from the Seseña near to Madrid (Spain) after a massive fire in the biggest illegal tire dump in Europe due to harmful emissions of polycyclic aromatic hydrocarbons and other hazardous compounds.[10,11] Furthermore, disposed tires can allow stagnant rain water pools to gather in the curved structures of tires. Such conditions creates ideal places for breeding grounds for rodents and mosquitoes, which carry various diseases.[12,13] Another issue is the complex structure of end-of-life tires, which can affect the characteristics of waste tire recycling products and the repeatability of their properties. The differences in the composition of passenger car tires and truck tires are shown in the Table 3.1. It is commonly known that truck tires contain more natural rubber and less carbon black than passenger car tires, which is related to different performance properties of passenger car tires (*e.g.* low rolling resistance, improved skid resistance and good wear).[14]

Table 3.1 Typical composition of passenger and truck tires in Europe.[14]

Composition	Passenger car tire	Truck tire
Natural rubber	22%	30%
Synthetic rubber	23%	15%
Carbon black	28%	20%
Other additives (*e.g.* curing system, fibers, processing aids, *etc.*)	14%	10%
Steel	13%	25%
Estimated average weight of new tire	8.5 kg	65 kg

To prevent the aforementioned issues, current legal and economic regulations require producers to recycle waste tires.[15] Nowadays, energy recovery is commonly used to utilize waste tires as an alternative to fuel in cement kilns and power plants. The calorific values of waste tires are comparable, or even higher than those for coal. Therefore, waste tires can be used as a cheap alternative fuel and are commonly used for up to 75% of the fuel required in cement kilns.[16] On the other hand, the energy equivalent to the production of 1 kg of passenger car tire is around 128 MJ, while burning of waste tires allows recovery of only *ca.* 30 MJ kg^{-1}. For comparison, the size reduction method by mechanical disintegration of 1–1.5 kg of worn tires requires around 6.8 MJ of additional energy.[17] This confirms that searching for new methods for the potential application of ground tire rubber (GTR) in industry, which allow more environmental-friendly "up-cycling" of waste tires is reasonable.

At present, many attempts have focused on the application of GTR as a filler or modifier in rubber compounds, thermoplastic composites, thermoplastic elastomers, polyurethanes, epoxy resins and other polymer matrices, which was comprehensively described in recent review papers published by Karger-Kocsis *et al.*, Ramarad *et al.* and Sienkiewicz *et al.*[18–20] GTR was also incorporated into construction and building materials based on cement, concrete and bitumen. Comprehensive literature reviews in this field were presented in works by Lo Presti, Shu and Huang and Wang *et al.*[21–23]

Another interesting solution is valorization of GTR by using pyrolysis or reclaiming/devulcanization. Pyrolysis is a process conducted under an inert atmosphere, in which GTR is subjected to a high temperature. These conditions allow thermal decomposition of GTR into simpler components, such as gaseous products, pyrolytic oils and pyrolytic carbon, which have found diverse industrial applications.[24,25]

Reclaiming (sometimes termed in the literature as the devulcanization process) involves transformation of waste rubber using thermal, mechanical or chemical energy. During the reclaiming process, the three-dimensional network of the cross-linked rubber is destroyed and compared to GTR, the obtained reclaimed rubber can be easily processed, shaped and vulcanized. Specific conditions (*e.g.* high temperature, high shear forces, *etc.*) during reclaiming strongly affect the final properties of the obtained reclaimed GTR.[26–28]

Among the above mentioned methods, the application of GTR during preparation of thermoplastic elastomers (TPEs) seems to be a very promising route, suitable for implementation at the industrial scale. According to ISO 18064, the description of a thermoplastic elastomer is "consisting of a polymer or blend of polymers that has properties at its service temperature similar to those of vulcanized rubber but can be processed and reprocessed at elevated temperature like a thermoplastic". Additionally, in the case of TPEs the important parameters are elongation at break and compression set. The requirement for TPEs is elongation at break of at least 100%, while permanent set should be lower than 50%.[29]

In comparison to cross-linked rubbers the main advantage of TPEs is their easy processing and recycling using commonly applied methods such as injection molding or extrusion, which affect the beneficial ratio of price to product quality. Furthermore, TPEs can be prepared by physical or reactive blending of semi-crystalline plastomers and amorphous elastomers. This allows a wide range of possible selections of components in the polymer blends composition and thus their final performance properties. Additionally, manufacturing of TPEs can be performed on mixing devices commonly used in polymer technology (*e.g.* extruders, batch mixers, *etc.*), which does not enforce a change of apparatus and reduces the final cost of TPEs production. The main disadvantage of TPEs is their cost, which is still higher than for conventional chemically cross-linked rubber compounds. Additionally, TPEs are characterized by lower chemical resistance and thermal stability compared to chemically cross-linked elastomers.

The clear advantage of TPEs means that in the last few years research studies on novel thermoplastics/elastomer blends with unique properties have received growing interest among science and industry representatives worldwide. Many of currently conducted research studies are focused on the optimization of processing conditions and further compatibilization of blends based on polyolefins (polyethylene, polypropylene) and different rubbers, such as natural rubber,[30] styrene-butadiene rubber,[31] ethylene-propylene-diene rubber,[32] butyl rubber,[33] acrylonitrile butadiene rubber[34] and other elastomeric phases (*e.g.* ethylene vinyl acetate copolymer,[35] styrene-butadiene-styrene block copolymer,[36] polyethylene-octene elastomer[37] or polyurethane[38]).

One of the first papers about TPE with reclaimed rubber was published by Al-Malaika and Amir in 1989.[39] Since then, many attempts have focused on industrial application of waste rubber as an elastomeric phase during the preparation of TPEs.[40] For example, in 2002 a research group from Toyota Central Research & Design Labs. Inc. started the mass production of TPEs (capacity of 350 000 kg per year) based on post-production waste ethylene propylene diene monomer (EPDM) generated during manufacturing of weather strips.[41,42] The produced TPEs were used in automotive parts (*e.g.* vacuum sensing hoses, door glass runs, *etc.*) in order to determine their practical uses. However, to the best of our knowledge, commercialization of thermoplastics/GTR blends is still very limited, although the promising industrial application of such products is bitumen modification.[43–47]

In this chapter the recent developments in extrusion processing and compatibilization of selected thermoplastics/GTR blends are presented and discussed. We hope that this work will provide new, useful information for students and researchers starting studies in this field.

3.2 Thermodynamics of Polymer Blends Containing GTR

Polymer mixtures containing GTR can be classified as polymer blends or as polymer composites depending on the size of the dispersed GTR phase. The thermodynamics of such systems is complex and depends mainly on morphology of the phase and on interactions at the interface of the phase. As the GTR forms a dispersed phase, the effective contact area at various surface activities and the GTR – polymer interactions contribute to a great extent to the static and dynamic properties of the composite. To obtain the optimal morphology of TPEs, the elastomer particles promoting elasticity, in our case GTR, should be finely dispersed in a relatively small amount of the plastic.[48] TPEs obtained with the use of GTR differ from the conventional thermoplastic vulcanizates (TPVs) in chemical and physical structure development. In the technology of conventional TPVs the elastomer phase undergoes breakup/coalescence and is cross-linked during the process. Obtaining GTR filled TPEs is based on processing of the plastics with already cross-linked rubber particles, in which partial devulcanization of their surfaces is desirable, leading to better physical interaction at their surfaces. The surface properties are responsible for the evaluation of a stable co-continuous structure *via* agglomeration of the elastomer particles.

Dey *et al.* studied TPVs based on natural rubber, ethylene octene copolymer and carbon black, with the absence of hydrogen bonding and acid–base interactions, that is also valid for the GTR filled TPEs.[49] The authors applied Wang theory[50] in order to evaluate a spontaneous agglomeration of elastomer particles, to determine what could be expected in the case of positive change in the adhesive energy (ΔW) before and after filler agglomeration. ΔW was calculated using eqn (3.1):

$$\Delta W = 2[(\gamma_f^{d})^{1/2} - (\gamma_p^{d})^{1/2}]^2 + 2[(\gamma_f^{p})^{1/2} - (\gamma_p^{p})^{1/2}]^2 \tag{3.1}$$

In which γ_f^{d} and γ_f^{p} are the dispersive and polar components of the filler; γ_p^{d} and γ_p^{p} are the surface energy of the filler and the dispersive and polar components of the surface energy of the polymer.

The dispersive and polar components of the surface energy are available experimentally from the contact angle measurements. The greater difference in free energies leads to a higher tendency for the GTR particles to agglomerate in the polymer matrix to facilitate the phase inversion. The modification of the surface energy of GTR has been performed by surface modification of the GTR particles with variable functional groups (*e.g.* maleic anhydride, allylamine, acrylamide or chloride). Upon obtaining thermoplastics/GRT blends with properties of TPEs and showing enhanced elastic recovery, the key aspect is an efficient degradation and devulcanisation of GTR, necessary for the formation of the co-continuous phase structure of TPEs.

3.3 Preparation of Thermoplastics/GTR Blends in Variable Conditions

Preparation of polymer blends *via* melt blending is a simple, solvent-free and cost-effective route to obtain new polymeric materials with the desired performance properties. Melt-compounding is usually performed by using the extrusion technique, which is one of the fastest growing methods of polymer processing.[51] This is due to the fact that it is commonly used in industry extruders to allow continuity of the process, short cycle times, high efficiency and suitable quality of products. Crucial factors affecting extrusion effectiveness are extruder construction (*e.g.* screw configuration) and extrusion parameters (*e.g.* temperature, screw speed, shear forces, residence time distribution, *etc.*), which have a significant effect on the homogeneity and consequently the final properties of the obtained materials. In this section, we have presented and discussed the results of our research studies regarding the preparation of thermoplastics/GTR blends in variable conditions.

3.3.1 Statistical Methods Used in Extrusion

Currently, the application of statistical methods combined with computer simulations has been gaining in importance, especially during extrusion of polymers.[52,53] This approach was successfully applied for thermoplastic/GTR blends. For example, Zhang *et al.* studied the properties of polypropylene/GTR blends compatibilized by bitumen and maleic anhydride grafted styrene-ethylene/butadiene-styrene copolymer.[54,55] During the studies statistical methods based on neural networks and genetic algorithms were applied to optimize the PP/GTR blends prepared in a co-rotating twin screw extruder, which allowed a reduction in the number of evaluated samples to only seven. The obtained results indicated that PP/GTR blends show good ductility with increasing content of bitumen and maleic anhydride grafted styrene-ethylene/butadiene-styrene copolymer as compatibilizers, while their optimal ratio was experimentally validated based on the mechanical property measurements.

da Costa *et al.* optimized the composition of polypropylene/ethylene-propylene-diene monomer/GTR blends using the response surface methodology (RSM), which reduced the number of studied samples to thirteen.[56] The authors proved that a higher content of GTR in studied ternary blends caused deterioration of their mechanical properties due to poor interfacial adhesion between the thermoplastic matrix and the cross-linked GTR. Additionally, it was observed that ethylene-propylene-diene rubber acts like a physical compatibilizier in polypropylene/GTR blends. As presented, the main advantage of the application of statistical methods is the significant reduction of the number of time-consuming and costly experimental works.

3.3.2 Importance of Extrusion Temperature

Formela and Bogucki examined the impact of extrusion parameters on selected properties of a 50/50 wt% blend based on post-production recycled high density polyethylene and GTR.[57] The authors used RSM to determine the effects of the barrel temperature, T (160–240 °C), screw speed, n (250–650 rpm) and throughput, Q (1–2.5 kg h^{-1}) on the processing and static mechanical properties of the obtained products. The experimental results are summarized in Table 3.2. Based on the obtained results empirical models for determining the relationship between the extrusion conditions and properties of the studied thermoplastic compositions were developed.

It was observed that in the studied extrusion conditions, the barrel temperature (T) had the strongest and most decisive impact on the mass flow rate of recycled high density polyethylene/GTR blends. A higher temperature during processing enhanced the thermo-mechanical degradation of polyethylene and ground tire rubber, which increased the mass flow rate of the obtained blends. Additionally, it was noticed that the changes in mass flow rate are correlated with other processing parameters, such as energy consumption, melt temperature or melt pressure. For example, increasing screw speed resulted in higher MFR values and simultaneously higher energy consumption and melt temperature (lower melt pressure). This is related to a tendency of thermo-mechanically degraded blends to stick to the screws, which causes problems with extrusion of dosed material and an increase of the energy consumption per 1 kg. On the other hand, the higher throughput causes a decrease in the mass flow rate of the obtained blends, which resulted in better mixing and transport during extrusion.

Table 3.2 The correlation between the extrusion parameters, processing and mechanical properties of recycled high density polyethylene/ground tire rubber blends (in ratio 50/50 wt%).[57]

Extrusion parameters									
n (rpm)	T (°C)	Q (kg h^{-1})	Energy consumption (kJ kg^{-1})	Melt temperature (°C)	Melt pressure (bar)	Melt flow rate, (g/10 min)	Tensile strength (MPa)	Elongation at break (%)	Hardness (°Sh D)
250	160	1.00	3240	167	15.3	7.0	10.0	8	46
650	160	1.00	7020	176	12.9	7.4	10.0	9	43
250	240	1.00	2484	252	8.2	9.5	10.0	8	44
650	240	1.00	5472	254	6.4	13.3	8.7	6	40
250	160	2.50	2002	172	24.0	5.1	8.8	9	40
650	160	2.50	3902	177	18.7	6.5	9.3	8	44
250	240	2.50	1354	252	13.2	6.8	8.8	8	42
650	240	2.50	2707	255	9.8	9.8	9.2	7	42
450	200	1.75	3621	214	13.6	6.0	8.7	8	42

It is well known that the molecular weight of polymers has a strong influence on their physico-mechanical properties. The melt flow rate is an indirect measure of the molecular weight. However, in the discussed research there is no simple correlation between increased MFR (decreased molecular weight) and deterioration of tensile properties. This observation could be explained by partial reclaiming on GTR surface which improved the interfacial adhesion between the polymer matrix and the GTR phase due to higher physical interactions at the phases boundary.[58,59] In this way, additional interactions between the reclaimed GTR and polyethylene matrix are able to compensate the negative effects related to thermo-mechanical degradation of polyethylene and the decrease of its molecular weight.

In the Figure 3.1 the morphologies of low density polyethylene/unmodified GTR (Figure 3.1A) and low density polyethylene/thermo-mechanically reclaimed GTR (Figure 3.1B) previously subjected to tensile tests (crosshead speed: 50 mm min^{-1}) are presented. The images show the surface perpendicular to the direction of the applied stress. As can be observed, the sample with reclaimed GTR was characterized by the jagged surface of the polyethylene, which confirmed the higher compatibility of the polyethylene matrix with partially reclaimed GTR.[60]

Similar observations were described by Scaffaro *et al.* who studied the impact of temperature, screw speed, ratio of used components, mixing procedure and formulation methods (injection molding/compression molding) on the mechanical properties of recycled polyethylene/GTR blends.[61] The obtained results indicated that an increase of the processing temperature resulted in the reclaiming/devulcanization of GTR which subsequently increased the interfacial compatibility with the polyethylene matrix. On the other hand, using too high a temperature of the extruder barrel (at the level of 300 °C) caused partial carbonization of GTR particles, which resulted in a higher viscosity of the studied blends and their lower elasticity. Furthermore, in the studied case, the residence time and mixing procedure had only a slight impact on the final properties of the polyethylene/GTR blends.

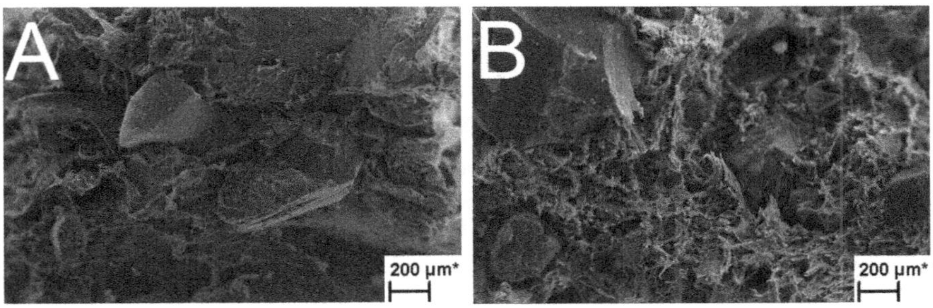

Figure 3.1 Scanning electron microscopy images of low density polyethylene modified with 50 wt%: (A) unmodified GTR and (B) GTR thermo-mechanically reclaimed using a co-rotating twin screw extruder. Adapted from ref. 60 with permission from PRZEMYSŁ CHEMICZNY.

3.3.3 Effect of Extrusion Settings

In work published by Formela *et al.* the effects of material dosing order and screw speed during extrusion on properties of high density polyethylene/GTR blends were investigated.[62] As presented in Table 3.3, the studied variables have a significant influence on the processing parameters (*e.g.* melt temperature, melt pressure or screw torque) and mass flow rate. It was found that the studied conditions have a negligible impact on the mechanical properties of the examined blends, which corresponded with the results of the dynamic mechanical analysis, thermogravimetric analysis and differential scanning calorimetry measurements. Moreover, it was observed that material dosing order affected changes to the surface morphology in the studied blends, which was due to plasticization of the thermoplastic matrix by low-molecular weight compounds formed during reclaiming of GTR.

Recently, de Sousa *et al.* studied the effects of extrusion conditions on the properties of dynamically vulcanized polyethylene/microwave treated GTR blends (in ratio 40/60 wt%).[63] The authors indicated that material dosing order and screw speed have a significant impact on the dynamic vulcanization efficiency of studied blends. It was observed that a higher screw speed resulted in finer GTR particles and a narrower particle size distribution in the polyethylene matrix. However, the finest morphology of the dispersed GTR phase and consequently the highest mechanical properties were obtained for blends obtained when treated GTR is added to the side feeder, while polyethylene is added to the extruder hooper. This could be related to partial plasticization of GTR by the thermoplastic matrix and consequently higher efficiency of the GTR reclaiming process.[64]

Table 3.3 Processing parameters and mechanical properties of HDPE/GTR blends obtained in different conditions.[62]

n (rpm)	Materials dosing order		Screw torque (Nm)	Melt pressure (bar)	Melt temperature (°C)	Melt flow rate (g/10 min.)	Tensile strength (MPa)	Elongation at break (%)	Hardness (°Sh D)
	HDPE	GTR							
50	Hooper	Hooper	30	5	182	9.8	6.7	12	42
250	Hooper	Hooper	30	4	184	9.1	6.9	11	43
50	Side feeder	Hooper	28	4	197	9.3	7.8	9	43
250	Side feeder	Hooper	22	5	195	9.1	6.4	7	43
50	Hooper	Side feeder	25	5	189	7.8	6.5	9	41
250	Hooper	Side feeder	30	5	187	8.1	6.7	9	43

3.3.4 Combined Impact of Thermoplastic Matrix Type and Screw Configuration

Formela *et al.* examined the effects of high density polyethylene type and screw configuration on the processing and mechanical properties of high density polyethylene/GTR blends.[65] HDPE/GTR blends were prepared using a twin screw extruder Leistritz ZSE model 27HP. The gear in this model of extruder allows a change to the direction of rotation of the screw from co-rotating to counter-rotating, which was also investigated. The results obtained during these studies are summarized in Tables 3.4 and 3.5. As could be observed, regardless of the mixing condition and the HDPE type used during experiments, the mechanical properties of HDPE blended with 50 wt% GTR were much lower compared to pure HDPE. Application of GTR as a modifier of the polyethylene matrix resulted in a four-fold decrease in the tensile strength values from *ca.* 28–33 MPa to around 7–9 MPa. The tensile strength of HDPE/GTR blends was comparable to values determined for rGTR. The hardness of the HDPE/GTR blends decreased to about 21–27 °Sh D compared to pure HDPE, which is related to the high elasticity of the three-dimensional network structure of GTR. HDPE type has a strong effect on the elongation at break of the studied blends. Dispersion of the GTR particles in the high density polyethylene matrix resulted in decreasing of its elongation at break. Comparing to determined tensile strength values, the elongation at break results do not correspond with values of elongation at break for rGTR, which is due to poor compatibility between the polyethylene and GTR. Additionally, it was found that screw configuration had a significant impact on the screw torque values, which is a parameter strongly affecting extrusion efficiency. On the other hand, in the studied conditions the effect

Table 3.4 Processing and mechanical properties of HDPE/GTR blends as a function of HDPE type and screw speed.[65]

Sample composition	HDPE type	Screw speed (rpm)	Screw torque (Nm)	Mass flow rate (g/10 min.)	Tensile strength (MPa)	Elongation at break (%)	Hardness (°Sh D)
rGTR	—	—	—	—	6.90	310	—
HDPE	rHDPE 6 g/10 min	—	—	6	24	26	63
	HDPE 8 g/10 min	—	—	8	33	650	69
	HDPE 30 g/10 min	—	—	30	28	200	68
HDPE/GTR 50/50 wt%	rHDPE 6 g/10 min	50	35	1.8	7.6	21	47
		250	23	1.8	8.0	27	47
	HDPE 8 g/10 min	50	37	2.7	8.5	29	49
		250	20	2.7	8.3	26	47
	HDPE 30 g/10 min	50	21	10.1	7.2	10	42
		250	21	10.2	6.9	9	43

Table 3.5 The combined impact of screw configuration and screw speed on processing and mechanical properties of HDPE/GTR blends.[65]

Sample composition	Screw configuration	Rotational direction of the screw	Screw speed (rpm)	Screw torque (Nm)	Mass flow rate (g/10 min.)	Tensile strength (MPa)	Elongation at break (%)	Hardness (°Sh D)
rGTR	—	—	—	—	—	6.9	310	—
HDPE	—	—	—	—	30	28.0	200	68
HDPE/GTR 50/50 wt%	Profile A	Counter-rotating	50	27	8.5	7.2	10	42
			250	17	10.4	7.4	12	42
	Profile B	Co-rotating	50	21	10.1	7.2	10	42
			250	21	10.2	6.9	9	43
	Profile C	Co-rotating	50	27	10.3	7.4	11	41
			250	33	10.4	7.5	12	41

of the screw configuration on mass flow rate and mechanical properties was negligible (see Table 3.5), which may result from high mass flow rate of the used HDPE and consequently its higher stability during thermo-mechanical processing.

It is highlighted in this section that variable mixing conditions affect (more or less) the residence time distribution of the thermoplastic/GTR blends during extrusion. Residence time distribution is the time that the temperature and shear forces are acting on the flowing components inside the barrel extruder. This parameter consequently affects the processing as well as the final properties of the obtained materials. Presented examples confirmed that the optimization of the processing conditions, particularly mixing and molding conditions, is one the simplest and low-cost routes to improve the performance properties of thermoplastic/GTR blends. However, it should be noted that none of above mentioned examples reached the property requirement characteristics for TPEs (elongation at break of at least 100%, while permanent set should be lower than 50%). This indicated that optimization of the processing conditions should be combined with further compatibilization of thermoplastic/GTR blends, which will be discussed in the next section.

3.4 Routes for Compatibilization of Thermoplastics/GTR Blends

In general, a higher content of GTR in polymer blends or composites results in deterioration of their mechanical properties. This is due to poor interfacial adhesion between the polymer matrix and cross-linked GTR phase with larger size. Higher compatibility in thermoplastics/GTR blends can be obtained by creating additional physical and/or chemical interactions at the interface between the used components, which allow improvement of the quality of the thermoplastics filled with GTR. Simultaneously, it should be noted, that not all compatibilization routes allow obtaining of TPEs containing GTR.

3.4.1 Cross-linking

Sonnier *et al.* studied the impact of dicumyl peroxide on the properties of polyethylene/GTR blends.[66] It was found that free radicals formed during dynamic curing of polyethylene/GTR blends cause a co-cross-linking of both phases, which creates interfacial adhesion between polyethylene and GTR particles. The formed interface promotes compatibilization and enhanced the mechanical properties of the obtained blends. Moreover, the authors proved that preparation of TPEs from polyethylene/GTR blends compatibilized by dynamic curing is not possible. This is due to the fact that the mean particle size requirement for rubber inclusions in TPEs (or thermoplastic vulcanizates) is very hard to reach with GTR particles. However, in the literature the term "thermoplastic elastomers" for thermoplastic composition filled with GTR is often incorrectly used, probably to catch the readers attention.

For better presentation of this issue, Figure 3.2 shows the stress–strain curves of low-density polyethylene (LDPE) melt-compounded with 50 wt% of different elastomers: ground tire rubber (GTR), partially cross-linked butyl rubber (IIR, Kalar® 5215), and styrene-butadiene-styrene block copolymer (SBS, Kraton® 1102).[67] The course of the curves for the LDPE/SBS copolymer blends suggest better compatibility of LDPE with a SBS copolymer than in the case of the partially cross-linked IIR and GTR. It was observed that incorporation of 50 wt% of GTR or partially cross-linked IIR into the LDPE resulted in a significant decrease of the tensile strength (from ~8 MPa to ~3.6–4.1 MPa). Despite the similar tensile strength of LDPE/GTR and LDPE/partially cross-linked IIR blends, significant differences in elongation at break were noticed. This is related to the chemical structure and viscosity of the elastomer (crosslink density, filler content), which affect its dispersion into the thermoplastic matrix and therefore the physical interaction in the examined blends.

He *et al.* pointed out that the above mentioned limitation for polyethylene/GTR blends could be resolved using mechanochemically desulfurized GTR and dynamic curing with a dicumyl peroxide/phenolic resin mixed system.[68]

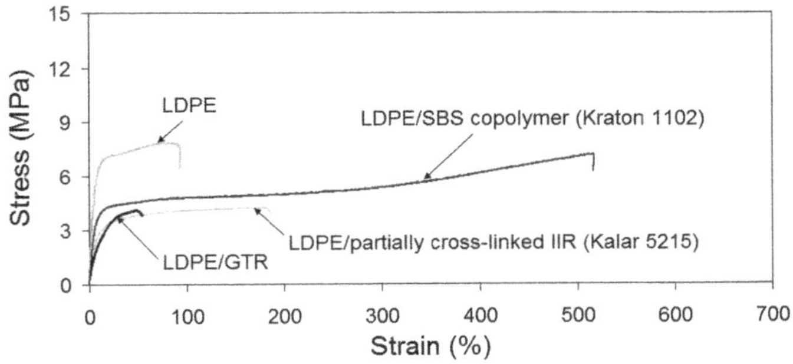

Figure 3.2 Stress-strain curves for LDPE/elastomer blends.[67]

Additionally, it is worth mentioning that an interesting alternative to dynamic curing is the application of electron beam irradiation to the thermoplastic/GTR blends, which was comprehensively studied by Mészáros *et al.*, Hassan *et al.*, Ramarad *et al.* and recently by Mali *et al.*[69–77]

3.4.2 Oxidization or Reclamation of GTR

Another solution is modification of GTR using controlled oxidation of the surface or the reclaiming/devulcanization process. These operations allow enhance flowability and processing of cross-linked rubber, which improve its dispersibility in the polymer matrix. Colom *et al.* treated GTR with various inorganic acids, such as perchloric acid, sulfuric acid (VI) or nitric acid (V).[78] The obtained modified GTR was then used as a reinforcing phase in the high density polyethylene. It was found that treatment of GTR by sulfuric acid (VI) or nitric acid (V) improved the interfacial adhesion between the polyethylene matrix and the GTR phase. On the other hand, application of perchloric acid did not enhance the compatibility and the final properties of the studied materials. Similar observations were made when sulfuric acid (VI) treated GTR was compared with GTR modified by a silane coupling agent or chlorination with trichloroisocyanuric acid.[79]

Colom *et al.* showed that the effect of the GTR particle size has a higher impact on the properties of HDPE/GTR blends than the acidic pre-treatment of GTR.[80] Moreover, Sonnier *et al.* indicated that GTR oxidation is a less effective strategy towards a non-polar polymer matrix compared to dynamic curing of thermoplastic/GTR blends.[81] Recently, work published by Gamez *et al.* indicated that the use sulfuric acid modified GTR during preparation of thermoplastic vulcanizates based on HDPE/EPDM blends is justified.[82]

Nevatia *et al.* showed that TPEs can be obtained by melt-compounding of waste plastics with reclaimed rubber.[83] On the other hand, other published works indicated that thermoplastics blended with reclaimed GTR did not receive the elongation at break required for TPEs.[84,85] These differences could be related to the diverse characteristics of the used components (*e.g.* thermoplastic matrix, source of waste rubber, curing system, *etc.*) and their processing conditions (*e.g.* temperature, shear forces, *etc.*). However, it should be pointed that a very important parameter affecting the final properties of thermoplastic/GTR blends is also the reclaiming degree of the GTR, which was recently described by Sripornsawat *et al.*[86]

3.4.3 Application of Additional Elastomer Phase

Another commonly applied method for the preparation of TPEs based on thermoplast/GTR blends is the addition of a fresh elastomer, which can be supported (depending on the quality requirements), by suitable compatibilization. Kumar *et al.* investigated the effects of elastomer type and dynamic curing on the properties of TPEs based on LDPE/GTR blends.[87] Based on the presented results, it can be stated that the dynamic curing of LDPE/GTR blends has a positive impact on the final products by increasing their elongation at break. Moreover, it was found that the used elastomers act as compatibilizers in LDPE/GTR blends, and that this follows an ordered sequence: ethylene propylene diene rubber > natural rubber > styrene-butadiene rubber. Similar observations were also reported elsewhere, showing the role of ethylene propylene diene rubber as an efficient compatibilizer of polyolefin/GTR blends.[88–90]

Abadchi *et al.* examined linear low-density polyethylene/natural rubber blends in which natural rubber was partially replaced with GTR.[91] It was observed that using 40 wt% GTR resulted in an increase of the tensile strength and hardness, while simultaneously decreasing the elongation at break in the tested blends. This is related to the possible migration of carbon black particles from GTR into the polymer matrix.[92] Additionally, the results show that use of maleic anhydride and dicumyl peroxide as reactive compatibilizers enhanced the interfacial interactions and so improved the mechanical properties of the studied blends. Similar observations were described by Formela *et al.*[93]

Formela and Haponiuk[94] evaluated the properties of low density polyethylene/GTR/partially cross-linked butyl rubber blends as a function of their composition. It was found that the mechanical properties of ternary blends with a higher content of elastomers (GTR and partially cross-linked butyl rubber) showed the same behavior, whereas for samples with only GTR as the elastomeric phase the respective parameters were worse.

Li *et al.* investigated the impact of the elastomer on the properties of HDPE/GTR blends.[95] During the studies two types of copolymers were used, a polar ethylene-vinyl acetate copolymer and a non-polar ethylene-octene copolymer. It was noticed that the non-polar ethylene-octene copolymer had a more beneficial impact on the mechanical properties of HDPE/GTR blends compared to the polar ethylene-vinyl acetate copolymer. This is due to the higher homogeneity of the HDPE/ethylene-octene copolymer blends and more efficient encapsulation of GTR particles by the non-polar elastomer, which significantly increased the ability of the polymer matrix to be deformed during mechanical stress.

In work by Formela *et al.* the mechanism of interaction in the ternary blends based on low density polyethylene, GTR and non-polar elastomer was described.[96] In this study two types of commercial non-polar elastomers were applied, styrene-butadiene-styrene block copolymers with different topologies and partially cross-linked butyl rubbers with different Mooney viscosities. It was found that the best mechanical properties (tensile strength and elongation at break) were determined for samples containing 50 wt% of LDPE as the main thermoplastic component. As presented in the Figure 3.3. The modification of LDPE/GTR blends by incorporation of non-polar elastomers occurs by encapsulation of GTR with the elastomer phase. The obtained results showed that the most critical factor affecting interfacial interactions in LDPE/GTR blends is the kind of elastomer, in particular its compatibility

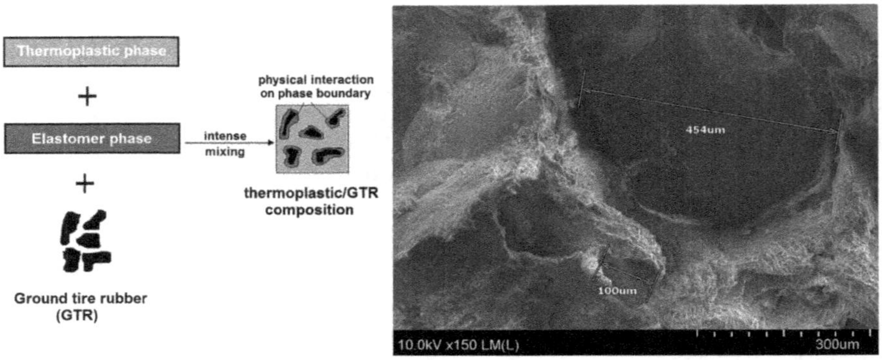

Figure 3.3 Compatibilization mechanism of thermoplastic/GTR using elastomer as a modifier. Reproduced from ref. 96 with permission from Elsevier, Copyright 2015.

with polyethylene and GTR. Another important factor is the melt viscosity of the thermoplastic matrix, which determines the compatibility with GTR particles. Moreover, the LDPE/GTR ratio and GTR particles size also had a significant impact on the compatibility of the investigated blends, which corresponded to observations made by Lima *et al.* and Wang *et al.*[97,98]

3.4.4 Grafted Polymers

Another commonly used method to enhance the compatibility of thermoplastic/GTR blends is the addition of grafted polymers. Phinyocheep *et al.* examined the impact of six different types of compatibilizers, styrene-butadiene-styrene block copolymer, styrene-ethylene/butadiene-styrene copolymer, maleic anhydride grafted styrene-ethylene/butadiene-styrene copolymer, maleic anhydride grafted polyethylene, maleic anhydride grafted polypropylene and glycidyl methacrylate grafted polypropylene on the mechanical properties of polypropylene/waste rubber blends.[99] In the studied conditions the most effective compatibilizers were the styrene-ethylene/butadiene-styrene copolymer and the maleic anhydride styrene-ethylene/butadiene-styrene copolymer, which is due to their good miscibility with the polypropylene matrix. Naskar *et al.* obtained maleated GTR, which was subsequently used in acrylated high-density polyethylene/ethylene propylene diene rubber blends.[100] The tensile properties of blends with maleated GTR were improved compared to blends with untreated GTR. Lee *et al.* investigated dynamic reactions inside the extruder during preparation of polypropylene/GTR and maleic anhydride grafted polypropylene/ground tire rubber blends compatibilized by maleic anhydride grafted styrene-ethylene/butadiene-styrene copolymer.[101] It was found that components with maleic anhydride groups allowed better dispersion of the GTR phase in thermoplastic matrix, which enhanced the interfacial adhesion and consequently the mechanical properties of the studied blends. Additionally, recyclabilty studies revealed that the tensile properties of the examined materials were retained even after 10 cycles.

Lu *et al.* evaluated the impact of maleic anhydride grafted polyethylene, styrene-butadiene-styrene block copolymer and ethylene-vinyl acetate copolymer on the properties of thermoplastic vulcanizate/GTR blends.[102] It was found that the mechanical parameters (tensile strength, tear strength, elongation at break) of the obtained blends decreased with a higher content of GTR and its increasing particle size. Among the examined compatibilizers the highest mechanical properties were determined for the blends modified with maleic anhydride grafted polyethylene.

Ramezani and Rodrigue proved that tensile and elastic recovery values of HDPE/GTR blends were significantly lower than those determined for maleic anhydride grafted polyethylene/GTR blends.[103]

Formela *et al.* investigated the effects of maleic anhydride grafted polyethylene and screw speed (shear forces) on the processing and mechanical properties of HDPE/GTR blends.[104] The obtained results are presented in the Table 3.6. It was noted that the increasing screw speed strongly affected the melt flow rate of blends modified with maleic anhydride grafted polyethylene, while its impact on mechanical properties was negligible. Moreover, application of 10 wt% of maleic anhydride grafted polyethylene into the HDPE/GTR blend decreased the melt flow rate and increased the tensile properties. Figure 3.4 shows the morphology of the HDPE/GTR blends without and with the anhydride grafted polyethylene previously subjected to the tensile test. It was observed that the thermoplastic phase in the HDPE/GTR 50/50 wt% blend was more ductile and showed a lower adhesion to the GTR phase compared to the HDPE/GTR/HDPE-*g*-MA 40/50/10 wt% blend. The surface of the HDPE/GTR blend compatibilized with the anhydride grafted polyethylene was characterized by higher amounts of shredded pieces of the thermoplastic phase formed during static tensile test.

Table 3.6 Processing parameters and mechanical properties of HDPE/GTR 50/50 wt% and HDPE/GTR/HDPE-*g*-MA 40/50/10 wt% blends.[104]

Sample composition	Screw speed (rpm)	Screw torque (Nm)	Melt temperature (°C)	Melt pressure (bar)	Melt flow rate, (g/10 min)	Tensile strength (MPa)	Elongation at break (%)	Hardness (°Sh D)
HDPE/GTR 50/50 wt%	50	30	182	5	9.8	6.8	12	42
	100	31	182	5	10.4	6.5	12	44
	150	30	183	4	10.0	6.4	10	44
	200	30	184	5	10.2	6.9	11	43
	250	30	184	4	9.1	6.7	11	43
HDPE/GTR/HDPE-*g*-MA 40/50/10 wt%	50	38	190	6	3.2	7.6	14	44
	100	40	190	6	3.3	7.1	13	42
	150	40	190	6	4.4	7.4	13	43
	200	41	192	6	5.1	7.1	11	43
	250	41	190	5	5.5	7.4	10	42

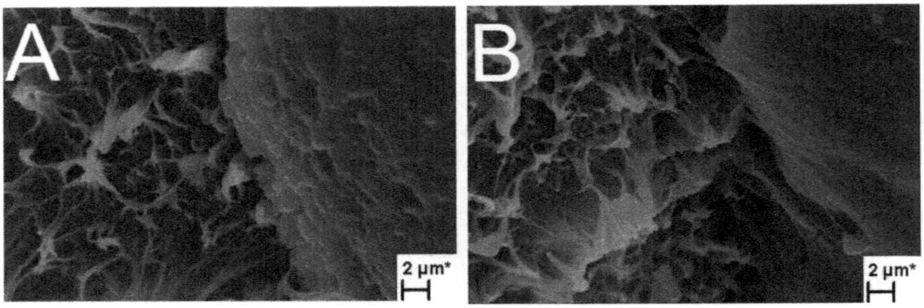

Figure 3.4 SEM images of samples (A) HDPE/GTR 50/50 wt% and (B) HDPE/GTR/HDPE-*g*-MA 40/50/10 wt% (magnification x5000). Adapted from ref. 104 with permission from PRZEMYSŁ CHEMICZNY.

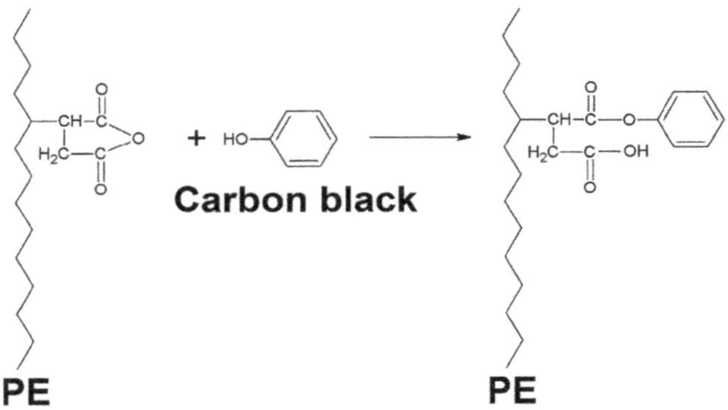

Figure 3.5 Chemical reaction between PE-*g*-MA and carbon black present in GTR.[105]

The scheme of the possible compatibilization mechanism of the polyethylene/GTR blends modified with maleic anhydride grafted polyethylene is shown in Figure 3.5. The higher compatibility of the above discussed systems is probably related to the reaction between the maleic anhydride and the hydroxyl group presented in carbon black.[105] Additionally, maleic anhydride groups can also react with the carboxyl group present on the surface of GTR after grinding or reclaiming (controlled oxidation).

3.4.5 Other Possibilities

Moreover, in the published papers there is a lot of information on other methods of thermoplastic/GTR blends compatibilization by using, for example, waxes and wetting agents, coupling agents, *in situ* polymerization on the GTR surface, reactive compatibilizers or a combination of diverse methods, for example: (i) devulcanization and graft polymers, (ii) particle size reduction and fresh elastomer phase or (iii) cross-linking and additional elastomer phase.[106–113]

Additionally, it should be pointed out that the low compatibility between the thermoplastic matrix and the GTR phase has a smaller impact on the properties of a the polymeric materials with a cellular structure. Therefore, the foaming of thermoplastic/GTR blends could be a promising route to produce new materials with unique properties.[114] However, in the last years, only a few research articles on the application of GTR in thermoplastics foams have been published.[115–120]

3.5 Conclusions

The information presented in this chapter indicates that the production of thermoplastics and TPEs filled with GTR will continue to grow in the future. In particular, the market is still searching for new solutions to GTR valorization, which will be an economic and ecological alternative for energy recovery. It seems that the current and future trends in this field of research will focus on reactive extrusion of thermoplastics/GTR blends. This technique allows preparation of novel TPEs based on GTR for which final performance properties can be successfully tailored by reactive additives (*e.g.* peroxides, functional monomers, *etc.*) or application of variable extrusion conditions (*e.g.* temperature, shear forces, *etc.*). Moreover, during extrusion partial reclaiming/devulcanization of GTR occurs, which also promotes the interfacial interaction between the thermoplastic matrix and the GTR phase.

The serious limitation of this method is that the obtained materials are characterized by weak mechanical properties (low compatibility) and usually by high stiffness, which limits their potential applications.

Additionally, during their processing, smelly volatile ingredients and degradation product are released. Low temperature "green devulcanization", more neutral for the environment and performed by advanced reactive extrusion is a promising technology in spite of enhancing the mechanical properties of thermoplastics/GTR blends when compared to the present state of art methods.

References

1. J. Datta and P. Kopczyńska, From polymer waste to potential main industrial products: actual state of recycling and recovering, *Crit. Rev. Environ. Sci. Technol.*, 2016, **46**(10), 905.
2. N. Singh, D. Hui, R. Singh, I. P. S. Ahuja, L. Feo and F. Fraternali, Recycling of plastic solid waste: a state of art review and future applications, *Composites, Part B*, 2016, **115**, 409.
3. PlasticsEurope raport, *Plastics - the Facts 2016: An Analysis of European Plastics Production, Demand and Waste Data*, 2016.
4. P. Kopczyńska and J. Datta, Rheological characteristics of oligomeric semiproducts gained via chemical degradation of polyurethane foam using crude glycerin in the presence of different catalysts, *Polym. Eng. Sci.*, 2017, **57**(8), 891.

5. M. Barczewski, D. Matykiewicz, J. Andrzejewski and K. Skórczewska, Application of waste bulk moulded composite (BMC) as a filler for isotactic polypropylene composites, *J. Adv. Res.*, 2016, **7**(3), 373.
6. K. Formela, M. Wołosiak, M. Klein and S. Wang, Characterization of volatile compounds, structural, thermal and physico-mechanical properties of cross-linked polyethylene foams degraded thermo-mechanically at variable times, *Polym. Degrad. Stab.*, 2016, **134**, 383.
7. Ł. Zedler, M. Przybysz, M. Klein, M. R. Saeb and K. Formela, Processing, physico-mechanical and thermal properties of reclaimed GTR and NBR/reclaimed GTR blends as function of various additives, *Polym. Degrad. Stab.*, 2017, **143C**, 186.
8. ETRMA – European Tyre & Rubber Manufacturers' Association (Belgium), *End-of-life Tyre Report, 2015*, 2016.
9. C. Former and E. Osen, State and prospects of rubber recycling, *Kautsch. Gummi Kunstst.*, 2003, **56**, 81.
10. S. J. Chen, H. B. Su, J. E. Chang, W. J. Lee, K. L. Huang, L. T. Hsieh, Y. C. Huang, W. Y. Lin and C. C. Lin, Emissions of polycyclic aromatic hydrocarbons (PAHs) from the pyrolysis of scrap tires, *Atmos. Environ.*, 2007, **41**(6), 1209.
11. M. Gągol, G. Boczkaj, J. Haponiuk and K. Formela, Investigation of volatile low molecular weight compounds formed during continuous reclaiming of ground tire rubber, *Polym. Degrad. Stab.*, 2015, **119**, 113.
12. J. Fiksel, B. Bakshi, A. Baral, E. Guerra and B. DeQuervain, Comparative life cycle assessment of beneficial applications for scrap tires, *Clean Technol. Environ. Policy*, 2011, **13**, 19.
13. F. D. B. de Sousa, in *Devulcanization of Elastomers and Applications in Elastomers*, ed. N. Cankaya, InTech, Rijeka, Croatia, 2017, p. 210.
14. A. Pehlken and D. H. Müller, Using information of the separation process of recycling scrap tires for process modeling, *Resour., Conserv. Recycl.*, 2009, **54**(2), 140.
15. V. Torretta, E. C. Rada, M. Ragazzi, E. Trulli, I. A. Istrate and L. I. Cioca, Treatment and disposal of tyres: two EU approaches. A review, *Waste Manage.*, 2015, **45**, 152.
16. J. Godlewska, Recovery and recycling of waste tires in Poland, *Procedia Eng.*, 2017, **182**, 229.
17. V. L. Schulman, *Post-consumer Tyres in the European Union*, DIK Workshop, Hannover, Germany, 2002.
18. J. Karger-Kocsis, L. Mészáros and T. Bárány, Ground tyre rubber (GTR) in thermoplastics, thermosets, and rubbers, *J. Mater. Sci.*, 2013, **48**(1), 1.
19. S. Ramarad, M. Khalid, C. T. Ratnam, A. L. Chuah and W. Rashmi, Waste tire rubber in polymer blends: a review on the evolution, properties and future, *Prog. Mater. Sci.*, 2015, **72**, 100.
20. M. Sienkiewicz, H. Janik, K. Borzędowska-Labuda and J. Kucińska-Lipka, Environmentally friendly polymer-rubber composites obtained from waste tyres: a review, *J. Cleaner Prod.*, 2017, **147**, 560.

21. D. Lo Presti, Recycled tyre rubber modified bitumens for road asphalt mixtures: a literature review, *Constr. Build. Mater.*, 2013, **49**, 863.
22. X. Shu and B. Huang, Recycling of waste tire rubber in asphalt and portland cement concrete: an overview, *Constr. Build. Mater.*, 2014, **67**, 217.
23. S. Wang, D. Cheng and F. Xiao, Recent developments in the application of chemical approaches to rubberized asphalt, *Constr. Build. Mater.*, 2017, **131**, 101.
24. N. Antoniou and A. Zabaniotou, Features of an efficient and environmentally attractive used tyres pyrolysis with energy and material recovery, *Renewable Sustainable Energy Rev.*, 2013, **20**, 539.
25. A. Quek and R. Balasubramanian, Liquefaction of waste tires by pyrolysis for oil and chemicals – a review, *J. Anal. Appl. Pyrolysis*, 2013, **101**, 1.
26. K. Formela, M. Cysewska and J. Haponiuk, Thermo-mechanical reclaiming of ground tire rubber via extrusion at low temperature: efficiency and limits, *J. Vinyl Addit. Technol.*, 2016, **22**(3), 213.
27. X. Colom, A. Faliq, K. Formela and J. Cañavate, FTIR spectroscopic and thermogravimetric characterization of ground tyre rubber devulcanized by microwave treatment, *Polym. Test.*, 2016, **52**, 200.
28. K. Formela, M. Klein, X. Colom and M. R. Saeb, Investigating the combined impact of plasticizer and shear force on the efficiency of low temperature reclaiming of ground tire rubber (GTR), *Polym. Degrad. Stab.*, 2016, **125**, 1.
29. W. M. Rzymski and H. J. Radusch, Thermoplastic elastomers manufactured of polymer blends, *Polimery*, 2002, **47**(4), 229.
30. E. E. Mastalygina, I. A. Varyana, N. N. Kolesnikova and A. A. Popov, Structural changes in the low-density polyethylene/natural rubber composites in the aqueous and soil media, *AIP Conf. Proc.*, 2016, **1736**, 020097.
31. S. Dubinin, Z. Hrdlička and V. Ducháček, Thermoplastic vulcanizates based on polyethylene and styrene-butadiene rubber. Part I: theoretical aspects, *Kautsch. Gummi Kunstst.*, 2011, **64**(6), 20.
32. A. B. Balaji, M. Khalid, C. T. Ratnam, F. M. Michael and R. Walvekar, Co-PP/EPDM blend optimization using D-optimal design for medical applications, *Polym.-Plast. Technol. Eng.*, 2017, **56**(2), 216.
33. R. Zhang, X. He, D. Yang and Z. Lai, Non-isothermal crystallization kinetics and segmental dynamics of high density polyethylene/butyl rubber blends, *Polym. Int.*, 2015, **64**(9), 1252.
34. H. Ismail, D. Galpaya and Z. Ahmad, Electron-beam irradiation of blends of polypropylene with recycled acrylonitrile-butadiene rubber, *J. Vinyl Addit. Technol.*, 2010, **16**(2), 141.
35. O. Y. Alothman, Processing and characterization of high density polyethylene/ethylene vinyl acetate blends with different VA contents, *Adv. Mater. Sci. Eng.*, 2012, **2012**, Article ID 635693.
36. V. Das, S. S. Gautam and A. K. Pandey, Effect of SBS content on low temperature impact strength, morphology and rheology of PP-cp/SBS Blends, *Polym.-Plast. Technol. Eng.*, 2011, **50**(8), 825.

37. P. Svoboda, K. Trivedi, D. Svobodova, P. Mokrejs, V. Vasek, K. Mori, T. Ougizawa and T. Inoue, Isothermal crystallization in polypropylene/ethylene–octene copolymer blends, *Mater. Chem. Phys.*, 2011, **131**, 84.
38. S. Jia, J. Qu, W. Liu, C. Wu, R. Chen, S. Zhai and Z. Huang, Thermoplastic polyurethane/polypropylene blends based on novel vane extruder: a study of morphology and mechanical properties, *Polym. Eng. Sci.*, 2014, **54**(3), 716.
39. S. Al-Malaika and E. J. Amir, Thermoplastic elastomers: Part III. Ageing and mechanical properties of natural rubber-reclaimed rubber/polypropylene systems and their role as solid phase dispersants in polypropylene/polyethylene blends, *Polym. Degrad. Stab.*, 1989, **26**, 31.
40. R. S. Rajeev and S. K. De, Thermoplastic elastomers based on waste rubber and plastics, *Rubber Chem. Technol.*, 2004, **77**(3), 569.
41. K. Fukumori and M. Matsushita, Material Recycling Technology of Crosslinked Rubber Waste, *R&D Review of Toyota CRDL*, 2003, **38**, 39.
42. K. Fukumori, M. Matsushita, M. Mouri, H. Okamoto, N. Sato, K. Takeuchi and Y. Suzuki, Dynamic devulcanization and dynamic vulcanization for recycling of crosslinked rubber, *Kautsch. Gummi Kunstst.*, 2006, **59**, 405.
43. S. Wang, Q. Wang, X. Wu and Y. Zhang, Asphalt modified by thermoplastic elastomer based on recycled rubber, *Constr. Build. Mater.*, 2015, **79**, 337.
44. S. Wang, C. Yuan and D. Jiaxi, Crumb tire rubber and polyethylene mutually stabilized in asphalt by screw extrusion, *J. Appl. Polym. Sci.*, 2014, **131**(23), 41189.
45. S. Wang and Y. Xie, Crumb tire rubber polyolefin elastomer modified asphalt with hot storage stability, *Prog. Rubber, Plast. Recycl. Technol.*, 2016, **32**(1), 25.
46. K. Formela, M. Sulkowski, M. R. Saeb, X. Colom and J. Haponiuk, Assessment of microstructure, physical and thermal properties of bitumen modified with LDPE/GTR/elastomer ternary blends, *Constr. Build. Mater.*, 2016, **106**, 160.
47. M. H. Karahrodi, O. M. Jazani, S. M. R. Paran, K. Formela and M. R. Saeb, Modification of thermal and rheological characteristics of bitumen by waste PET/GTR blends, *Constr. Build. Mater.*, 2017, **134**, 157.
48. J. George, K. T. Varughese and S. Thomas, Dynamically vulcanised thermoplastic elastomer blends of polyethylene and nitrile rubber, *Polymer*, 2000, **41**(4), 1507.
49. P. Dey, K. Naskar and G. B. Nando, Theoretical and morphological evaluation of dynamic viscoelasticity and thermo-mechanical characteristics of TPV composites, *Polymer*, 2015, **70**, 161.
50. M.-J. Wang, Effect of polymer-filler and filler-filler interactions on dynamic properties of filled vulcanizates, *Rubber Chem. Technol.*, 1998, **71**(3), 520.
51. T. Sakai, Screw extrusion technology – past, present and future, *Polimery*, 2013, **58**(11–12), 847.

52. J. Stasiek, K. Bajer, A. Stasiek and M. Bogucki, Co-rotation twin-screw extruders for polymer materials: a method for experimental studying the extrusion process, *Przem. Chem.*, 2012, **91**(2), 224.
53. K. Formela, M. Bogucki and P. Stączek, Effect of the conditions of thermomechanical reclaiming of ground tire rubber on the properties of reclaimed rubber materials, *Polimery*, 2015, **60**(1), 57.
54. S. L. Zhang, Z. X. Zhang, K. Pal, Z. X. Xin and J. K. Kim, Prediction of mechanical properties of polypropylene/waste ground rubber tire powder treated by bitumen composites via uniform design and artificial neural networks, *Mater. Des.*, 2010, **31**, 1900.
55. S. L. Zhang, Z. X. Zhang, K. Pal, Z. X. Xin and J. K. Kim, Prediction of mechanical properties of waste polypropylene/waste ground tire powder blends using artificial neutral networks, *Mater. Des.*, 2010, **31**, 3624.
56. H. M. da Costa, V. D. Ramos, W. S. da Silva and A. S. Sirqueira, Analysis and optimization of polypropylene (PP)/ethylene-propylene-diene monomer (EPDM)/scrap rubber tire (SRT) mixtures using RSM methodology, *Polym. Test.*, 2010, **29**(5), 572.
57. K. Formela and M. Bogucki, Use of response surface methodology in characteristics of properties of recycled high density polyethylene-ground tire rubber compositions, *Polimery*, 2014, **59**(6), 488.
58. X. Zhang, C. Lu and M. Liang, Properties of natural rubber vulcanizates containing mechanochemically devulcanized ground tire rubber, *J. Polym. Res.*, 2009, **16**(4), 411.
59. I. Mangili, M. Lasagni, M. Anzano, E. Collina, V. Tatangelo, A. Franzetti, P. Caracino and A. I. Isayev, Mechanical and rheological properties of natural rubber compounds containing devulcanized ground tire rubber from several methods, *Polym. Degrad. Stab.*, 2015, **121**, 369.
60. K. Formela, P. Stankiewicz, K. Kołacka, Ł. Piszczyk and J. Haponiuk, Effect of processing and devulcanization conditions on compatibility and properties of low density polyethylene/ground tire rubber compositions, *Przem. Chem.*, 2012, **91**(9), 1762.
61. R. Scaffaro, N. Tzankova Dintcheva, M. A. Nocilla and F. P. La Mantia, Formulation, characterization and optimization of the processing condition of blends of recycled polyethylene and ground tyre rubber: mechanical and rheological analysis, *Polym. Degrad. Stab.*, 2005, **90**, 281.
62. K. Formela, M. Cysewska and J. Haponiuk, The influence of materials dosing order and shear forces on the properties of high density polyethylene/ground tire rubber blends obtained in co-rotating twin screw extruder, *Polimery*, 2014, **59**(4), 324.
63. F. D. B. de Sousa, C. H. Scuracchio, G. H. Hu and S. Hoppe, Effects of processing parameters on the properties of microwave-devulcanized ground tire rubber/polyethylene dynamically revulcanized blends, *J. Appl. Polym. Sci.*, 2016, **133**(23), 43503.
64. Y. Zhang, J. Shen, T. Chen, Z. Xu, X. Chen and M. Zhang, Process for devulcanization of rubber, U.S. Patent Application 0082475, 2009.

65. K. Formela, M. Cysewska and J. Korol, Effect of compounding conditions on static and dynamic mechanical properties of high density polyethylene-ground tire rubber blends, *Int. Polym. Process.*, 2014, **29**(2), 272.
66. R. Sonnier, E. Leroy, L. Clerc, A. Bergeret, J. M. Lopez-Cuesta, A. S. Bretelle and P. Ienny, Compatibilizing thermoplastic/ground tyre rubber powder blends: efficiency and limits, *Polym. Test.*, 2008, **27**(7), 901.
67. K. Formela, Ł. Piszczyk, J. Haponiuk, K. Bajer and M. Cysewska, Interfacial adhesion evaluation in low-density polyethylene/elastomer blends, *J. Vinyl Addit. Technol.*, 2016, **22**, 492.
68. M. He, Y. Li, B. Qiao, X. Ma, J. Song and M. Wang, Effect of dicumyl peroxide and phenolic resin as a mixed curing system on the mechanical properties and morphology of TPVs based on HDPE/ground tire rubber, *Polym. Compos.*, 2015, **36**(10), 1907.
69. L. Mészáros, T. Bárány and T. Czvikovszky, EB-promoted recycling of waste tire rubber with polyolefins, *Radiat. Phys. Chem.*, 2012, **81**(9), 1357.
70. L. Mészáros, M. Fejős and T. Bárány, Mechanical properties of recycled LDPE/EVA/ground tyre rubber blends: effects of EVA content and post-irradiation, *J. Appl. Polym. Sci.*, 2012, **125**(1), 512.
71. M. M. Hassan, R. O. Aly, S. E. Abdel Aal, A. M. El-Masry and E. S. Fathy, Mechanochemical devulcanization and gamma irradiation of devulcanized waste rubber/high density polyethylene thermoplastic elastomer, *J. Ind. Eng. Chem.*, 2013, **19**(5), 1722.
72. M. M. Hassan, N. A. Badway, M. Y. Elnaggar and E. A. Hegazy, Synergistic effect of gamma radiation and peroxide on dynamic vulcanization of thermoplastic vulcanizes based on (devulcanized tubber)/polypropylene, *J. Vinyl Addit. Technol.*, 2014, **20**(3), 168.
73. M. M. Hassan, N. A. Badway, M. Y. Elnaggar and E. A. Hegazy, Effects of peroxide and gamma radiation on properties of devulcanized rubber/polypropylene/ethylene propylene diene monomer formulation, *J. Appl. Polym. Sci.*, 2014, **131**(16), 40611.
74. M. M. Hassan, N. A. Badway, M. Y. Elnaggar and E. A. Hegazy, Physical properties of irradiated thermoplastic elastomeric olefins based on (devulcanized rubber)/(high crystalline polypropylene) formulations, *J. Vinyl Addit. Technol.*, 2015, **21**(1), 33.
75. S. Ramarad, C. T. Ratnam, M. Khalid and A. L. Chuah, Improving the properties of reclaimed waste tire rubber by blending with poly(ethylene-co-vinyl acetate) and electron beam irradiation, *J. Appl. Polym. Sci.*, 2015, **132**(11), 41649.
76. S. Ramarad, C. T. Ratnam, M. Khalid, A. L. Chuahd and S. Hanson, Improved crystallinity and dynamic mechanical properties of reclaimed waste tire rubber/EVA blends under the influence of electron beam irradiation, *Radiat. Phys. Chem.*, 2017, **130**, 362.

77. M. N. Mali, A. A. Arakh, K. A. Dubey and S. T. Mhaske, Influence of triallyl cyanurate as co-agent on gamma irradiation cured high density polyethylene/reclaimed tire rubber blend, *Radiat. Phys. Chem.*, 2017, **131**, 66.
78. X. Colom, F. Carrillo and J. Cañavate, Composites reinforced with reused tyres: surface oxidant treatment to improve the interfacial compatibility, *Composites, Part A*, 2007, **38**(1), 44.
79. X. Colom, J. Cañavate, F. Carrillo, J. I. Velasco, P. Pages, R. Mujal and F. Nogues, Structural and mechanical studies on modified reused tyres composites, *Eur. Polym. J.*, 2006, **42**(10), 2369.
80. X. Colom, J. Cañavate, F. Carrillo and J. J. Suñol, Effect of the particle size and acid pretreatments on compatibility and properties of recycled HDPE plastic bottles filled with ground tyre powder, *J. Appl. Polym. Sci.*, 2009, **112**(4), 1882.
81. R. Sonnier, E. Leroy, L. Clerc, A. Bergeret and J. M. Lopez-Cuesta, Polyethylene/ground tyre rubber blends: Influence of particle morphology and oxidation on mechanical properties, *Polym. Test.*, 2007, **26**(2), 274.
82. J. F. H. Gamez, E. H. Hernandez, R. I. N. Cespedes, M. G. N. Velazquez, S. G. S. Rosales, F. S. Corral, P. G. Morones, S. F. Tavizon, R. D. de Leon and L. F. Cepeda, Mechanical reinforcement of thermoplastic vulcanizates using ground tyre rubber modified with sulfuric acid, *Polym. Compos.*, 2018, **39**(1), 229.
83. P. Nevatia, T. S. Banerjee, B. Dutta, A. Jha, A. K. Naskar and A. K. Bhowmick, Thermoplastic elastomers from reclaimed rubber and waste plastics, *J. Appl. Polym. Sci.*, 2001, **83**(9), 2035.
84. T. Luo and A. I. Isayev, Rubber/plastic blends based on devulcanized ground tire rubber, *J. Elastomers Plast.*, 1998, **30**(2), 133.
85. C. K. Hong and A. I. Isayev, Plastic/rubber blends of ultrasonically devulcanized GRT with HDPE, *J. Elastomers Plast.*, 2001, **33**(1), 47.
86. B. Sripornsawat, S. Saiwari, S. Pichaiyut and C. Nakason, Influence of ground tire rubber devulcanization conditions on properties of its thermoplastic vulcanizate blends with copolyester, *Eur. Polym. J.*, 2016, **85**, 279.
87. C. R. Kumar, I. Fuhrmann and J. Karger-Kocsis, LDPE-based thermoplastic elastomers containing ground tire rubber with and without dynamic curing, *Polym. Degrad. Stab.*, 2002, **76**(1), 137.
88. O. P. Grigoryeva, A. M. Fainleib, A. L. Tolstov, O. M. Starostenko, E. Lievana and J. Karger-Kocsis, Thermoplastic elastomers based on recycled high-density polyethylene, ethylene–propylene–diene monomer rubber, and ground tire rubber, *J. Appl. Polym. Sci.*, 2005, **95**(3), 659.
89. Z. Hrdlicka, A. Kuta and J. Hajek, Thermoplastic elastomer blends based on waste rubber and low-density polyethylene, *Polimery*, 2010, **55**(11–12), 832.
90. J. Cañavate, P. Casas, X. Colom and F. Nogues, Formulations for thermoplastic vulcanizates based on high density polyethylene,

ethylene-propylene-diene monomer, and ground tyre rubber, *J. Compos. Mater.*, 2011, **45**(11), 1189.
91. M. R. Abadchi, A. J. Arani and H. Nazockdast, Partial replacement of NR by GTR in thermoplastic elastomer based on LLDPE/NR through using reactive blending: its effects on morphology, rheological, and mechanical properties, *J. Appl. Polym. Sci.*, 2010, **115**(4), 2416.
92. K. Formela and J. Haponiuk, Curing characteristics, mechanical properties and morphology of butyl rubber filled with ground tire rubber (GTR), *Iran. Polym. J.*, 2014, **23**(3), 185.
93. K. Formela, J. Haponiuk, P. Stankiewicz, J. Ryl and K. Końska, Effect of slightly crosslinked butyl rubber and compatibilizer additives on properties of recycled polyethylene-ground tire rubber blends, *Przem. Chem.*, 2013, **92**(4), 444.
94. K. Formela and J. Haponiuk, Characterization and properties of LDPE/ground tire rubber/cross-linked butyl rubber blends, *J. Vinyl Addit. Technol.*, 2014, **20**(4), 237.
95. Y. Li, Y. Zhang and Y. X. Zhang, Morphology and mechanical properties of HDPE/SRP/elastomer composites: effect of elastomer polarity, *Polym. Test.*, 2014, **23**(1), 83.
96. K. Formela, J. Korol and M. R. Saeb, Interfacially modified LDPE/GTR composites with non-polar elastomers: from microstructure to macro-behavior, *Polym. Test.*, 2015, **42**, 89.
97. P. Lima, J. Oliveira and V. Costa, Partial replacement of EPDM by GTR in thermoplastic elastomers based on PP/EPDM: Effects on morphology and mechanical properties, *J. Appl. Polym. Sci.*, 2014, **131**(8), 40160.
98. L. Wang, F. Lang, S. Li, F. Du and Z. Wang, Thermoplastic elastomers based on high-density polyethylene and waste ground rubber tire composites compatibilized by styrene–butadiene block copolymer, *J. Thermoplast. Compos. Mater.*, 2014, **27**(11), 1479.
99. P. Phinyocheep, F. H. Axtell and T. Laosee, Influence of compatibilizers on mechanical properties, crystallization, and morphology of polypropylene/scrap rubber dust blend, *J. Appl. Polym. Sci.*, 2002, **86**(1), 148.
100. A. K. Naskar, S. K. De and A. K. Bhowmick, Thermoplastic elastomeric composition based on maleic anhydride–grafted ground rubber tire, *J. Appl. Polym. Sci.*, 2002, **84**(2), 370.
101. S. H. Lee, M. Balasubramanian and J. K. Kim, Dynamic reaction inside co-rotating twin screw extruder. II. Waste ground rubber tire powder/polypropylene blends, *J. Appl. Polym. Sci.*, 2007, **106**(5), 3209.
102. X. Lu, W. Wang and L. Yu, Waste ground rubber tire powder/thermoplastic vulcanizate blends: preparation, characterization, and compatibility, *J. Appl. Polym. Sci.*, 2014, **131**(3), 39868.
103. A. K. Ramezani and D. Rodrigue, Highly filled thermoplastic elastomers from ground tire rubber, maleated polyethylene and high density polyethylene, *Plast., Rubber Compos.*, 2013, **42**, 115.

104. K. Formela, J. Korol, M. Cysewska and J. Haponiuk, Effect of addition of HDPE modified with maleic anhydride and shear forces on properties of HDPE-ground tire rubber blends, *Przem. Chem.*, 2013, **92**(4), 512.
105. S. H. Lee, M. Balasubramanian and J. K. Kim, Dynamic reaction inside co-rotating twin screw extruder. I. Truck tire model material/polypropylene blends, *J. Appl. Polym. Sci.*, 2007, **106**(5), 3193.
106. J. Cañavate, F. Carrillo, P. Casas, X. Colom and J. J. Suñol, The use of waxes and wetting additives to improve compatibility between hdpe and ground tyre rubber, *J. Compos. Mater.*, 2010, **44**(10), 1233.
107. Y. Zhou and M. Fan, Recycled tyre rubber-thermoplastic composites through interface optimization, *RSC Adv.*, 2017, **7**, 29263.
108. R. Sulcis, L. Lotti, S. Coiai, F. Ciardelli and E. Passaglia, Novel HDPE/ground tyre rubber composite materials obtained through in-situ polymerization and polymerization filling technique, *J. Appl. Polym. Sci.*, 2014, **131**, 40313.
109. O. Grigoryeva, A. Fainleib, O. Starostenko, A. Tolstov and W. Brostow, Thermoplastic elastomers from rubber and recycled polyethylene: chemical reactions at interphases for property enhancement, *Polym. Int.*, 2004, **53**(11), 1693.
110. A. Tolstov, O. Grigoryeva, A. Fainleib, I. Danilenko, A. Spanoudaki, P. Pissis and J. Grenet, Reactive compatibilization of polyethylene/ground tire rubber inhomogeneous blends *via* interactions of pre-functionalized polymers in interface, *Macromol. Symp.*, 2007, **254**(1), 226.
111. P. Song, S. Li and S. Wang, Interfacial interaction between degraded ground tire rubber and polyethylene, *Polym. Degrad. Stab.*, 2017, **143**, 85.
112. Z. Hrdlička, P. M. M. Cebriá, V. Štefan and A. Kuta, Thermoplastic elastomeric blends based on waste tires and polyethylene: the role of rubber particle size, *Prog. Rubber, Plast. Recycl. Technol.*, 2016, **32**(3), 129.
113. Y. Lu, Y. Yang, P. Xiao, Y. Feng, L. Liu, M. Tian, X. Li and L. Zhang, Effect of interfacial enhancing on morphology, mechanical, and rheological properties of polypropylene-ground tire rubber powder blends, *J. Appl. Polym. Sci.*, 2017, **134**(40), 45354.
114. J. Karger-Kocsis, Waste tyre rubber – what to do next? *eXPRESS Polym. Lett.*, 2013, **7**(5), 406.
115. Z. X. Xin, Z. X. Zhang, K. Pal, D. J. Kang, S. H. Lee and J. K. Kim, Effects of waste ground rubber tire powder (WGRT) and chemical blowing agent content on the cell morphology and physicomechanical properties of injection-molded polypropylene/WGRT foams, *J. Vinyl Addit. Technol.*, 2009, **15**, 275.
116. Z. X. Xin, Z. X. Zhang, K. Pal, K. J. Kim, D. J. Kang, J. K. Kim and B. S. Bang, Microcellular structure of PP/waste rubber powder blends with supercritical CO_2 by foam extrusion process, *J. Cell. Plast.*, 2009, **45**, 499.
117. Z. X. Xin, Z. X. Zhang, B. S. Zhang, K. Pal, X. Deng, S. H. Lee and J. K. Kim, Expanded waste ground rubber tire powder/polypropylene composites: processing-structure relationships, *J. Compos. Mater.*, 2009, **43**, 3003.

118. Z. X. Xin, Z. X. Zhang, K. Pal, J. U. Byeon, S. H. Lee and J. K. Kim, Study of microcellular injection-molded polypropylene/waste ground rubber tire powder blend, *Mater. Des.*, 2010, **31**, 589.
119. J. Zhu, X. Zhang, M. Liang and C. Lu, Enhancement of processability and foamability of ground tire rubber powder and LDPE blends through solid state shear milling, *J. Polym. Res.*, 2011, **18**, 533.
120. D. Tian, X. Zhang, J. Zhu and C. Lu, Morphology and physical properties of composite foams based on low-density polyethylene and ground tire rubber, *J. Vinyl Addit. Technol.*, 2013, **19**, 105.

CHAPTER 4

Waste Rubber Based Composite Foams

MAPOLOKO MPHO PHIRI[a], MOTSHABI ALINAH SIBEKO[a], SHANGANYANI PERCY HLANGOTHI[a] AND MAYA JACOB JOHN*[a,b]

[a]Department of Chemistry, Faculty of Science, Nelson Mandela University, PO Box 1600, Port Elizabeth 6031, South Africa; [b]CSIR Material Science and Manufacturing, Polymers and Composites Competence Area, PO Box 1124, Port Elizabeth 6000, South Africa
*E-mail: mjohn@csir.co.za

4.1 Introduction

An estimated 22 million tonnes of rubber are processed per year throughout the world, the majority of which goes into production of tyres.[1] At the end of their useful life, massive amounts of the waste rubber are created. Waste rubber may be in the form of waste tyre rubber, waste latex gloves, and waste conveyor belts. The increase in stockpiles of waste tyres is a serious problem to the environment and the economy, hence most industrialised countries are in need of innovative solutions to address the issue.[2,3] Tyres, being highly engineered materials, can withstand most chemical

Green Chemistry Series No. 59
Rubber Recycling: Challenges and Developments
Edited by Jin Kuk Kim, Prosenjit Saha, Sabu Thomas, Józef T. Haponiuk and M. K. Aswathi

Published by the Royal Society of Chemistry, www.rsc.org

and biological attacks, making their disposal into landfill highly impractical and, in some countries, illegal.[4,5] In addition, the accumulation of waste tyres in massive stockpiles is a fire hazard that can result in toxic gases being released into the environment.[6] The well-known and adopted waste management strategy for waste materials advocates for prevention and re-use of the waste, preceding recycling, energy recovery and disposal.[7] For tyres, re-use is mainly limited to re-treading, which can only be done once, or at most twice for all pneumatic car tyres depending on the strength of the carcass. This means re-using waste tyres is neither adequate nor capable of eliminating the waste tyres problem. Alternative approaches often adopted for the valorisation of waste tyres include material recovery (*e.g.* rubber, carbon and metals) from tyres, through processes such as crumbing and devulcanization, and energy recovery through processes such as pyrolysis.

The use of waste ground tyre rubber (GTR) in composites is well established in the literature and has been used for decades.[8–11] In most of these applications, GTR served mainly as a damping additive and for improved impact in both concrete and in polymer based composites.[12–14] Studies have also reported that incorporation of up to 20% of fine crumb rubber aggregate and up to 15% of coarse GTR or 15% of a mixture of fine and coarse GTR particles are suitable for structural applications. Up to 30% GTR modified concrete was found to be the most suitable for non-structural applications.[14,15] However; other techniques such as foaming have also been widely used to attain similar properties such as sound damping.

Composite foams are lightweight materials formed when blowing agents/air-entrapping agents (AEA) are used to disperse air pockets throughout the composite structure. Blowing agents used in the past were mainly halogenated hydrocarbon based, which are harmful to the environment.[16,17] However, greener options have since been adopted to align the process with environmental legislation.[18–21] Foaming has been used in cement applications since ancient history, mainly to impart thermal insulation, shock and sound absorption properties.[22–28] Other advantages of foaming include low production costs, which has attracted the attention of a number of industries such as the construction[26] and light metal application[27] industries. This technique has also been used for polymers such as polystyrene,[29,30] polycarbonate,[31] nylon 6,[32] polyolefins,[33,34] polymethylmethacrylate,[35,36] polyurethanes[20] and rubber.[37–39]

A well-known disadvantage of foamed composites is the reduced mechanical properties as opposed to unfoamed versions especially for concrete.[20] As a result, research trends in this field have inclined towards combining both foaming and addition of waste rubber in one composite material to take advantage of the properties each technique offers. Such composites were reported to possess superior thermal and acoustic damping properties as compared to composites where each technique was used independently.[10,40,41]

4.2 Processing of Rubber Foam Composites

Foamed rubber products can be produced by using a chemical or physical blowing agent that forms bubbles in the sample during the compounding process. Common processing technologies of rubber foams include foam extrusion, foam injection moulding, compression moulding and micro-foaming. Of all these processing technologies, injection moulding is highly preferred due to the associated reduced weight, improved aesthetics, faster production cycle time, high stiffness to weight ratio and the possibility of manufacturing large complex shaped products with high accuracy. In the physical blowing process, an inert gas is injected into the polymer melt where it dissolves under high pressure. Physical blowing agents are advantageous to chemical blowing agents in terms of economy, ecology, and efficient processing and they include the use of nitrogen, carbon dioxide and water.[42] Another physical process is mechanical foaming, which is exclusively used for latex emulsions. In this case, the degree of foaming depends on the design and engineering of suitable mixing equipment.[43] In all processes, varying the amount of blowing agent and additives results in changes in the formation of bubbles and bubble sizes during foaming. This in turn affects the foam density, cell size and cell shape, which eventually influences the structural behaviour and various properties of the polymer foam product.

In the case of solid rubbers, azodicarbonamide and modified azodicarbonamide compounds are the commonly used chemical blowing agents along with the use of additives such as zinc oxide (ZnO) and stearic acid, which are used as cell stabilizers or cell nucleating agents.[44,45] At high temperatures and in the presence of water, azodicarbonamide compounds decompose in an exothermic reaction releasing gases such as nitrogen, carbon monoxide and ammonia.[21] In this process, uniform cell structures and surfaces are obtained through even distribution of the blowing agents. Table 4.1 presents the most common exothermic blowing agents.

4.2.1 Processing of Foamed Composites with GTR

Compounding of foamed composites containing GTR is achieved through two main techniques, namely mechanical stirring and through mechanical milling methods, where the latter includes pan-milling and two-roll mill mixing processes. Mechanical stirring processes are usually used for compounding of foamed polyurethane (PU)/GTR and foamed concrete/GTR composites. These processes entail either mixing the dehydrated polyol, with GTR for a few minutes at speed ranges around 1200–2000 rpm. Subsequently, other ingredients such as the blowing agent (water or pentane), the catalyst or mixture of, and surfactant are added with further stirring for an additional minute using an overhead stirrer fitted with special dispersion discs for a specified amount of time. This solution mixture, referred to as A, is then

Table 4.1 Common exothermic blowing agents.[43]

Blowing agent	Decomposition temperature [°C]	Gas yield in ml g^{-1},air, blowing agents	Comments
Azodicarbonamide (AZO ADCA)	205–215	270 N_2, CO, NH_3, CO_2	Used for elastomers and plastics
OXYbis (benzenesulphonyl hydrazide)	155–165	160 N_2, H_2O	Used for elastomers and plastics
p-Toluenesulphonyl hydrazide	120–130	120 N_2, H_2O	
Toluenesulphonyl semicarbazide	230–250	140 N_2, H_2O	Used for high temperature thermoplastics
5-Phenyltetrazole	215–225	220 N_2	Used for high temperature thermoplastics

mixed with polyisocyanide (B) and placed in the mould where foaming is allowed to take place. The foam is finally cured at a high temperature for a prescribed amount of time.[46,47] In other cases, the foam is prepared separately and then mixed with the wet concrete mixture containing GTR, placed in the moulds and then cured.[25,27] Alternatively, GTR is mixed with the foam prior to moulding and curing[48] or mixed with B instead of A as previously highlighted.[49]

Mechanical milling methods are employed for solid state mixing of rubber or thermoplastic matrices with GTR followed by foaming. These methods entail milling the matrix with GTR in a pan milling setup and then the powder from the pan-milling process is melt mixed with the blowing agent, crosslinking agent, and blowing co-agent. This melt is then placed in a mould and press cured at a high temperature.[50,51] A typical formulation for a foamed linear low density polyethylene (LDPE)/GTR composite is shown in Table 4.2.

Zhu *et al.* prepared foamed linear low density polyethylene (LDPE)/GTR composites by co-milling GTR powder with LDPE pellets in the pan-mill at 30 rpms for 20 cycles at ambient temperature. The pan-mill was equipped with a jack of circulating water to remove excess heat generated by the milling process. The product was then melt mixed for 4 min in a Brabender Plasti-Corder at 130 °C, at the same rotation speed while adding the blowing agent, crosslinking agent and blowing co-agent for an additional 6 minutes. The mixture was placed in a mould and pressed at 180 °C in a hydraulic press at 9 MPa for 15 min. Foaming occurred during rapid depressurization, and the obtained foams were allowed to cool down at room temperature.[51] In another study, the mechanochemically pan-milled GTR was blended with ethylene propylene diene monomer (EPDM) on a two-roll mixing mill (diameter, 150 mm; working distance, 320 mm; speed of the slow roll, 15 rpm; friction ratio, 1 : 1.4) at about 50 °C. Initially, the mixture of powdered GTR and virgin EPDM were masticated followed by

Table 4.2 Compound recipe used for foaming LDPE/GTR composites.[a] Redrawn with permission from the *Journal of Vinyl & Additive Technology*.[53]

	Formulation code										
Ingredients (phr)	1	2	3	4	5	6	7	8	9	10	11
LDPE	100	100	100	100	100	100	100	100	100	100	100
Milled GTR	—	10	20	30	40	50	—	—	—	—	—
Raw GTR	—	—	—	—	—	—	10	20	30	40	50
AC	5	5	5	5	5	5	5	5	5	5	5
DCP	2	2	2	2	2	2	2	2	2	2	2
Zinc oxide	4	4	4	4	4	4	4	4	4	4	4
Stearic acid	1.5	1.5	1.5	1.5	1.5	1.5	1.5	1.5	1.5	1.5	1.5

[a]The tabulated amounts are parts by weight.

mixing with other compounding ingredients for 10 minutes. The blends were placed in a mould followed by pressing in a hydraulic press at 160 °C and 8 MPa for 15 min. Post pressurization, the foam expanded almost immediately and the foam was allowed some time to cool down to room temperature.[41]

4.3 Properties of Foamed/GTR Composites

4.3.1 Morphological Properties

The microcellular and cellular structure of the foamed/GTR composites are usually studied *via* robust surface techniques such as scanning electron microscopy (SEM), transmission electron microscopy (TEM) and atomic force microscopy (AFM). Studies have found that the microstructure and morphology of foamed composites with GTR is controlled by a number of factors, such as the type and concentration of the AEA agent used,[10,21] concentration[52] and particle size of the GTR in the composite[46,57] and finally by the processing conditions.[21,31,53]

It is well known that the three-dimensional cross-linked structure of GTR results in poor interaction when added to polymer matrices.[31,47,54] As such, studies have been conducted to modify the structure of GTR by thermo-mechanical,[55,56] and mechano-chemical[46,48,57] devulcanization methods. Solid-state thermo-mechanical milling with a pan-mill reactor is widely used to break the cross-linked structure of GTR and achieve partial devulcanization due to its high capacity, relatively low cost, energy efficiency, and the fact that chemicals need not be used.[48] The pan-mill has a pair of three-dimensional scissors that apply strong shear forces to the material during milling. In addition, the technique also has multi-functions such as pulverization, dispersion, and mixing, as well as activation.[58] Generally, for polymer composites, the particle size of filler plays a significant role in the final properties of the material. During the pan-milling process, the particle size of GTR is controlled by varying the number of milling cycles. Zhang *et al.* investigated the particle size of GTR before pan milling, after 5 cycles and after 20 cycles of milling. Before milling, large particle sizes with the average size of 300 μm and smooth surface were observed, while; after milling, particle sizes decreased with the increase in the number of milling cycles. In addition, SEM images reported after 5 cycles of milling, showed irregular particles with rough surface, while a fluffy and ultrafine powder with a particle size of about 35 μm was observed after 20 cycles of milling.[57]

The foamability of polymer/filler composites and their resultant properties depends on factors such as foaming pressure, foaming elasticity and adhesion between the polymer and the filler.[57,59] For example, in the case of foamed EPDM/GTR composites, the presence of untreated GTR decreased the foaming pressure due to poor adhesion between the two components. This was attributed to the lack of compatibility that enhanced the formation of defects during foaming, thus allowing the gas to escape and inhibit bubbles from growing. On the other hand, increased foaming pressure was observed with devulcanized GTR and this was ascribed to the improved interfacial adhesion between the polymer and GTR.[41,57]

Polyurethane (PU) foams can form either open-cell or closed-cell structures with different properties such as density. SEM images of open and closed-cell structures are shown in Figures 4.1 and 4.2, respectively. Normally, a closed-cell structure is formed by using low amounts of physical blowing agent while an open-cell structure requires a high amount of blowing agent, especially water.[60,61] The disruption of cell structures was observed in the presence of untreated GTR particles in both closed and open cells. This resulted in less regular structures with various cell sizes as can be seen in Figures 4.1b and 4.2b. This behaviour was ascribed to the poor interfacial interaction between the PU and the large GTR particle sizes, which readily formed agglomerates and disrupted foamability. Nonetheless, foamed PU/GTR composites prepared with devulcanized GTR (shown in Figures 4.1c and 4.2c) showed more well-defined cells.[46] Similar observations were reported in the case of foamed LDPE/GTR composites.[50,52] The

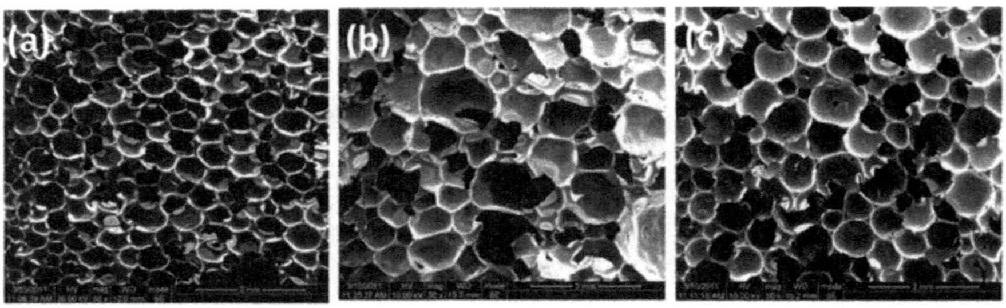

Figure 4.1 SEM images showing open cell of pristine PU form (a) and PU/GTR foamed composites (b) 10% untreated; (c) 10% devulcanized GTR. Reproduced from ref. 46 with permission from John Wiley and Sons, © 2012 Wiley Periodicals, Inc.

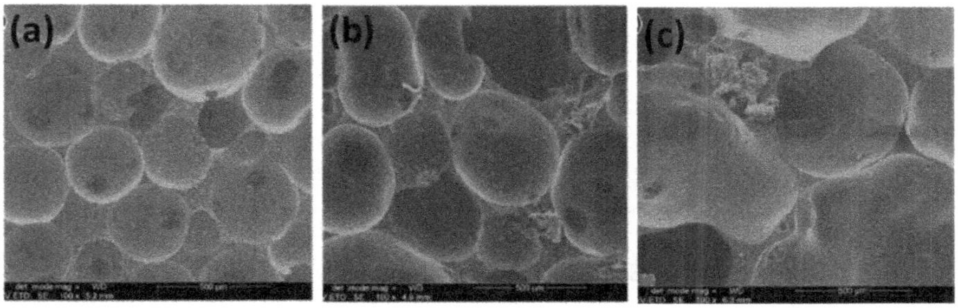

Figure 4.2 SEM images showing closed cell of pristine PU foam (a) and PU/GTR foamed composites with (b) 10% untreated; (c) 10% devulcanized GTR. Reproduced from ref. 55 under the terms of the CC BY 4.0 Licence, https://creativecommons.org/licenses/by/4.0/, © The Author(s) 2014.

influence of GTR particle size on the cell morphology of foamed PU/GTR composites was also investigated and found to be more uniform with the smallest GTR particle size.[46]

In the case of polypropylene (PP)/waste ground tyre rubber (WGTR) foamed composites, the average cell size decreases with the increase in WGTR loading compared to the pristine PP foam while the cell density and relative density increases.[62–64] The decreases in cell size was mainly attributed to the high viscosity of the composite which inhibited cell nucleation and growth. Uniform cell distribution was observed at 20–25 wt% WGTR loading, however, at 40 and 50 wt% WGTR loading, the cell distribution was non-uniform.[63,64] This was brought about by the weak surface adhesion between the PP and WGTR, aiding channel formation through which CO_2 escaped from the composite. Compatibilizers such as PP-g-MA[65] and SEBS-g-MA[66] have been introduced in PP/WGTR foamed composites. The addition of SEBS-g-MA at 5 and 20 wt% loading further decreased the cell size of the PP/WGTR foamed composites compared to the composites without the compatibilizer. Thus, SEBS-g-MA improved the interfacial interaction between PP and WGTR and hence prevented channel formation.

In cement, the morphological changes in the rubberised composites showed two types of porous structures. Firstly, the porosity typical of the cement matrix and secondly, the porosity induced by the presence of GTR in the matrix.[67] In a similar notion, it was shown that in foamed GTR cement composites, the higher GTR content promoted air entrapment in the composites even at higher AEA concentration. Thus, GTR helped stabilize air bubbles in the composites regardless of the aeration level.[10] In addition, the SEM images revealed that around the GTR aggregates, hydration products were of a different nature from the ones distributed elsewhere in the composite matrix. It was therefore suggested that the calcium aluminium sulfate hydration product tended to migrate to the places of low pressure, which are located near to cavities and large pores.[10]

4.3.2 Physical Properties

Physical properties such as the rebound resilience of polymer foams are important for applications such as sealing and packaging. The increase in rebound resilience for foamed composites of EPDM/GTR[41] and foamed LDPE/GTR[50] were reported upon the addition of devulcanized GTR, this was attributed to the stress transfer from the EDPM to GTR. As GTR is elastic, it enhanced the rebound resilience. Similar observations were reported in the case of foamed LDPE/GTR composites. In the case of foamed EPDM/GTR composites with untreated GTR, a decrease in rebound resilience was observed and ascribed to the lack of stress transfer due to poor interfacial adhesion between untreated GTR and EPDM foam. Furthermore, the presence of GTR aggregates serve as defect points that diminish the rebound resilience properties of the composites.

The incorporation of GTR into polymer foams influenced the density of the foamed polymer composites.[48] Generally, the increase in density of foamed PU/GTR composites was observed with the addition of both untreated[68,69] and devulcanized GTR.[55,58] This was ascribed to the high density of GTR compared to that of PU foam. However, foamed PU/GTR composites containing devulcanized GTR particles showed an even higher density increase, due to the improved interfacial interaction between the components, which improved the foamability and formed a well-defined cell structure.

The process of foaming and partially replacing sand with GTR in cement and concrete applications has been found to result in the reduced density of composites. This has been ascribed to the low density of GTR (1.1 g cm^{-3}) as opposed to that of sand (2.7 g cm^{-3}).[9] On the other hand, air pockets in the matrix of the composite also contributed to the lightweight property of the said composites. These composites were also reported to have decreased viscosities and as such were easily workable as compared to the neat composites. Furthermore, the reduced workability of the concrete/GTR foams was observed by a reduction in the compaction factor[10,70] or slump height of the mortars.[71,72] However, these composites had significantly higher water absorption due to high porosity.[10,73]

4.3.3 Mechanical Properties

The weak interfacial linkage between the GTR and concrete has been reported to result in upto a 90% reduction in the mechanical properties of the foamed cement composites, such as the mechanical, flexural, and compression strength. Compressibility tests are usually studied according to ASTM C39[74] standards where the maximum load carried and compressive strength by the specimen is recorded and calculated using a formula similar to the one shown in eqn (4.1).

$$f'c = \frac{P}{A} \tag{4.1}$$

where $f'c$ is the compressive strength in MPa, P is the maximum load carried by the specimen, N, A (width × thickness) is the surface area of the specimen carrying the load in mm^2. The compressibility of foamed cement composites was found to decrease as a function of increasing GTR content in the composites. However, it increased with aging of the composite for 7, 21 and 28 days.[10,70] Similar results were previously reported for cement/GTR composites in which the detrimental decline of mechanical properties was assigned to the air void formation which was promoted by the incorporation of GTR.[15] Thus, the air pockets in composites exacerbate the weakened interfacial bond between the two matrices. The same reasoning was acknowledged for the decline in flexural and tensile strength.

It has been established that the shape of cells and the density of the foam have a significant influence on the mechanical properties. In several studies, it was found that the compression strength of foamed PU increases with the addition of both untreated[68,69,75] and devulcanized GTR.[48,55] However, the addition of untreated GTR decreased the compression strength of foamed PU/GTR composite at higher loading, and the maximum strength was observed at 6%[75] and 1.4%[70] loadings. The decrease in composites strength was associated with the presence of agglomerates and their tendency to tear the walls of the cell structure. In the case of devulcanized GTR, an increase in compression strength of foamed PU/GTR composites was reported with the increase in GTR loading. This behaviour was related to the improved interaction between the PU foam and GRT.[48,55]

The particle size of a filler has a significant influence on the mechanical properties of polymer composites, and as such, the influence of the GTR particle size on the mechanical properties of polymer foams has been reported.[46,52,57] Without milling, the tensile strength and elongation at break of foamed EPDM/GTR composites were reported as 0.27 MPa and 290.7% respectively, and after 20 cycles of milling, the tensile strength increased to 0.72 MPa and elongation at break increased to 748.2%. The enhanced compatibility between the devulcanized GTR and EPDM foam was attributed to the large surface area and activated GTR surface enhanced stress transfer.[57] In another study, it was reported that the addition of untreated GTR also increased the tensile strength of foamed PU/GTR composites, however, the elongation at break only increased at low GTR loading and decreased significantly as the loading increased.[68] It was observed that the addition of GTR increased the thickness of the cell wall from 91.95 to 102.15 μm, which ultimately increased the stiffness of the foamed composites and decreased the ability of the foamed composites to elongate.[68]

4.3.4 Damping Properties

Damping is described as a measure of energy loss in the form of friction as the material is perturbed by vibration. This property is measured *via* ultrasonic pulse velocity measurements based on the model proposed by Albano *et al.*[71] or through vibration measurements. In cement, the ultrasonic pulse velocity and resistance indicated an increase in damping as a function of the increase in AEA content and the increase in the percentage of GTR in the composites. The vibration damping properties on the other hand were found to increase linearly with increasing GTR and AEA contents in the composites.[10] The increase in damping was attributed to the increase in the number of discontinuities in the material. Thus, vibration damping was found to be higher in foamed GTR composites when compared to standard cement mortar. Additionally, an increase of between 2–4 times the damping coefficient of the reference mortar was observed for rubberised foamed composites. These can be observed on the time histories obtained from the fundamental transverse frequency test for mortars containing 40% GTR as a function of the increasing AEA content (Figure 4.3).[10] These results corroborated the findings of Najim and Hall[76] on rubberized concrete, however, in foamed GTR composites the damping was found to be much more pronounced.

Foams have been used in thermal insulation and acoustic absorption applications, but, the ability of the material to be used as an absorbent depends on its sound absorption efficiency.[77,78] As such, materials that are

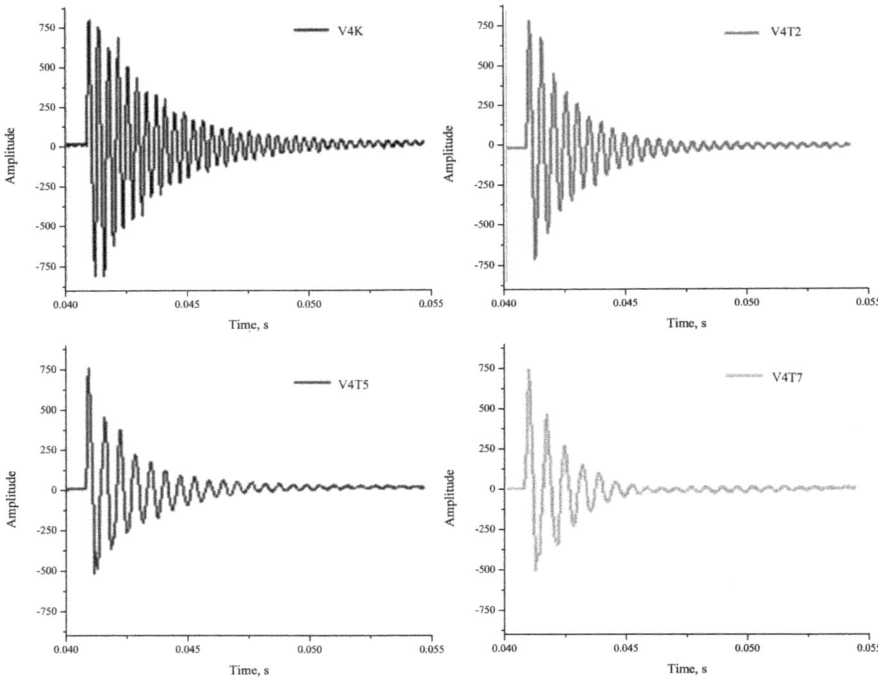

Figure 4.3 Typical signal responses in time domain for mortars V4K (40% GTR and 0% AEA), V4T2 (40% GTR and 0.250% AEA), V4T5 (40% GTR and 0.0500% AEA) and V4T7(40% GTR and 0.0750% AEA) at 25 kHz. Reproduced from ref. 10 with permission from Elsevier, Copyright 2014.

able to absorb sound waves in wide frequency regions are desired. Gayathri *et al.*[49] studied the sound absorption efficiency of foamed composites over the frequency range of 100–200 Hz while Zhang *et al.*[46] focused on the middle frequency range of 500–1500 Hz. In both studies it was found that the sound absorption efficiency of PU foam increased upon the addition of GTR. Nonetheless, foamed PU/GTR composites prepared with devulcanized GTR showed higher sound absorption properties compared to untreated GTR. In the presence of devulcanized GRT, the sound absorption coefficient increased from 0.15 to 0.35 at 30 wt% loading and this phenomenon was attributed to the increase in the viscoelastic properties of GTR, as the devulcanization of GTR facilitated the movement of rubber chains. Viscoelastic polymer foams are reported to have a good performance in acoustic absorption due to their ability to weaken certain vibrations and absorb sound energy.[79]

4.3.5 Thermal Properties

Thermal properties of composites are usually studied by employing differential scanning calorimetry (DSC), thermogravimetric analysis (TGA), dynamic mechanical analysis (DMA) and thermal conductivity measurements. However, the thermal behaviour of foamed polymer composites is limited.[55,68] For instance, the thermal stability of foamed PU/GTR composites prepared with untreated (R) and devulcanized (D) GTR slightly increased when compared to pristine PU foam (Figure 4.4), and this was ascribed to the presence of carbon black in GTR.[55,68] Also, the addition of pan-milled or untreated GTR was found to increase the glass transition temperature of PU foams, as can be seen in Figure 4.5. This increase in the glass transition temperature was attributed to the decreased mobility of the polymer chains which resulted in the stiffening of the PU foams.[55]

Thermal conductivity studies of foamed concrete showed a significant decrease in thermal conductivity with the incorporation of GTR (Figure 4.6). These observations were ascribed to the lower thermal conductivity of GTR relative to that of sand particles, as well as the facilitation of air absorption by GTR.[80]

4.4 Studies of Waste Rubber Foams

The microstructure of foamed reclaimed natural rubber from waste latex gloves (r-NRG) was studied and the effect of varying the sodium bicarbonate (SBC) concentration on the compressive and water contact behaviour of foams was studied. The blowing agent concentration was varied from 4, 6, 8, 10 and 12 phr and then melt mixed at 60 °C with r-NRG in an internal mixer. Higher concentrations of blowing agent subsequently generated more gaseous media and hence reduced the relative foam density, and cell structures with large pore sizes were formed. This phenomenon was described by

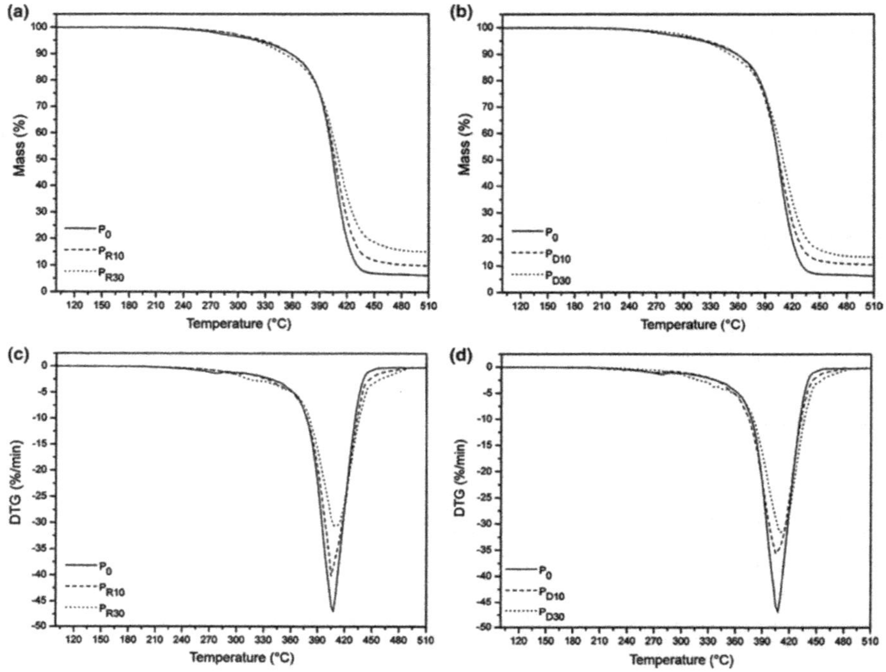

Figure 4.4 Mass loss ((a) untreated (R) and (b) devulcanized (D) GTR) and differential thermogravimetric curves ((c) untreated (R) and (d) devulcanized (D) GTR) as a function of temperature for polyurethane foams filled with ground tire rubber. Reproduced from ref. 55 under the terms of the CC BY 4.0 Licence, https://creativecommons.org/licenses/by/4.0/, © The Author(s) 2014.

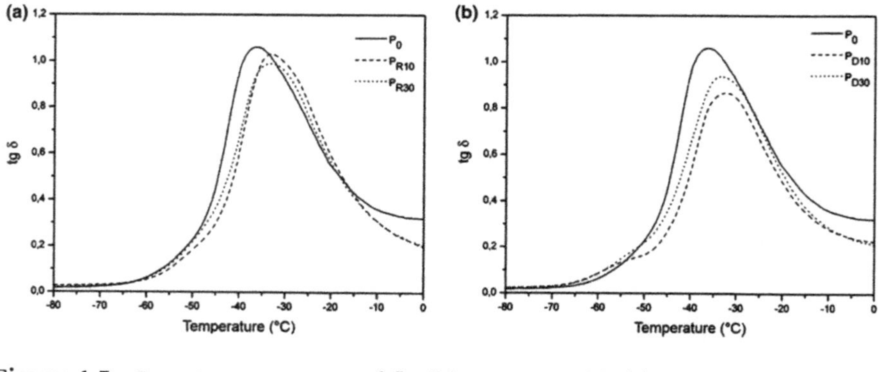

Figure 4.5 Loss tangent curves of flexible PU/GTR with (a) untreated (R) and (b) treated (D) GTR. Reproduced from ref. 55 under the terms of the CC BY 4.0 Licence, https://creativecommons.org/licenses/by/4.0/, © The Author(s) 2014.

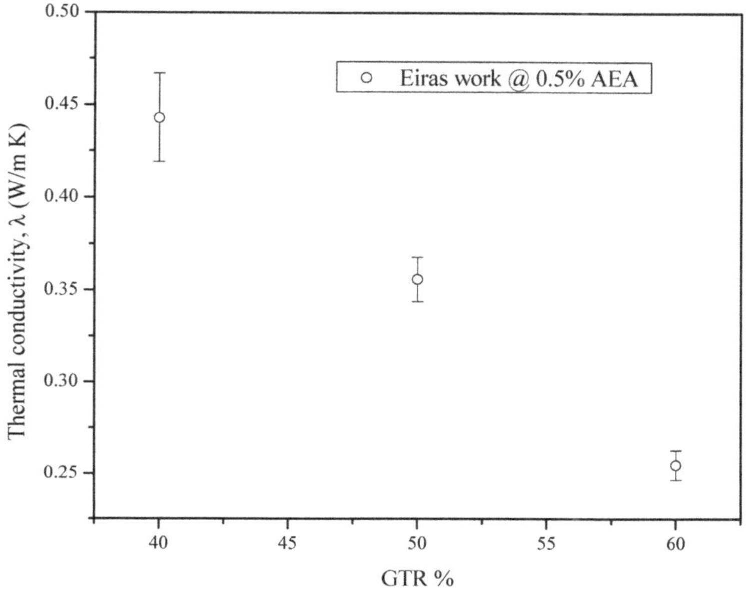

Figure 4.6 Thermal conductivity with GTR content. Reproduced from ref. 10 with permission from Elsevier, Copyright 2014.

Zakaria,[81] who reported that high blowing agent concentrations shorten the growth time of the foam. Thus, minimal gas bubbles are escaping through the foam surface allowing more foam expansion and consequently resulting in reduction of the foam relative density. Additionally, due to the open cell structure, more water absorption was realised while a reduction in compressive strength was observed upon increasing the blowing agent concentration in the foams. In this study, a blowing agent at a concentration of 4 phr was found to have the highest compressive strength, highest relative density and lowest water absorption.[82] Chin *et al.* studied the foamed blends of reclaimed NR from latex gloves (r-NRG) with Malaysian Natural Rubber (SMR 20). The blends were prepared by varying the r-NRG content from 20–95% using a two roll mill compounded with ADC as a blowing agent. The cure characteristics and scotch time of the blends decreased with the increase of r-NRG in the blends due to the semi cross-linked nature of r-NRG. In addition, a detrimental decline in the tensile properties and elongation at break was observed with an increase of r-NRG in the blends. This indicated poor compatibility between the SMR 20 and r-NRG phases.[83] Blends of epoxidised natural rubber (ENR) and reclaimed rubber glove blends were also prepared using SBC as a blowing agent using melt compounding. It was shown in this study that the ENR/RR with the blend ratio of 90/10 yielded the rubber foam with the highest relative density of 0.85, the lowest water absorption rate of 0.1 g h^{-1} and an improved energy absorption behaviour, amongst other properties.[84]

Composites of foamed LDPE/waste EPDM (w-EPDM) were prepared by mixing LDPE pellets and waste EPDM powder (0, 30 and 40%) using a twin-screw extruder.[85] The influence of foaming conditions such as pressure, temperature and the amount of w-EPDM on the foamed microcellular structure and mechanical properties were reported. The addition of w-EPDM decreased the microcellular size of the foam LDPE, and at 30% w-EPDM loading the cell pattern and distribution was mostly uniform. However, at 40% the microcellular structure was uneven due to presence of w-EPDM agglomerates. The same behaviour was reported in the case of foamed LDPE/waste GTR composites.[50] The most uniform microcell patterns with smaller dimensions were observed at optimum conditions of 30% w-EPDM, 100 °C and 200 bar. The density and the compression strength of the composites increased with an increase in w-EDPM loading. On the other hand, the tensile strength and elongation at break of the foamed LDPE composites decreased with the increase in w-EPDM loading. This was attributed to the lack of compatibility of highly cross-linked w-EPDM and LDPE which resulted in inferior mechanical properties. However, the tensile strength of the foamed composites was improved by incorporating DCP at 1.50 phr loading. The presence of DCP induced crosslinking resulting in finer and more uniformly arranged microcells, ultimately increasing the tensile strength and compression strength even more. Furthermore, the increase in crystallinity, and the number of crosslinking sites between w-EPDM and LDPE decreased the elongation at break.[85]

Various amounts of recycled EPDM (r-EDPM) (0, 35, 50 and 65 wt%) was blended with virgin PP using twin-screw extrusion.[86] The samples were pelletized and foamed by injection moulding, using ACD as a blowing agent. The tensile strength, tensile modulus, flexural modulus and hardness of the foamed PP decreased with the increase in r-EPDM loading. This was due to the elastic nature of r-EPDM which lowered the stiffness of the composites, and as a result, elongation at break increased. The toughness of the composites also increased with r-EPDM loading, hence, the soft EDPM absorbed energy through elastic deformation.

4.5 Applications of Waste Rubber Foam Composites

4.5.1 Non-structural Applications

Foamed concrete/GTR composites lack mechanical strength and as such have limited use in structural applications. Eiras *et al.* found that the mechanical strength of these composites ranged between 1–10 MPa making them suitable for non-structural applications. The authors also found that composites containing 40–60% GTR at AEA concentrations ranging between (0.125–0.500%) met the requirements[10] for application in masonry units. Another application which has been patented is in the production of composite boards for re-roofing a roof deck using foamed polyurethane filled with GTR.[87]

4.5.2 Lightweight Applications

Flexible foamed polyurethane/GTR composites have been investigated for use as floating trays and application as compression-absorbing buoys. For these applications, it is required that the apparent density of the composite be lower than that of water to achieve the floatability in water, as well as good compressive stress absorption.[47] For the floating trays, the foam with a bulk apparent density of 89 kg m^{-3} was reported to be ideal and was composed of 150% of rubber, 4% of water, 0.6% of Tegostab B8225, 0.086% of DABCO (1,4-diazabicyclo[2,2,2] octane, 33LV, Air products) catalyst, 0.25% of DBTL (dibutyltin dilaurate, T12, Air Products) catalyst, and an isocyanate index of 105%. From the compression tests, the foam capable of absorbing the highest compressive stress was composed of 200% of rubber, 3% of water, 0.6% of Tegostab B8225, 0.086% of DABCO catalyst, 0.25% of DBTL catalyst, an isocyanate index of 105%, with an apparent density of 121 kg m^{-3}. This foam absorbed a compressive stress of 0.07 MPa when subjected to deformation of up to 60%.[47]

4.5.3 Sound and Vibration Absorption

Foamed concrete/GTR and foamed polyurethane/GTR composites have been shown to display excellent acoustic absorption, and as such they find applications in sound proofing rooms.[37] Mechano-chemical treated foamed PU/GTR composites were also reported to have better sound absorption, especially in the frequency region near 1000 Hz.[46]

4.5.4 Insulation and Impact Isolation

The low thermal conductivity of the foamed concrete/GTR composites makes them ideal for thermal energy saving in buildings. Thus, these materials can be used as a base for roofing to lower the heat transfer in and out of the building.[37] These materials also possess high damping properties suitable for use in impact isolation.

4.5.5 Drainage Systems

As foamed composites have high porosity, it has been suggested that they could be ideal for use in drainage systems to capture storm water and in pavements to replenish underground water.[37,88]

4.6 Concluding Remarks

This chapter gives an overview of waste rubber composite foams, their processing techniques and applications in various sectors. The use of waste rubber from different sources, for example waste GTR, waste latex glove and waste EPDM, in composite foams have been discussed. As waste rubber foams are lightweight and have insulating properties, they are commonly used for thermal insulation, energy absorption and non-structural applications. Future research will focus on improving the strength properties of the rubber based foams.

References

1. European-Commission, Converting waste rubber into quality products Europe 2012. Available from http://ec.europa.eu/research/infocentre/article_en.cfm?id=/research/headlines/news/article_12_11_30_en.html&item=Infocentre&artid=28453, Cited 18 Oct 2017.
2. CIWMB, *California Waste Tire Generation, Markets, and Disposal*, 2006 CIWMB Staff Report, Sacramento, Califonia, 2003.
3. A. Uruburu, E. Ponce-Cueto, R. J. Cobo-Benita and J. Ordieres-Meré, *Waste Manage.*, 2013, **33**, 679.
4. C. Hammer and T. A. Gray, "*Designing building products made with recycled tires*", California Integrated Waste Management Board, California: 2004 June 2004. Report No.
5. K. C. Baranwal, "*Akron Rubber Development Laboratory*" ASTM Standards and Testing of Recycle Rubber San Francisco, California 2003 [updated 29 Apr 2003].
6. S. Perez, Spanish tyre dump fire triggers evacuation of 9,000 people Spain, *The Guardian*, 2016[updated 13 May 2016], Available from https://www.theguardian.com/world/2016/may/13/spanish-tyre-dump-fire-triggers-evacuation-of-9000-people.
7. European-Commission, Directive 2008/98/EC on waste (WasteFramework Directive) 2008 [updated 09 Jun 2016 cited 08 Sep 2017], Available from http://ec.europa.eu/environment/waste/framework/.
8. N. I. Fattuhi and L. A. Clark, *Constr. Build. Mater.*, 1996, **10**, 229.
9. L. He, Y. Ma, Q. Liu and Y. Mu, *Constr. Build. Mater.*, 2016, **120**, 403.
10. J. N. Eiras, F. Segovia, M. V. Borrachero, J. Monzó, M. Bonilla and J. Payá, *Mater. Des.*, 2014, **59**, 550.
11. S. L. Zhang, Z. X. Xin, Z. X. Zhang and J. K. Kim, *Waste Manage.*, 2009, **29**, 1480.
12. R. Sulcis, L. Lotti, S. Coiai, F. Ciardelli and E. Passaglia, *J. Appl. Polym. Sci.*, 2014, **131**, 40313.
13. R. Maderuelo-Sanz, M. Martín-Castizo and R. Vílchez-Gómez, *Appl. Acoust.*, 2011, **72**, 823.
14. M. R. Hall, K. B. Najim and C. J. Hopfe, *Appl. Therm. Eng.*, 2012, **33–34**, 77.
15. K. B. Najim and M. R. Hall, *Constr. Mater.*, 2012, **166**, 7.
16. K. Ashida and K. Ashida, Pat. 4898893 A, 1990.
17. H. P. Doerge, Pat. 4960804 A, 1990.
18. B. Krause, R. Mettinkhof, N. F. A. van der Vegt and M. Wessling, *Macromolecules*, 2001, **34**, 874.
19. R. Butler, I. Hopkinson and A. I. Cooper, *J. Am. Chem. Soc.*, 2003, **125**, 14473.

20. X. Dai, Z. Liu, Y. Wang, G. Yang, J. Xu and B. Han, *J. Supercrit. Fluids*, 2005, **33**, 259.
21. Z. X. Xin, Z. X. Zhang, K. Pal, B. X. Lu, X. Deng and S. H. Lee, *J. Vinyl Addit. Technol.*, 2009, **15**, 266.
22. C. P. Park, M. Brucker, G. Eschenlauer, M. E. Schaller and J. F. Koenig, Pat. 6583193 A, 2003.
23. N. M. Zahari, I. A. Rahman, A. Zaidi and A. Mujahid, Malaysian Technical Universities Conference on Engineering and Technology (MUCEET); 2009, MS Garden, Kuantan, Pahang, Malaysia, 20–22 June 2009.
24. A. P. Proshin, V. A. Beregovoi, A. M. Beregovoi and A. I. Eremkin, *Use of Foamed Concrete in Construction*, 2005, pp. 113–120.
25. A. Giannakou and M. R. Jones, *Innovations and Developments in Concrete Materials and Construction*, 2002, pp. 533–544.
26. B. B. Quist and J. J. M. Zuiderwijk, Pat. 3887385 A, 1975.
27. F. García-Moreno, *Materials*, 2016, **9**, 85.
28. M. T. Ergene, Pat. 3867159 A, 1975.
29. J. B. Sisson, Pat. 2917217 A, 1959.
30. D. Wiley, Pat. 3814779 A, 1974.
31. H. Chen, R. Guan, J. Zhao, S. Jiang, Z. Ke and S. Zha, *J. Wuhan Univ. Technol. Mater. Sci. Ed.*, 2013, **28**, 235.
32. H. Yousefian and D. Rodrigue, *J. Cell. Plast.*, 2017, **53**, 253.
33. Y. W. Shin, Pat. 4746564, 1988.
34. E. H. Tejeda, C. Z. Sahagún, R. González-Núñez and D. Rodrigue, *J. Cell. Plast.*, 2005, **41**, 417.
35. S. Siripurapu, J. M. DeSimone, S. A. Khan and R. J. Spontak, *Macromolecules*, 2005, **38**, 2271.
36. J. Pinto, D. Morselli, V. Bernardo, B. Notario, D. Fragouli, M. A. Rodriguez-Perez and A. Athanassiou, *Polymer*, 2017, **124**, 176.
37. L. Shao, J. Z. Ji, J. Z. Ma, C. H. Xue, Z. L. Ma and J. Zhang, *Nature*, 2016, **6**, 36931.
38. N. N. Najib, Z. M. Ariff, A. A. Bakar and C. S. Sipaut, *Mater. Des.*, 2011, **32**, 505.
39. L. Oliveira-Salmazo, A. Lopez-Gil, F. Silva-Bellucci, A. E. Job and M. A. Rodriguez-Perez, *Ind. Crops Prod.*, 2016, **80**, 26.
40. K. A. Paine and R. K. Dhir, *Constr. Mater.*, 2010, **163**, 7.
41. D. Zhang, W. Tian, J. Zhang, H. Zhu and C. Lu, *Prog. Rubber, Plast. Recycl. Technol.*, 2013, **29**, 81.
42. W. Michaeli, K. Westermann and S. Sitz, *J. Cell. Plast.*, 2011, **47**, 483.
43. J. Stehr, *Gummi, Fasern, Kunstst.*, 2015, 812.
44. Z. X. Xin, Z. X. Zhang, K. Pal, J. U. Byeon, S. H. Lee and J. K. Kim, *Mater. Des.*, 2010, **31**, 589.
45. H. J. Tai and J. B. Wang, *J. Cell. Plast.*, 1997, **33**, 304.
46. X. Zhang, Z. Lu, D. Tian, H. Li and C. Lu, *J. Appl. Polym. Sci.*, 2013, **127**, 4006.
47. A. G. Cachaço, M. D. Afonso and M. L. Pinto, *J. Appl. Polym. Sci.*, 2013, **129**, 2873.

48. L. Piszczyk, A. Hejna, M. Danowska, M. Strankowski and K. Formela, *J. Reinf. Plast. Compos.*, 2015, **34**, 708.
49. R. Gayathri, R. Vasanthakumari and C. Padmanabhan, *Int. J. Sci. Eng. Res.*, 2013, **4**, 301.
50. D. Tian, X. Zhang, J. Zhu and C. Lu, *J. Vinyl Addit. Technol.*, 2013, **19**, 105.
51. J. Zhu, X. Zhang, M. Liang and C. Lu, *J. Polym. Res.*, 2011, **18**, 533.
52. D. Tian, X. Zhang, J. Zhu and C. Lu, *J. Vinyl Addit. Technol.*, 2013, **19**, 105.
53. S. V. Patel, *The Effect of Processing Conditions and Composition on Starch Microcellular Foam Properties*. 2009.
54. X. Zhang, X. Zhu, M. Liang and C. Lu, *J. Appl. Polym. Sci.*, 2009, **114**, 1118.
55. L. Piszczyk, A. Hejna, K. Formela, M. Danowska and M. Strankowski, *Iran. Polym. J.*, 2015, **24**, 75.
56. L. Piszczyk, A. Hejna, K. Formela, M. Danowska and M. Strankowski, *Cell. Polym.*, 2015, **34**, 45.
57. X. Zhang, D. Tian, W. Zhang, J. Zhu and C. Lu, *Prog. Rubber, Plast. Recycl. Technol.*, 2013, **29**, 81.
58. C. Lu, X. Zhang, M. Liang, *Mechanical Recycling and Processing of Waste Crosslinked Polymers: Waste Tire Rubber and Waste XLPE from Cable Scraps, the 5th ISFR*. 2009.
59. S. Datta, A. Bhattacharya, S. De, E. Kontos and J. Wefer, *Polymer*, 1996, **37**, 2581.
60. S. N. Singh, J. S. Costa Jr. and R. E. Berrier, Pat. 1999061504 A, 2001.
61. K. Yuge and H. Muramatsu, Pat. 5457128 A, 1995.
62. Z. X. Xin, Z. X. Zhang, B. S. Zhang, K. Pal, X. Deng and S. H. Lee, *J. Compos. Mater.*, 2009, **43**, 3003.
63. Z. X. Xin, Z. X. Zhang, K. Pal, K. J. Kim, D. J. Kang and J. K. Kim, *J. Cell. Plast.*, 2009, **45**, 499.
64. Z. X. Zhang, S. L. Zhang, Z. X. Xin and J. K. Kim, *e-Polymers*, 2007, **7**, 1556.
65. Z. X. Zhang, S. L. Zhang, S. H. Lee, D. J. Kang, D. Bang and J. K. Kim, *Mater. Lett.*, 2008, **62**, 4396.
66. Z. X. Zhang, S. H. Lee, J. K. Kim, S. L. Zhang and Z. X. Xin, *Macromol. Res.*, 2008, **16**, 404.
67. M. Turki, E. Bretagne, M. Rouis and M. Quéneudec, *Constr. Build. Mater.*, 2009, **23**, 2715.
68. R. Gayathri, R. Vasanthakumari and C. Padmanabhan, *Int. J. Sci. Eng. Res.*, 2013, **4**, 301.
69. K. Formela, A. Hejna, L. Zedler, M. Przybysz, J. Ryl and M. R. Saeb, *Ind. Crops Prod.*, 2017, **108**, 844.
70. S. N. Shah, P. D. Jadhao and S. Dumne, *Am. J. Eng. Res.*, 2014, **3**, 93.
71. C. Albano, N. Camacho, J. Reyes, J. Feliu and M. Hernández, *Compos. Struct.*, 2005, **71**, 439.
72. H. A. Toutanji, *Constr. Build. Mater.*, 1999, **13**, 171.
73. A. A. Hilal, *J. Eng. Sci.*, 2011, **4**, 1.
74. A. Standard, *Standard Test Method for Compressive Strength of Cylindrical Concrete Specimens, C39-86*, 1997, pp. 20–24.

75. S. K. Subramaniyan, S. Mahzan, M. I. bin Ghazali, M. Ismon and A. M. A. Zaidi, *Appl. Mech. Mater.*, 2013, **315**, 861.
76. K. B. Najim and M. R. Hall, *Constr. Build. Mater.*, 2012, **27**, 521.
77. C. H. Sung, K. S. Lee, K. S. Lee, S. M. Oh, J. H. Kim and M. S. Kim, *Macromol. Res.*, 2007, **15**, 443.
78. J. G. Gwon, S. K. Kim and J. H. Kim, *Mater. Des.*, 2016, **89**, 448.
79. S. Chen, Y. Jiang, J. Chen and D. Wang, *Adv. Mater. Sci. Eng.*, 2015, **2015**, 1.
80. M. R. Hall, K. B. Najim and C. J. Hopfe, *Appl. Therm. Eng.*, 2012, **33**, 77.
81. Z. Zakaria, *Characterisation of Polyethylene Foam and Its Structure-properties Relationship in Shock Absorbing Application*: Universiti Sains Malaysia; 2007.
82. N. Mohamad, M. Mazliah, Z. N. Sharafina, M. A. Asyraf, M. Abdollah and A. H. Effendy, *Proceedings of Malaysian International Tribology Conference*; Malaysian Tribology Society, 2015, p. 86.
83. K. P. Chin, N. Y. Wan and C. S. M. Saad, *Pertanika J. Sci. & Technol.*, 2011, **19**, 171.
84. M. A. Mahamood, N. Mohamad, A. R. Jeeffeerie, A. H. Mohd Zain, M. I. S. Iqbal and H. E. A. Maulod, *J. Mech. Eng.*, 2016, **1**, 113.
85. K. M. Jeong, Y. J. Hong, P. Saha, S. H. Park and J. K. Kim, *Waste Manage. Res.*, 2014, **32**, 1113.
86. P. Mahallati and D. Rodrigue, *Cell. Polym.*, 2014, **33**, 233.
87. J. Geary, B. Douglas, T. O'Connor, G. Brandt and J. Letts, Pat. 20030082365, 2001.
88. P. Tennis, M. Leming and D. Akers, *Pervious Concrete Pavements, EB302*, Portland Cement Association, Skokie, Illinois, and National Ready Mixed Concrete Association, Silver Spring, Maryland, 2004.

CHAPTER 5

Recycling of Tire Rubbers and Their Re-usability

PARTHEBAN MANOHARAN AND KINSUK NASKAR*

Rubber Technology Centre, Indian Institute of Technology, Kharagpur 721302, India
*E-mail: knaskar@rtc.iitkgp.ernet.in, naskark73@gmail.com

5.1 Introduction

During the last few decades, the generation of waste/used tire rubbers and other rubber products has increased dramatically due to the increased number of automobiles. Millions of tons of waste/used tires are dumped in natural environments. The terms waste/used tire rubber correspond to tires at the end of their life cycle. Waste tires are nearly non-degradable and resides in a large number of land spaces and represents a severe environmental pollution issue worldwide.[1,2] Many attempts have been made to safely dispose of unavoidable waste tire rubbers by developing a new recycling technique and through re-use of materials, for both ecological and economic reasons.[3–5] Europe, Japan and the United States obligate the automobile industry to recycle automotive rubber materials. In order to protect the earth's atmosphere, this gives rise to a demand for products that produce

Green Chemistry Series No. 59
Rubber Recycling: Challenges and Developments
Edited by Jin Kuk Kim, Prosenjit Saha, Sabu Thomas, Józef T. Haponiuk and M. K. Aswathi

Published by the Royal Society of Chemistry, www.rsc.org

the smallest amount of environmental pollution impact.[6] Proper waste tire management *modus operandi* such as recycling, reclaiming, retreading and reuse are widely accepted processes to discard used/waste tires. The life cycle and the reusability of tire rubbers are displayed in Figure 5.1. The tire is fabricated by using different materials such as rubber, filler, fiber, antioxidants, accelerator and curative. Since 1839, the sulfur (curative) based vulcanization technique has been the traditional method to vulcanize the tire rubbers. Sulfur will react with active sites in the rubber chain to form the chemical connections or crosslinks between rubber chains. Sulfur vulcanization creates poly-sulfide, di-sulfide and mono-sulfide crosslinking in rubber chains.[7] Mono-sulfide crosslinks are thermally highly stable as compared with di-sulfide crosslinks and poly-sulfide crosslinks. Styrene butadiene rubber (SBR) polybutadiene rubber (PBD), isobutylene-isoprene rubber (IIR), brominated (B-IIR), chlorinated (C-IIR), natural rubber (NR) and ethylene–propylene–diene rubber (EPDM) are the commonly used raw rubbers in tire composites.[8–10] Polymers can be divided into two groups: thermoplastics and thermosetting materials. Thermoplastics soften when heated, making it possible to (re-)shape them at higher temperatures. When the rubber chains are crosslinked with sulfur; it becomes a thermoset material and therefore cannot be softened or remodeled by increasing temperature. The three-dimensional network (crosslinked chain network) has to be destroyed in order to achieve successful recycling of tire rubbers. This can be obtained by either separation of the crosslinks or by breaking of the rubber chains. Reclamation and de-vulcanization are defined as essential terms for rubber network breakdown.[11–13] Thus, different de-vulcanization methods have been developed to destroy the three-dimensional structure of rubber. Reclaiming is the process that breaks down the vulcanized rubber and plasticizes it by heat in the presence of chemicals or a high-pressure stream. The

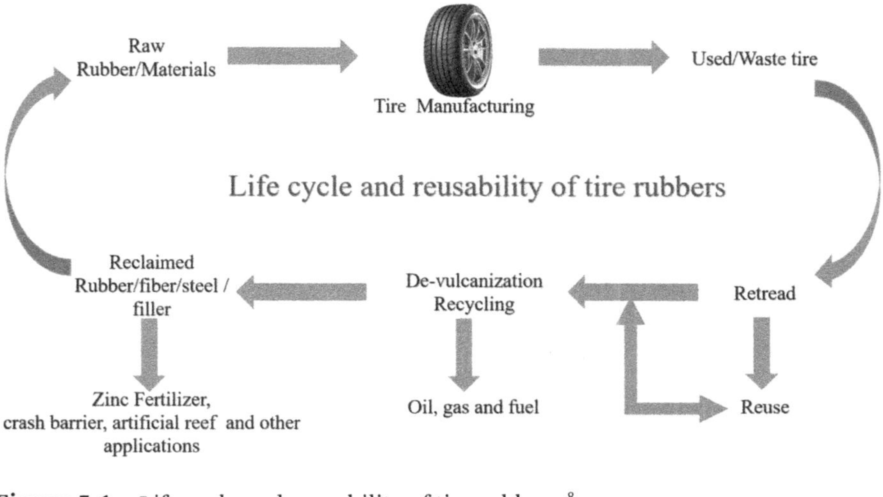

Figure 5.1 Life cycle and reusability of tire rubbers.[8]

reusability of the reclaimed rubber is of prime interest to the scientific and technological domain, as an alternative way of protecting the environment. Retreading is another method of recycling for used tires. In retreading, the old tire tread is replaced with a new tire tread. Retreading is another efficient way to utilize used tires and a more economically viable method, as it saves 30–50% of the expense of obtaining a new tire. Entirely worn-out tires can be reused in many applications, such as crash barriers, artificial reefs and playground equipment. Reusability of waste/used tire rubber signifies a potential "green" source of rubber and is the preferred substitute for raw natural and synthetic rubbers. On the other hand, waste/used tire rubbers have been converted into oil, char, gas and fuel by the incineration recycling method. Waste/used tire rubbers have a high-value of fuel, approximately 3.3×10^4 kJ kg^{-1} calorific value, which is similar to that of the calorific value of coal, 2.8×10^4 kJ kg^{-1}.[14] Thus, this chapter explores the role of reusability and end-of-life tires, various types of recycling methods, reclamation processes and reclamation recipes are described with special reference to tire rubbers. De-vulcanization and reclamation are reviewed in detail by their reaction mechanism and chemistry.

5.2 Tire Composition, Tire Parts and End-of-life Tires

When automobiles were invented, rubber became the material of choice for manufacturing tires. Automobile tires are the most prominent rubber articles in regard to volume and importance. More than seventy percent of the synthetic rubbers and natural rubber produced in the world are consumed by the tire industry. In 1845, the world's first pneumatic tire was invented by Scottish engineer Robert William Thomson in Great Britain. These tires were commercially unsuccessful due to having less resistance to the irregularities of road surfaces, because the rubber tubes were filled with air protected by a canvas cover with leather treads. More than forty years later, in 1888 John Boyd Dunlop introduced the pneumatic tire. This invention and its first applications were for bicycle tires.[15,16] In the following years, numerous improvements and developments have been made on tires. A tire is a much more complex object than it looks. Actually, a tire is many things: geometrically, it is a torus; mechanically, a flexible-membrane pressure container; and structurally, a high-performance composite. A tire is made up of 40–50% rubbery materials, such as raw rubber (NR, SBR and so on), 20–30% filler (carbon black, silica and so on), 5–10% fiber (rayon, polyester, steel and so on) and finally a few percent of vulcanizing agents.[4,17] An overview of tire components is depicted in Figure 5.2. Since 1920, tire research has been influenced by a new generation of filler, chemicals, and fibers that give rise to a significant improvement in tire quality. Moreover, the commercial introduction of the radial tire in 1948 had the advantage of a high traction, lower rolling resistance of the tire, improved handling and longer

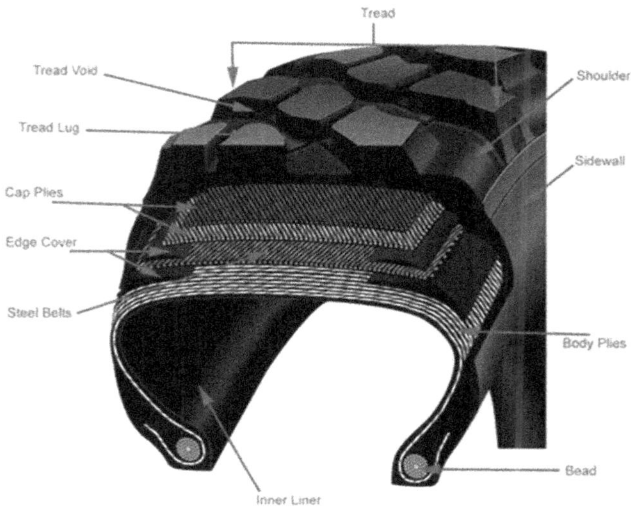

Figure 5.2 Tire components in cutaway view.[9]

Table 5.1 Rubbers used in tire components and their applications.[9]

Component	Applications	Rubbers
Tread	The tread is designed and compounded for resistance to abrasion, traction, low rolling resistance, and durability.	NR, SBR, BR
Inner liner	Designed to maintain the air impermeability in tire.	IIR, BIIR, CIIR
Belt	The belts are layers of rubber coated with either textile fiberglass or steel, which wrap circumferentially around plies.	NR
Sidewall	Thinnest area of tire; resistance to weathering, ozone and fatigue life.	NR, BR, EPDM
Carcass	Structural part of tire which supports tread block.	NR,BR,SBR
Bead	Acts as a load transfer between the rim and the tire.	NR
Cap-plies	Plies provide tensile resistance to inflation pressure along cord direction.	NR, BR

service life. Ever since 1980, the tire industry has continued improving the rolling resistance for better fuel economy, high wet traction for winter tires and durability of tires.[18] Each part of a tire has its own specific formulation with a variety of constituents for better tire performance.[9,10] Table 5.1 summarizes the most important tire rubbers with respect to tire components. Tire components are prominent resources and most of them can be reused, recycled and recovered. However, tires suffer from wear and have a limited lifetime due to their use. After their replacement, the unusable tires are known as end-of-life tires. In particular, end-of-life tires are considered to be an important factor for sustainable material recovery. Overall two billion waste/used tires are discarded globally per year. The waste/used tires are collected by tire service centers in China and European countries.[19] This collection is aimed at reducing the environmental pollution and making high-quality reclaim rubber materials, which might make adequate substitutes for the raw materials. In recent years, intensive research and development efforts have been better directed towards making end-of-life tires into useful products instead of polluting the environment. Amongst all possible ways of handling post-consumer tires, one of the most commonly applied approaches is recycling and de-vulcanization.

5.3 Why Recycle Tire Rubbers?

Waste/used tires were widely disposed of in landfill. Whole tires were not added to landfills because they could float to the surface, break the cover in the landfill, and expose once buried waste to the environment. The major disadvantage of landfilling with waste tire is the contamination of neighboring groundwater sources from the undesirable leachate constituents. However, currently landfill space for waste rubber has become scarcer, more expensive and restrictive.[19–22]

Rubber recycling is almost as old as rubber manufacturing itself. In 1820, Charles Macintosh and his partner, Thomas Hancock, developed a machine called a "masticator" to grind and recycle the scrap rubber produced during the rain coat-making process. Charles Goodyear patented a process for recycling scrap rubber by grinding it as fine as possible and blending it into a new rubber as an elastomeric filler. In 1858, Hall patented the "heater process" whereby lightly crosslinked waste rubber from boots and shoes was partially de-vulcanized at a higher temperature in the presence of steam to obtain a re-vulcanizable product, known as "reclaim" which could be blended in suitable proportions with fresh raw rubbers. In order to make a high quality "reclaim" the fiber must be removed from the tire rubber scrap. During World War II, reclaimed rubber was extensively used, as the supply of raw natural rubber from Malaysia was stopped. Annually, two billion waste/used tires are discarded into the environment worldwide. However, in recent years, these tires have been recycled in a cradle-to-cradle approach or incinerated for energy recovery. The concept of waste tire rubber recycling has been followed encompassing two main approaches, these are: (i) improvement in compound design and methods of tire rubber compounding to increase the product life, thereby reducing the waste generation and (ii) to develop technologies for efficient recycling. The advantages of the recycling/de-vulcanization are a reduction of CO_2 emissions, reuse of tires in tires, cost reduction and a reduction in the need for primary raw materials.[7,8,14] Research and development in the recycling area has resulted in the following technologies: incineration, pyrolysis, reclaiming, de-vulcanization, and pulverization.[2,4,8,23–25] Figure 5.3 shows the flowchart of the utilization of waste/used tire rubbers.

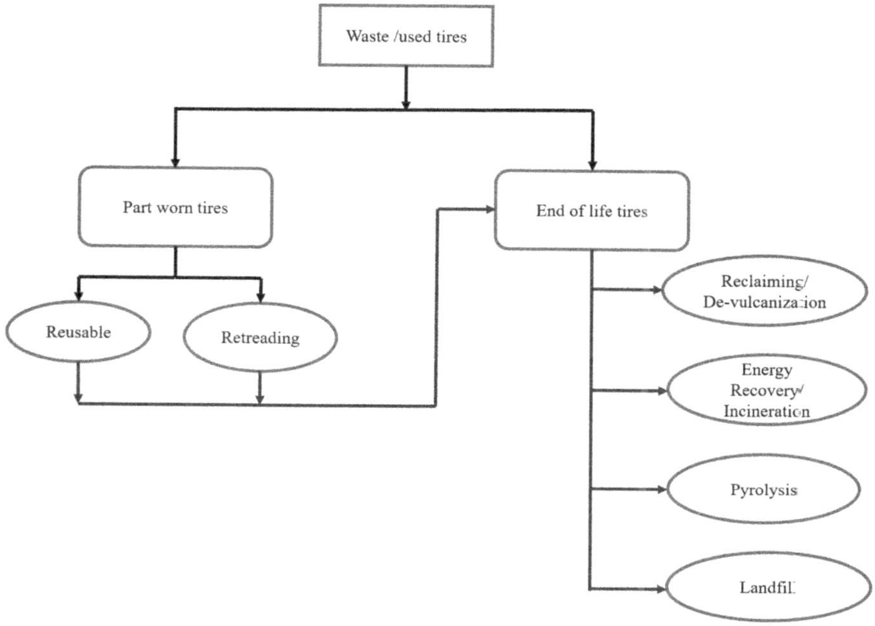

Figure 5.3 Flowchart of waste/used tire utilization.[2]

- **Incineration**: To completely burn the waste tires for energy generation. Waste/used tire are used as a fuel source in cement kilns, for production of steam, electrical energy, pulp, paper, lime and steel.
- **Pyrolysis**: Decomposing of the rubber component in the presence of heat and the absence of oxygen, thereby avoiding oxidation.
- **Reclaiming**: This process breaks down the vulcanized rubber together with main chain scission by heat in the presence of chemicals or high-pressure steam.
- **De-vulcanization**: This method aims to selectively break the sulfide crosslinks in the vulcanizates with minimum main chain scission.

5.4 Recycling of Waste/Used Tire Rubbers

Figure 5.4 represents the scheme of vulcanization, de-vulcanization and reclaiming while recycling the waste rubbers. In vulcanization, the rubber chains are crosslinked using a curing agent. In reclaiming, the waste tire rubbers are normally ground and then treated thermally, chemically and then intensely worked mechanically to partially de-vulcanize the rubber component. This partially de-vulcanized rubber material is widely known as reclaimed rubber, which can be compounded, processed and re-vulcanized, like virgin rubber. Prior to de-vulcanization, size reduction

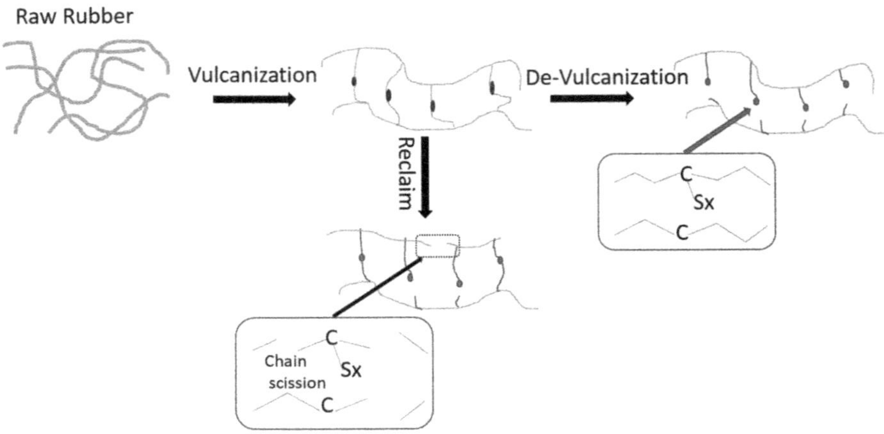

Figure 5.4 Plausible scheme of vulcanization, de-vulcanization and reclaiming.[4]

is the most important process to increase the surface area. Cutting and chipping of waste tire rubber is very difficult, firstly due to the presence of steel wire in the tire, and secondly due to the toughness and high resilience of tire rubber. The type of machinery and the process conditions determine the size of the chopped tires. Several successive chopping or grinding operations must be conducted to arrive at the final product. The most popular machines consist of arrays of spaced circular blades on pairs of counter-rotating shafts. Blades containing hooks, projected from the periphery counter-rotate in an overlapping fashion. Cryogenic grinding, wet grinding and ambient grinding are the most familiar grinding methods. Nowadays, several technologies have been developed for shredding and grinding of the wastetires prior to recycling. The de-vulcanization of a sulfur-cured rubber is the process of cleaving, totally or partially, the poly-, di- and mono-sulfide crosslinks, which are formed during vulcanization. Complete de-vulcanization of crosslinked rubber is the primary goal for a commercially viable process. Practically however, total de-vulcanization of sulfur-cured rubber is difficult and sometimes de-polymerization (breakage of C–C linkage) occurs along with cleavage of C–S or S–S bonds *via* oxidation of the rubber by the combination of heat, pressure, chemical agents and mechanical shearing action. There are numerous methods and process that have been developed to reuse waste/used tire rubbers by reclaiming and de-vulcanization. However, there are more efficient de-vulcanization methods, namely, the chemical method, mechanical method, thermal method, microwave method, ultrasonic method and biological method.[4,7,8,26–28] All of these methods and process are discussed in the following sections with respect to the tire rubbers. Table 5.2 shows the typical characteristics of different types of de-vulcanization method with regard to the tire rubbers.

5.4.1 Chemical De-vulcanization Method

The chemical de-vulcanization method is the process of cleaving the sulfur–sulfur linkages in the vulcanizate using chemicals, leaving the main chain intact. The chemical de-vulcanization method is the most commonly used de-crosslinking method in the rubber recycling industry. As the fundamental objective of reclamation is to break the crosslinks, all kinds of reclamation method can be regarded as a chemical method. In recent years, Myhre and Saiwari have published extensive reviews on chemical de-vulcanization of tire rubbers.[8–10] They described that triphenylphosphine is suitable for converting the poly-sulfide linkages to mono-sulfide and to a lesser extent, to di-sulfide links. The typical energies required to cleave sulfur or carbon bonds are tabulated in Table 5.3. Sodium-di-*n*-butyl phosphite, hexane-1-thiol and a binary mixture of thiols and amines have been reported to cleave di-and polysulfide crosslinks but not mono-sulfide crosslinks. Propane-thiol/piperidine mixture has been reported to be responsible for cleavage of the poly-sulfide linkage only, but di and mono-sulfide

Table 5.2 Various types of de-vulcanization methods and the basis of processing according to the tire rubbers.[7,9]

De-vulcanization method	Basis of process	Rubbers
Thermal[b]	Heat-induced cross-link scission	NR SBR BR
Thermal with chemicals[b]	Targeted chemical reactions at elevated temperature	SBR NR BR
Mechanical[a]	Shear-induced cross-link scission	NR SBR, EPDM IIR
Mechanical with chemicals[a]	Shear/chemical de-vulcanization	SBR, EPDM IIR
Thermo-mechanical with chemicals	Combination of heat, shear and chemical reactions to break crosslinks	EPDM SBR NR
Ultrasonic	Ultrasound energy used to break crosslinks	NR,SBR,IIR and BR
Microwave	Energy generated by microwaves	NR SBR BR
Microbiological	Microorganisms	NR SBR BR

[a]Either operated near ambient or heat introduced, for example, by heated extruder.
[b] Not commercialized: used only for research purpose.

Table 5.3 Energy required for cleaving sulfur or carbon bonds.[8]

Bond type	Bond energies (kJ mol)	Elastic constants of bonds (N m^{-1})	Location in rubber
S–S	270	~3	Di-sulfide and polysulfide
C–S	310	Intermediate value	Mono-sulfide crosslinks
C–C	370	~100	Main-chain bonds in rubber molecules

links remained unaffected. Dithio-threitol quantitatively cleaves disulfide links. Methyl iodide has been known to be useful for cleavage of mono-sulfide links. Some researchers reported breakage of the polysulfide crosslinks in ground tire rubber particles by phase transfer catalysis. They used onium ions to promote the transport of hydroxide ions from water into swollen ground tire rubber particles, allowing breakage of polysulfide crosslinks with only a little or no breakage of main chains. However, methylated mono-sulfide crosslinks cannot be broken by this process. Therefore, thiols or a thiol-amine mixture along with methyl iodide are good chemical agents for total de-vulcanization of rubber. However, in the early 1990s, methyl iodide was reported as being carcinogenic.[20,21,29–31] Hence, utilization of methyl iodide, for rubber reclamation, is restricted nowadays. Utilization of diallyl disulfide, aryl disulfides (diphenyl disulfide, dibenzyl disulfide), diamyl disulfide and bis (alkoxy aryl) disulfide as an efficient de-vulcanizing agent has been reported in the chemical de-vulcanization method.[5,32] This de-vulcanization agent is supposed to selectively break the sulfur crosslinks in the rubber network and prevent the recombination of molecules. Adhikari and coworkers reported suitability of a renewable resource material (*i.e.*, a vegetable product) containing diallyl disulfide as a major constituent as a reclaiming agent for the de-vulcanization of NR and SBR vulcanizates. High molecular weight mercaptans were also found to be useful for de-vulcanization of rubber.[30] Diphenyl disulfide is the prominent de-vulcanizing reagent for NR and IIR when using supercritical carbon dioxide as the reaction medium. De-vulcanization by organic disulfides is generally associated with de-polymerization of rubber.[29] Mangaili *et al.* investigated diphenyl disulfide together with supercritical carbon dioxide as a de-vulcanizing agent for ground truck tire rubber.[33] Figure 5.5 shows the reaction scheme for diphenyl disulfide and natural rubber.[33] During the milling or extrusion of scrap rubber free radicals are also generated by oxidative or thermal cleavage of C–C bonds. These radicals combine with radicals generated by the cleavage of disulfides, leading to degraded or depolymerized polymer chains. Olefin metathesis degradation is a stimulating chemical approach for the degradation of waste/used tire rubber. Rearrangement of double bonds between the carbon atoms of olefins takes place during metathesis because of the rupturing and reassembly of C=C bonds. The efficiency of recycling is calculated by the sol content, gel content, crosslink density, sulfur content, green strength and physico-mechanical properties of the recovered material. The thermo-chemical method is a mixture of thermal power and chemical agents. In thermo-chemical de-vulcanization, several factors affect the properties of the ground tire rubber. The predominant problems are inhomogeneous de-vulcanization, which is caused by poor diffusion of the de-vulcanization aid into the rubber particles. Saiwari *et al.* studied the effect of diphenyl disulfide in a thermo-chemical method for passenger car tire rubbers. They stated the importance of a homogenous breakdown of the crosslinked network with optimum de-vulcanization conditions for the thermo-chemical method.[12]

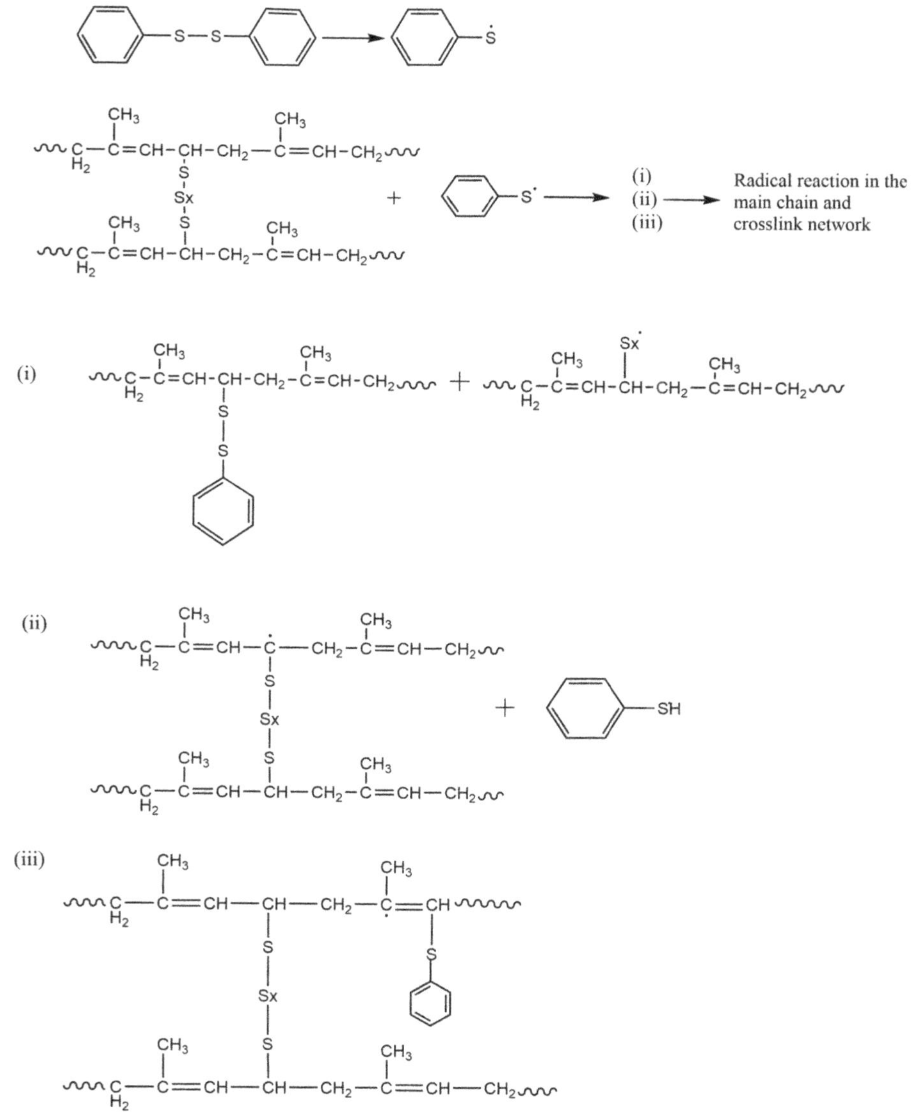

Figure 5.5 Reaction scheme for the diphenyl disulfide and Natural rubber. Reproduced from ref. 33 with permission from Elsevier, Copyright 2014.

5.4.2 Mechanical Method

Generally, the mechanical de-vulcanization method is the most prominent method for industries to recycle waste rubber because it is considered to be a low-cost and environmentally friendly method. High-shear mechanical milling technology can be used to reclaim ground rubber, and partial

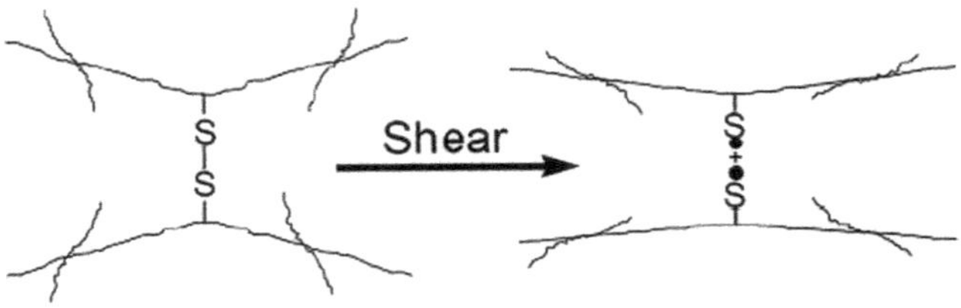

Figure 5.6 Plausible mechanism for rubber network rupture. Reproduced from ref. 9 with permission from S. Saiwari.

de-crosslinking can also be achieved by this technique.[20] This de-vulcanization takes place without any apparent requirement of chemicals at ambient temperature, but the high mechanical force can lead to breakage of the main chains of macromolecules. It can be carried out in a two roll mill, intermixer (batch), extruder (continuous) or specialist equipment (*e.g.* high-stress mixing (HSM) process). A plausible mechanism for the shear forces breaking the crosslinks of vulcanizates is shown in Figure 5.6.[9] However, poor physical properties are obtained in reclaimed rubber because of the high shear mechanical process which facilitates main chain scission. Another shortcoming is the difficulty of achieving consistent properties for reclaimed rubber when using the mechanical method.[34]

5.4.2.1 Mechanochemical Method

Later, mechanical de-vulcanization was successfully carried out with a multifunctional de-vulcanizing agent at ambient temperature. The multifunctional de-vulcanizing agents are disulfides, for example, aryl disulfides or diphenyl sulfides, tetramethyl thiuram disulfide (TMTD), thiophenols and their zinc salts and mercaptanes. Organic disulfides are widely used as a multifunctional de-vulcanizing agent for the de-vulcanization of rubbers. During mechanical processing (two roll mill or extrusion), shear force cleaves the S–S bonds of crosslinks in rubber to generate polymer radicals. Disulfides also break into radicals when they are processed with rubber. These radicals combine with the broken polymer chain radical and thereby prevent their recombination, resulting in de-crosslinking. The concentration of the de-vulcanization agent is maintained around 0.5 to 4 wt% during the mechanochemical de-vulcanization process. Ghorai *et al.* reported the utilization of bis(3-triethoxysilyl propyl) tetrasulfide (TESPT) as a novel de-vulcanizing agent in an open two roll mixing mill for ground NR. They concluded that the superior de-vulcanization occurred at 6 mL concentration of TESPT and the corresponding time of de-vulcanization was 40 min.[35] As a general rule, TESPT is used as a silane coupling agent to match the polarity between the non-polar rubbers (NR, SBR) to the polar filler (silica). Gosh *et al.* also investigated TESPT as a de-vulcanizing agent in SBR based vulcanizates in the mechanochemical method.[32] The efficiency of de-vulcanization has been

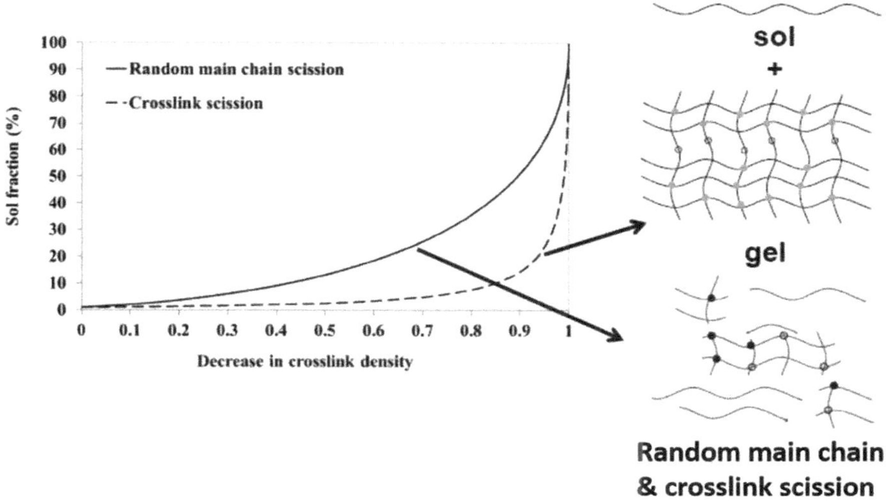

Figure 5.7 Scheme and Analytics of Horikx plot. Reproduced from ref. 9 with permission from S. Saiwari.

concluded by the random main chain scission *versus* crosslink scission curves in a Horikx plot as shown in Figure 5.7. This plot correlates the sol content of the de-vulcanizates to the crosslink density and theoretically predicts their change for two cases (i) main chain scission (ii) crosslink scission. The experimental data for the de-vulcanized rubber was found to lie near the theoretically predicted crosslink scission. Hence, it can be concluded that the method has an excellent mill processability and extrusion processability coupled with the de-vulcanizing agent.

5.4.2.2 Thermo-mechanical Method with De-vulcanizing Agents

The thermo-mechanical method is a combination of a mechanical shearing action and heating of the material. Main chain degradation takes place during the mechanical processes due to the strong rise in material temperature. Carbon-to-carbon bonds and sulfur crosslinks are cleaved together in the thermo-mechanical recycling method and it can be identified through the formation of soluble branched structures known as fragments of gel. Nowadays, most of these methods attempt a thermo-mechanical de-vulcanization method with a chemical agent (di-sulfides, thiols, amines and unsaturated compounds) at a higher temperature for material recycling.[36,37] This method explores the reclaimed materials properties at 70–80% of the recovery, as compared to the original gum vulcanizate. Around 1 to 2 wt% quantities of the chemical agent are added during thermo-mechanical recycling. It acts as a softener and reduces the thermal degradation resistance of the vulcanizate by weakening the interaction between the filler and rubber chains.

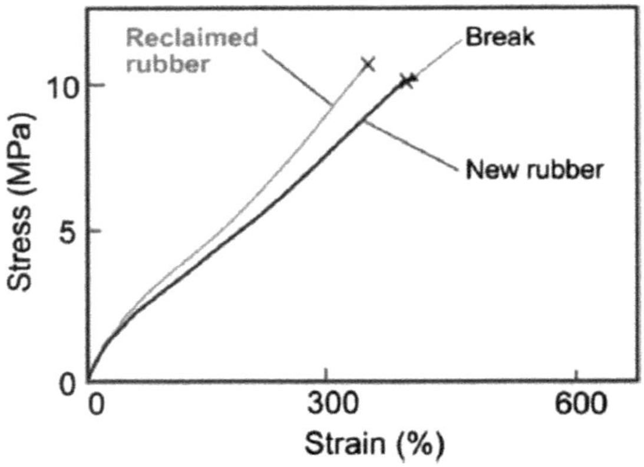

Figure 5.8 Stress *versus* strain curves of reclaimed EPDM vulcanizates compared with raw EPDM.[7,9] Reproduced from ref. 9 with permission from S. Saiwari.

One of the first continuous rubber reclaiming processes involved a single screw extruder called a reclaimator and used a feedstock of 0.6 mm ground, fiber-free rubber scrap. In order to introduce the high shearing action to the bulk of the material at elevated temperatures, the clearances of the flights of the screw to the wall are set at optimized minimal values. EPDM vulcanizates are treated by a thermo-mechanical de-vulcanization method under optimal conditions and exhibit superior tensile properties as compared with raw EPDM (Figure 5.8). Recently the tread part of waste tires was de-vulcanized by a twin-screw extruder. The effects of the processing conditions, such as the barrel temperature and screw speed were studied. The de-vulcanized rubbers were blended with virgin rubber and re-vulcanized successfully. The results showed that the mechanical properties of the virgin rubber with de-vulcanized compounds were slightly inferior to the pure virgin compound.[11]

5.4.3 Energy Recovery Method

The energy recovery method is widely considered to be a thermo-chemical method. In this context, it is important to ensure reusability of used tires more efficiently by retrieving resources and energy from them. Thermo-chemical treatment of tires is recognized as the most advanced and significant disposal solutions through which energy and resources can be recovered. Thermal treatment is generally subdivided into:

- Combustion or incineration (with air atmosphere): scrap rubbers are converted into flue, gas and ash.
- Gasification (limited air atmosphere): scrap rubbers are converted into gas, ash and tar.
- Pyrolysis (no air): scrap rubbers are converted into gas, oil and char.

Incineration and gasification involve mild and severe oxidation, respectively, while pyrolysis is performed under inert conditions. These methods have the following advantages:

- Volume reduction of waste by more than 90%.
- Net energy recovery with possible material recovery.
- Nonpolluting and capable of destroying most organic substances.

There are also some problems associated with the thermal treatment, such as disposal of ash, generation of toxic gases, soot generation, and so forth.

5.4.3.1 Incineration

Incineration is defined as the process of oxidizing combustible material to give inert residuals. It is a spontaneous, highly exothermic process occurring above 400 °C and once initiated, it becomes self-supporting. The conversion temperature is around 1000 °C, and thus the reactor should be designed to be resistant to high temperatures for a high yield and low soot content. Waste tires with a calorific value of 4500–8000 KCal/Kg are employed as the incinerating fuel. The generated heat produced during the process can be utilized for industrial purposes as well as for the generation of electricity. Oriaku *et al.* reported the recovery of carbon black through incineration by burning the tire in a limited air supply.[38] The thick smoke enriched with carbon black particles were allowed to precipitate onto the filter surface placed in its way.

5.4.3.2 Gasification

Gasification is a partial oxidation process in which the oxidant reacts with the solid char in an endothermic reaction giving gaseous carbon monoxide and hydrogen along with carbon dioxide and light hydrocarbons. The composition of gas produced is dependent on the operating conditions and the concentration of the oxidizing agent. The gas produced from gasification is specifically known as syngas and is employed as a fuel for gas turbines and in fuel cells. Machin *et al.* have reported the generation of electrical and thermal energy *via* the gasification process.[39]

5.4.3.3 Pyrolysis

Scrap tire pyrolysis was the subject of several studies by the rubber, oil, and carbon black industries. Pyrolysis, also known as thermal cracking, is a process in which polymer molecules are heated, in the partial or total absence of air, until they fragment into several smaller, dissimilar, random-sized molecules of alcohols, hydrocarbons, and others substances. The pyrolysis process is considered to be an efficient process that converts waste tire scrap into useful energy sources: Solid char, liquid oil and gases. This method involves the decomposition of tire rubber at a temperature of around 200–900 °C in the absence of oxygen.[24] The complexity of the rubber composition in tires makes the pyrolysis process more complicated, involving many complex chemical reactions converting rubber into liquids, gases and char, as shown in Table 5.4. Temperature plays an important role in this process, as it determines the stability of the product formed and also the rate of thermal decomposition.[8–10]

The production of gas is favored at a higher temperature as a higher temperature leads to the cracking of heavier hydrocarbons, giving lighter products such as methane and hydrogen gas. Low-temperature pyrolysis leads to the production of heavier hydrocarbon oils, that are mainly aromatic. The solid tire pyrolytic product, for example tire-derived char, is often mesoporous in nature with a very low surface area and are often employed as adsorbents after the activation process. After subsequent treatment of these char, they can be applied instead of carbon black. In order to use pyrolytic char, the sulphur content should be much below 1%. As the sulphur is chemically bound, steam activation is often performed to reduce the sulphur content. The pyrolysis process is also influenced by the presence of additives such as plasticizers, stabilizers and so forth. The yield depends on the particular characteristics of the reactor, such as type, size, the efficiency of heat transfer, and residence time. Table 5.5 compares different pyrolysis techniques.

Table 5.4 Pyrolysis product of used tires.[9]

Primary products	wt%	Content	Products derived from primary products
Gas	10–30	Hydrogen, CO_2, CO, methane, ethane, propane and butane other hydrocarbons sulfur	
Oil	38–55	Aromatic compounds, high in sulfur content (0.3–1.0%), alkanes, alkenes, ketones and aldehydes	Carbon black
Char	33–38	>15% of ash (ZnO), 3–5% of sulfur	Activated carbon

Table 5.5 Comparison of various pyrolysis techniques.[9]

Process	Processing conditions	Complexity	Disadvantages
Vacuum pyrolysis	High	High temperatures and energies	High risk
Plasma pyrolysis	High		High cost
Microwave pyrolysis	Medium		Low heat transfer
Catalytic pyrolysis	Temperature app. 450 °C	Medium	Low oil yield
Liquefaction	Mild	Low	Oil contamination of the char

The important technologies employed are based on vacuum, plasma, microwave, and catalysis plasma.

5.4.4 Microwave Method

In the microwave de-vulcanization method, a controlled dose of microwave energy at a specified frequency and energy level is applied to cleave all carbon–sulfur and sulfur–sulfur bonds substantially, but the amount of microwave energy is insufficient to sever the carbon–carbon bonds. In general, 300 MHz to 300 GHz microwave energy are used to cleave the crosslinked bonds, without breaking the main chain of the polymers (carbon–carbon chemical bond).[40] In other words, poly-sulfide crosslinks are easily cleaved by the microwave method as compared with mono- and di-sulfide crosslinks due to the difference in bond energy. The energy required for cleaving sulfur or carbon bonds is presented in Table 5.3. During microwave de-vulcanization, volatile organics with low molecular weights are released in the form of sulfur dioxide (SO_2) due to the sulfur bonds being broken, as well as the sulfoxide and sulfide groups.[8,40,41] The novelty of this method is that the elastomer waste can be de-vulcanized without de-polymerization to a material capable of being re-compounded and re-vulcanized to give a physical property essentially equivalent to the original vulcanizate. This process leads to this method being economical and eco-friendly. A highly polar rubber absorbs a greater amount of microwave energy and generates sufficient heat to cut down the network structure as compared with non-polar rubbers. This is attributed to that fact that the molecular interaction of rubber is increased with respect to the applied electromagnetic field and raises the temperature of the material.[42] Figure 5.9 shows the process diagram of a microwave de-vulcanization method. This method reduces the processing time and energy in order to maintain the degradation, continuity, and handling of a bulk amount of materials. Garcia conducted microwave de-vulcanization of ground tire rubber (GTR) by using different microwave exposure times. Figure 5.10 shows the preliminary observations of GTR under different microwave exposure periods. Long microwave exposure periods lead to a drastic change in the texture of the GRT and also structural changes, such as the decrease of the insoluble phase content, breaking of chemical bonds and an increase in the fluidity of rubber.[41] SBR is a widely used rubber for tire tread applications, scrap rubber can be efficiently de-vulcanized by using the microwave method in combination with the mechanical

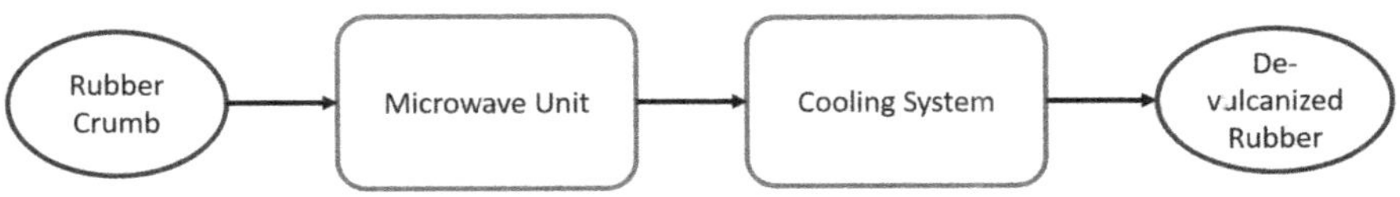

Figure 5.9 Schematic diagram of a microwave de-vulcanization method.[4]

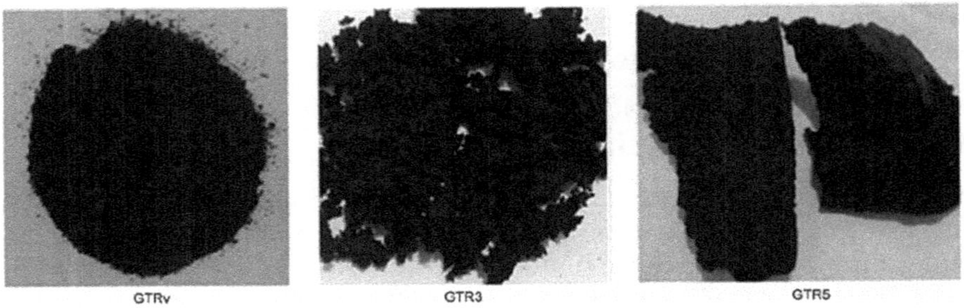

Figure 5.10 Photos of GTR at different microwave exposure time: untreated (GTRv), 3 min (GTR3) and 5 min (GTR5) respectively. This observation was noted only after the sheet out process in two roll mill.2 Reproduced from ref. 41 with permission from Express Polymer Letters, © BME-PT.

processing technique. Microwave de-vulcanized SBR is a promising, sustainable, and cheap filler material in polymer-based composites.[43] In recent years microwave irradiation de-vulcanization has been carried out using a chemical agent. Diphenyl disulfide (DPDS), hexadecyl amine (HDA), N-cyclohexyl-2-benzothiazole sulfenamide (CBS) and 2-mercaptobenzothiazole disulfide (MBTS) chemicals are widely used as de-vulcanization agents, among these DPDS is the most useful.[23]

5.4.5 Ultrasonic Method

Ultrasonic energy based de-vulcanization for crosslinked rubbers was introduced and patented by Panofsky. In the patent, he claimed that when solid rubber articles such as tire chips were immersed into a hydrocarbon solvent and an ultrasonic energy of 20 kHz (which is beyond the range of human hearing) at power intensity of greater than 100 W was applied, bulk rubber effectively disintegrated and dissolved into the solvent. Okuda and Hatano also reported ultrasonic reclamation of NR vulcanizates. They subjected the NR vulcanizate to ultrasonic energy of 50 kHz for 20 min of de-vulcanization followed by re-vulcanization and obtained reclaimed rubber with similar properties to those of original rubber.[4] This ultrasonic method is widely accepted for industrial applications due to the non-requirement of solvents and any chemical agents. Ultrasonic chain scission is characterized as non-random, which means that the cleavage of polymeric chains preferentially takes place near the middle of the chain compared to random scission under the thermo-chemical process. The ultrasonic de-vulcanization process creates a greater intensity of ultrasonic waves in the presence of pressure and heat which helps to breakdown the rubber network by generating cavitation. This ultrasonic method is widely accepted as a green de-vulcanization process. There are two types of ultrasonic reactors *viz.* the co-axial reactor and the barrel reactor. The de-vulcanization zone in the first reactor (coaxial reactor) is located at the exit from an extruder without the imposition of additional shearing, whereas, in the second reactor (barrel reactor), it is located in the barrel where additional shearing takes place due to the screw rotation. The barrel reactor permits continuous de-vulcanization with an output as high as 2.53 g s^{-1}.[44,45] Gel fraction and crosslink density show that the GRT in both the reactors is partially de-vulcanized. A correlation between the gel fraction and crosslink density obtained in both reactors indicated that additional shearing has a positive influence on improving the efficiency of de-vulcanization. Various synthetic rubbers, GTR and NR are de-vulcanized by using an ultrasonic single-screw extruder. Guyale rubber vulcanizate has also been found to undergo ultrasonic de-vulcanization. To improve the ultrasonic de-vulcanization process twin screw extruders were connected to make de-vulcanization more efficient. Figure 5.11 shows an ultrasonic twin screw micro-extruder. Recently Mangili *et al.* reported the influence of process variables (ultrasonic amplitude, temperature, gap size, screw speed and flow rate) on ultrasonic de-vulcanization of GTR in a co-rotating twin-screw extruder.[46] This co-rotating twin screw extruder ultrasonic de-vulcanization was applied to NR, GTR, PBR, SBR, EPDM and polyurethane rubber. The decrease of crosslink density and gel fraction indicated that the increasing ultrasonic amplitudes and decreasing flow rate led to better de-vulcanization. Ranking of the degree of de-vulcanization is found to be NR > SBR > EPDM. De-vulcanized IIR exhibits poor cure characteristics and physical properties due to the reduction of the molecular weight during ultrasonic de-vulcanization. Overall, ultrasound amplitude was found to be a governing factor for the properties of the de-vulcanized material.

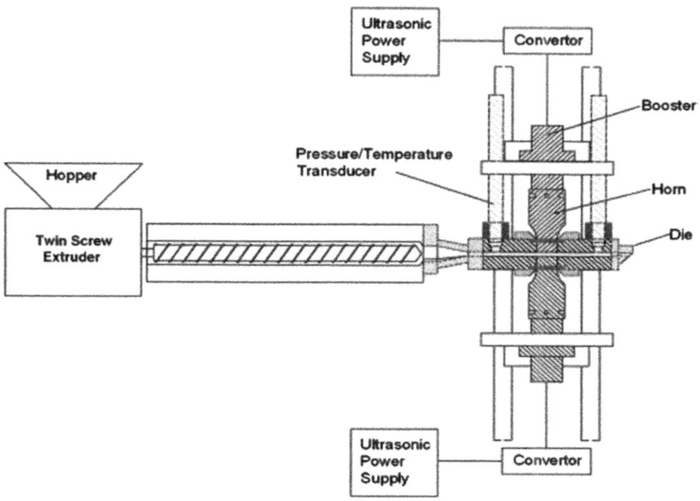

Figure 5.11 The ultrasonic twin screw micro-extruder. Reproduced from ref. 44 with permission from Elsevier, Copyright 2009.

5.4.6 Biological Method

The main motivation for the biological de-vulcanization method is to develop an eco-friendly process and eliminate the chemical agent in the existing de-vulcanization process. Biological de-vulcanization of waste tire rubber was first reported using sulfur consuming microorganisms belonging to the bacterial genera *thiobacillus*, *sulfolobus* and *rhodococcus*. These bacteria attack the sulfur–sulfur bonds only, without affecting the rubber hydrocarbon chains, so that the rubber can be re-vulcanized using fresh sulfur. The most widely used microorganisms for the desulfurization of tire rubbers are *T. ferrooxidans*, *T. thiooxidans*, *T. thioparus*, and *Sulfolobus acidocaldarius*. Microbial desulfurization is the most prominent biological method used for rubber recycling. A biological attack on natural rubber latex is very easy, as it contains other biodegradable ingredients such as proteinous materials. However, biodegradation of NR obtained from latex was difficult because of the removal of degradable materials and the hindered growth of microorganisms. However, the toxic substances usually present in compounded and vulcanized NR were reported to hinder the growth of microorganisms and kill them. In the case of synthetic rubber vulcanizates the problem of biodegradation is more acute. Beckman *et al.* discussed in their review the susceptibility of elastomeric articles as the source of the microbial attack.[55] Warner discussed the use of microorganisms to remove sulfur from coal, but they also mentioned the difficulty in removing organically bound sulfur. They reported that some thermophilic bacteria transform sulfur, present in coal, into hydrogen sulfide at a temperature above 100 °C.[47] This indicated the possibility of microbial desulfurization of the bound sulfur present in rubber vulcanizates. Many researchers have reviewed the possibilities of microbial de-vulcanization of waste rubber.[48,49] One of the difficulties with bacterial treatment is suppressing the antibacterial effects on curing and anti-aging agents in the rubber products. The curing agents, such as tetramethyl thiuram monosulfide (TMTM) and TMTD, and an anti-aging agent such as N-(1,3-dimethyl butyl)-N-phenyl-*p*-phenylenediamine (Dusantox 6PPD), have inhibitory effects on the growth of desulfurizing-bacteria, namely *Thiobacillus Ferrooxidans* and *Rhodococcus rhodochrous*. Therefore, cryo-grinding and the successive removal of toxic additives by extraction with ethanol were applied before the bacterial treatments. Sato *et al.* reported that NR vulcanizate was directly de-vulcanized by a lignin-degrading and wood rotting basidiomycete, *Ceriporiopsis subvermispora*.[56] They also analyzed the products of fungal de-vulcanization using various meticulous characterization methods. Li *et al.* used *Thiobacillus* sp. (a kind of bacillus isolated from the soil samples of a sulfur-rich coal mine) with a strong sulfur-oxidizing capacity to devulcanize the surface of GTR biologically. The properties of the devulcanized GTR/NR compounds and virgin GTR/NR compounds were compared. A better interface coherence between the devulcanized GTR and NR matrix was found by DMA and SEM studies, and explained the improvement in the mechanical properties for the devulcanized GTR/NR compounds.[48] Corynebacterium, Nocardia, Mycobacterium (CNM) is one of the most effective rubber-degrading strains within the group of prokaryotes. Tatangelo *et al.* recently reported the importance of Gordonia de-sulfuricans DSM 44462^T (rubber-degrading strains) in the bio-de-vulcanization process. This paper reports the influence of process parameters on the bacterial growth, such as the glucose concentration (C), dibenzothiophene concentration (DBT), and initial biomass (optical density, OD).[49] This biological de-vulcanization represents the an uncomplicated, minimal equipment approach with no elimination of toxic chemicals required, low energy consumption and rupturing of the sulfur crosslinked bonds on the surface without any major damage to the main chain. These characteristics make the process very promising for tire rubber recycling. The biological de-vulcanization of tire rubber leads to effective recycling with technological and economic advantages.

5.5 Reusability and Application of Tire Rubbers

In recent years, reusability of de-vulcanized/reclaimed tire rubbers has found numerous applications. It can be used as road paving, asphalt modification, supplementary fuel for cement kilns, acoustic barriers, paper mills, in energy-generating processes, rubber mats, fuel and oil absorbent rubber/ground rubber composites and thermoplastics/ground rubber composites.

5.5.1 Civil Engineering Applications

A large amount of reclaimed rubber is used in civil engineering applications such as playground surfaces, parking lots, bank stabilization, filling for under the road surface and as a asphalt modifier. Tire shreds have the following properties that civil engineers require, light weight (1/3 soil), low earth pressure (1/2 soil), good thermal insulation (8 times better), good drainage (10 times better), compressibility and vibration damping. Whole tires can be used to build highway barriers and construct artificial reefs, factory floor mats, and so forth. Many researchers have reported waste tire management in the civil engineering market and studied the properties of rubberized bitumen prepared by physical blending of bitumen with reclaimed rubber powders obtained from rejected rubber gloves and scrap tires.[21,50] The softening point, tenacity and energy to break increased progressively whilst the penetration number decreased with increasing rubber content. Rubberized bitumen prepared using glove crumbs produced overall better properties than that prepared using tire shavings. Moreover, rubberized bituminous mixes produced a higher resistance to permanent deformation and dynamic cracking compared with ordinary bituminous mixes. Rubberized asphalt has been developed to help overcome a few drawbacks posed by bitumen, and is used as the binder material in asphalt, like low heat resistance, the formation of ruts when exposed to heavy traffic and high-temperature dependency. These problems have been dealt with by blending the bitumen with rubber materials. It has also increased the durability, thinner lift, and increased the skid resistance. In recent years, a new type of pavement material using scrap rubber called elastic and permeable pavement material (EPPM) was proposed by Chung and Hong.[51] EPPM has been used to overcome water impermeability and brittleness in conventional rubber asphalt cement concrete. The key factors affecting the compressive strength and the water permeability of EPPM were the size of the ground scrap rubber and aggregate, the amounts, and the property of the binder. The optimum level of the binder was about 8 wt%. The compressive strength and water permeability of EPPM were 1.1 and 1.16 times higher than those of the conventional porous cement concrete, respectively.[52]

5.5.2 Commercial Application of De-vulcanized/Reclaimed Rubber

Partially or fully de-vulcanized reclaimed rubber is widely used in rubber formulations for partial replacement of virgin rubber. A large quantity of reclaimed rubber is used along with neat rubbers for manufacturing low-cost rubber goods such as bicycle tires, conveyer belts, ebonite, sponge rubbers, cables and so forth due to economic and processing advantages. In recent years, thermoplastics have gained much interest as a potential matrix for blending ground rubber as the resulting composite can be processed easily in conventional thermoplastic processing equipment.[53] Rubber crumb and tire shavings can be re-bonded to make excellent carpet under-cushion products. Reclaimed rubber is utilized as walkway covers for its non-slip properties and surface protection. Blending with small amounts of virgin rubber is a common practice. Good quality floor tiles can be produced from scrap crumb rubber by incorporating it into conventional rubber formulations or by bonding it to the rubber matrix using a binder. Such tiles are used in sports arenas and locker rooms. Recently, floor tiles have been manufactured in which the crumb rubber particles are bonded together by polyurethane adhesives and can be cured at room temperature. Crumb rubber is used for running and jogging tracks and sports fields and golf courses as crumb rubber provided resiliency and long-term performance at an acceptable cost. Rubber chips are used as a bulk surface cushion under children's playground equipment such as swings, slide and so forth.

5.5.3 Energy Production and Zinc Fertilizer

Generating energy by using waste/used tire rubber is an essential and motivating process of recycling. Scrap rubber is treated as a complementary material to the coal due to its similar calorific value. Waste rubber produces oil, char, and gas during pyrolysis. The pyrolysis process is regulated

Table 5.6 Energy balances for the production and recovery of rubber products. Reproduced from ref. 21 with permission from Elsevier, Copyright 2012.

Amount of energy required to produce a tire	87–115 MJ kg^{-1}
Amount of energy required to produce rubber goods	80–90 MJ kg^{-1}
Amount of energy obtained from the combustion of tires	32 MJ kg^{-1}
Amount of energy needed to comminute tires to a rubber recycle of grain size < 1.5 mm	1.8–4.3 MJ kg^{-1}

by the design of the reactor, temperature, and raw materials. High usage of diesel fuel is minimized by consuming a tire rubber derived fuel. Waste tire rubber has found good application in the generation of electricity and in the concrete industry. Table 5.6 clearly shows that the range of energy recovered is only 30–38% as compared with the energy initially invested for producing the tire rubbers.[21] Zinc (Zn) uptake is a necessity for healthy plant species, and ground rubber and/or waste tire rubber contains a greater amount of Zn and is also much cheaper than the commercially available $ZnSO_4$ fertilizer. Purified Zn has been used during the manufacturing of tires. Moreover, soil is an acidic media which can easily uptake the Zn from ground rubbers. Hence, ground rubber is an alternative harmless fertilizer source for Zn and some other nutrients for crops. Similarly, ground rubber helps plants to grow in the absence of lead and/or cadmium contamination.

5.5.4 Sound-proof Barriers

In general sound-roofing barriers on roads and acoustical sound-sealant is able to reduce environmental noise issues. Ground rubber can be a good acoustic material with a broadband absorption spectrum. Vázquez *et al.* developed a sound barrier material by incorporating crumb rubber through a wet process. Similarly, crumb rubber is used as a modifier of bituminous binders to reduce rolling noise on pavements and also has excellent acoustic durability properties.[54]

5.6 Advantages of Reclaimed/De-vulcanized Rubber

The major benefit of using reclaimed rubber is its lowered cost compared to virgin rubber. In addition to cost reduction, reclaimed rubber also imparts some improvements in processing. Reclaimed rubber has a much lower nerve than virgin polymers. As a result, compounds containing reclaim have much more uniform and lower die swell. It also increases calendar rates and in general, reduces calendar shrinkage. Due to the presence of crosslinked gel in reclaimed rubber, it is less thermoplastic than raw rubber compounds. Thus, extrudates have better dimensional stability and retain their shape during curing in hot air or open steam. Compounds containing reclaimed rubber have better air venting properties, thus blister formation while calendaring is reduced. Many workers observed that utilization of reclaim in a rubber formulation improved the aging characteristics of the product, and it exhibits less tendency to revert.

5.7 Disadvantages of Reclaimed/De-vulcanized Rubber

The inadequacies of reclaim are that it lowers the green strength and tensile strength of the compounds in which it is used. Reclaimed rubber can only be added to a compound in low percentages, and higher loadings generally result in severe degradation in physical properties of the cured compound. Nowadays, tire industries have drastically reduced the utilization of reclaimed rubber in tire formulations, as radial tires require high green strength for processing. Moreover, reuse of a large quantity of de-vulcanized whole tire rubbers containing a mixture of rubbers is not recommended as different rubbers are used in different components (*e.g.*, casing, tread, sidewall, shoulder) of the tire. Utilization of reclaim in large quantity is only possible in some rubber products of low-performance requirement such as mats, irrigation hose, conveyer belts, and so forth.

5.8 Conclusion

In summary, elastomers serve many functions in automobiles owing to their unique properties and performance. The tire is the foremost product of the automobile industries and waste/used tires are discarded in the environment after their limited lifetime. This represents a serious environmental pollution for the sustainable development of the world due to the non-degrading characteristics of crosslinked rubber. Enabling recycling loops for waste/used tires is both a challenge and an opportunity. In this chapter, several recycling methods and processes are described with regard to waste/used tire rubbers. De-vulcanization can be very effective with modernized technologies. Recycled materials will always have to compete with virgin raw materials on cost, availability, and quality. The use of reclaimed rubber in rubber compounds is also limited because of its detrimental effect on the properties of the compounds. However, the re-use of tire rubber makes a substantial contribution to reducing carbon emissions.

References

1. Z. Derakhshan, M. T. Ghaneian, A. H. Mahvi, G. O. Conti, M. Faramarzian, M. Dehghani and M. Ferrante, *Environ. Res.*, 2017, **158**, 462–469.
2. S. Ramarad, M. Khalid, C. Ratnam, A. L. Chuah and W. Rashmi, *Prog. Mater. Sci.*, 2015, 72, 100–140.

3. A. K. Naskar, Z. Bi, Y. Li, S. K. Akato, D. Saha, M. Chi, C. A. Bridges and M. P. Paranthaman, *RSC Adv.*, 2014, **4**, 38213–38221.
4. S. K. De, A. Isayev and K. Khait, *Rubber Recycling*, CRC Press, 2005.
5. D. De, A. Das, D. De, B. Dey, S. C. Debnath and B. C. Roy, *Eur. Polym. J.*, 2006, **42**, 917–927.
6. A. Pedram, N. B. Yusoff, O. E. Udoncy, A. B. Mahat, P. Pedram and A. Babalola, *Waste Manage.*, 2017, **60**, 460–470.
7. K. A. J. Dijkhuis, *Recycling of Vulcanized EPDM Rubber*, PhD thesis, University of Twente, The Netherlands, 2008.
8. R. Francis, *Recycling of Polymers: Methods, Characterization and Applications*, John Wiley & Sons, 2016.
9. S. Saiwari, *Post-consumer Tires Back into New Tires: De-vulcanization and Re-utilization of Passenger Car Tires*, PhD thesis, University of Twente, The Netherlands, 2013.
10. M. Myhre, S. Saiwari, W. Dierkes and J. W. M. Noordermeer, *Rubber Chem. Technol.*, 2012, **85**, 408–449.
11. H. Yazdani, M. Karrabi, I. Ghasmi, H. Azizi and G. R. Bakhshandeh, *J. Vinyl Addit. Technol.*, 2011, **17**, 64–69.
12. S. Saiwari, W. Dierkes and J. W. M. Noordermeer, *KGK, Kautsch. Gummi Kunstst.*, 2013, **66**, 20–25.
13. Z. X. Zhang, S. H. Lee, J. K. Kim, S. L. Zhang and Z. X. Xin, *Macromol. Res.*, 2008, **16**, 404–410.
14. V. V. Rajan, *Devulcanisation of NR Based Latex Products for Tyre Applications: Comparative Investigation of Different Devulcanisation Agents in Terms of Efficiency and Reaction Mechanism*, PhD thesis, University of Twente, The Netherlands, 2005.
15. X. Colom, X. Cañavate, P. Casas and F. Carrillo, in *Nanocomposites with Unique Properties and Applications in Medicine and Industry*, InTech, Editon edn., 2011.
16. A. ten Brinke, *Silica Reinforced Tyre Rubbers*, PhD thesis, University of Twente, The Netherlands, 2002.
17. G.-L. Zhuang, M.-Y. Wey and H.-H. Tseng, *J. Membr. Sci.*, 2016, **520**, 314–325.
18. P. Manoharan and K. Naskar, *J. Polym. Environ.*, 2017, 1–16.
19. V. Torretta, E. C. Rada, M. Ragazzi, E. Trulli, I. A. Istrate and L. I. Cioca, *Waste Manage.*, 2015, **45**, 152–160.
20. X. Guo, D. Xiang, G. Duan and P. Mou, *Waste Manage.*, 2010, **30**, 4–10.
21. M. Sienkiewicz, J. Kucinska-Lipka, H. Janik and A. Balas, *Waste Manage.*, 2012, **32**, 1742–1751.
22. A. K. Naskar, S. De, A. Bhowmick, P. Pramanik and R. Mukhopadhyay, *Rubber Chem. Technol.*, 2000, **73**, 902–911.
23. M. Molanorouzi and S. O. Mohaved, *Polym. Degrad. Stab.*, 2016, **128**, 115–125.
24. X. Wu, K. Formela, R. tur Rasool and S. Wang, *Polym. Degrad. Stab.*, 2017, **136**, 48–57.

25. A. M. Joseph, B. George and R. Alex, *Rubber Chem. Technol.*, 2016, **89**, 653–670.
26. S. Rooj, G. C. Basak, P. K. Maji and A. K. Bhowmick, *J. Polym. Environ.*, 2011, **19**, 382–390.
27. G. Jana, R. Mahaling and C. Das, *J. Appl. Polym. Sci.*, 2006, **99**, 2831–2840.
28. I. Mangili, M. Lasagni, M. Anzano, E. Collina, V. Tatangelo, A. Franzetti, P. Caracino and A. I. Isayev, *Polym. Degrad. Stab.*, 2015, **121**, 369–377.
29. Y. Ikeda, *Chemistry, Manufacture and Applications of Natural Rubber*, 2014, p. 436.
30. B. Adhikari, D. De and S. Maiti, *Prog. Polym. Sci.*, 2000, **25**, 909–948.
31. F. Sadaka, I. Campistron, A. Laguerre and J.-F. Pilard, *Polym. Degrad. Stab.*, 2012, **97**, 816–828.
32. J. Ghosh, S. Ghorai, S. Bhunia, M. Roy and D. De, *Polym. Eng. Sci.*, 2017, **58**, 74–85.
33. I. Mangili, E. Collina, M. Anzano, D. Pitea and M. Lasagni, *Polym. Degrad. Stab.*, 2014, **102**, 15–24.
34. Ł. Zedler, M. Przybysz, M. Klein, M. R. Saeb and K. Formela, *Polym. Degrad. Stab.*, 2017, **143**, 186–195.
35. S. Ghorai, S. Bhunia, M. Roy and D. De, *Polym. Degrad. Stab.*, 2016, **129**, 34–46.
36. K. Formela, M. Cysewska and J. T. Haponiuk, *J. Vinyl Addit. Technol.*, 2016, **22**, 213–221.
37. K. Formela, M. Formela, S. Thomas and J. Haponiuk, *Macromol. Symp.*, 2016, 64–72.
38. E. Oriaku, C. Agulanna, J. Odenigbo and N. Nnoruka, *Int. J. Multidiscip. Sci. Eng.*, 2013, **4**, 30–36.
39. E. B. Machin, D. T. Pedroso and J. A. de Carvalho Júnior, *Waste Manage.*, 2017, **68**, 412–420.
40. Y. J. Hong, K. M. Jeong, P. Saha, J. Suh and J. K. Kim, *Polym. Eng. Sci.*, 2015, **55**, 533–540.
41. P. Garcia, F. de Sousa, J. de Lima, S. Cruz and C. Scuracchio, *eXPRESS Polym. Lett.*, 2015, **9**, 1015–1026.
42. M. Khavarnia and S. O. Movahed, *J. Appl. Polym. Sci.*, 2016, **133**, 43363.
43. S. Seghar, N. A. Hocine, V. Mittal, S. Azem, F. Al-Zohbi, B. Schmaltz and N. Poirot, *eXPRESS Polym. Lett.*, 2015, **9**, 1076–1086.
44. A. Isayev, R. Kumar and T. M. Lewis, *Polymer*, 2009, **50**, 250–260.
45. B. Maridass and B. Gupta, *Polym. Compos.*, 2008, **29**, 1350–1357.
46. I. Mangili, M. Lasagni, K. Huang and A. I. Isayev, *Chemom. Intell. Lab. Syst.*, 2015, **144**, 1–10.
47. W. C. Warner, *Rubber Chem. Technol.*, 1994, **67**, 559–566.
48. Y. Li, S. Zhao and Y. Wang, *J. Polym. Res.*, 2012, **19**, 9864.
49. V. Tatangelo, I. Mangili, P. Caracino, M. Anzano, Z. Najmi, G. Bestetti, E. Collina, A. Franzetti and M. Lasagni, *Appl. Microbiol. Biotechnol.*, 2016, **100**, 8931–8942.

50. N. N. Eldin and A. B. Senouci, *Cem. Concr. Compos.*, 1994, **16**, 287–298.
51. K. H. Chung and Y. K. Hong, *J. Appl. Polym. Sci.*, 1999, **72**, 35–40.
52. A. M. Rashad, *Int. J. Sustainable Built Environ.*, 2016, **5**, 46–82.
53. Z. X. Zhang, S. L. Zhang, S. H. Lee, D. J. Kang, D. Bang and J. K. Kim, *Mater. Lett.*, 2008, **62**, 4396–4399.
54. V. Vázquez, J. Luong, M. Bueno, F. Terán and S. Paje, *Sci. Total Environ.*, 2016, **542**, 223–230.
55. J. A. Beckman, G. Crane, E. L. Kay and J. R. Laman, *Rubber Chem. Technol.*, 1974, **47**, 597–624.
56. S. Sato, Y. Ohashi, M. Kojima, T. Watanabe, Y. Honda, and T. Watanabe, *Chemosphere*, 2009, **77**, 798–804.

CHAPTER 6

Testing and Industrial Characterization of Waste Rubber

SUPRABHA BANDYOPADHYAY[†a], MD. MINHAJUR RAHMAN[†a], SOUMANTI HAZRA[a], JIN KUK KIM[b] AND PROSENJIT SAHA*[a]

[a]Indian Institute of Engineering Science and Technology, Shibpur, Dr. M.N. Dastur School of Materials Science and Engineering, Howrah 711103, India; [b]Gyeongsang National University, Department of Materials Engineering and Convergence Technology, Jinju 660701, South Korea
*E-mail: prosenjit@matsc.iiests.ac.in

6.1 Introduction

Characterization techniques for rubber evolved with the inadequate supply of a natural rubber source. However, the growing demand for rubber and rubber-based products has compelled researchers to find suitable synthetic elastomer compounds in order to meet the productivity demands. The behaviour of rubbers under exposure to chemicals has been set as the benchmark for determining several properties of rubber

[†]Both the authors have contributed equally.

Green Chemistry Series No. 59
Rubber Recycling: Challenges and Developments
Edited by Jin Kuk Kim, Prosenjit Saha, Sabu Thomas, Józef T. Haponiuk and M. K. Aswathi

Published by the Royal Society of Chemistry, www.rsc.org

before the advancement of such testing methods. The effect of heat and light on rubber based material has already been established before the discovery of vulcanization chemistry by Goodyear. Compositional analysis, conductivity testing, physical testing and so forth of rubbers gradually developed with time when researchers had faced obstacles in the development process of rubber vulcanizate based products.[1] The diversified application fields of rubbers have been explored by researchers over the years and henceforth, all sorts of possible characterization techniques are applicable for rubber based products. Microscopy analysis, rheological behaviour, anti-corrosive and anti-abrasive behaviour, and so forth are some of the characterization methods for rubber products. However, the set of characterization techniques are fixed according to the end applications. Bulk characterization techniques are designed for industrial applicability of the products.

With time another technological advancement has forced researchers to reuse the resources directly or through recycling for sustainable development and greener environment. In this scenario reuse of rubber stock piles and waste tyre tread compounds came to light for the development of advanced materials.[2] The properties of waste rubber depend on its reclamation and recycling techniques and hence the necessity lies for characterization of waste rubbers particles.[3] The grinding of rubbers to obtain powder was the first development in rubber recycling history. Soon after the development of the grinding technology, the devulcanization method was invented with the urge to obtain modified properties with appropriate miscibility of rubber crumb powder in rubber blends. In order to evaluate the properties of reclaimed rubber, identification of different components in waste rubber and their effect on the reclamation recipe had to be established first. The characterization techniques helped to establish the reaction mechanism and chemistry of reclamation technology.[4] To cite an example, thermogravimetric analysis (TGA) is often used for estimating the basic components of vulcanizates due to the process simplicity and rapidity. Specifically, differential thermal gravimetric analysis (DTGA) curves of single elastomers are treated as fingerprints for identification of the components.[5,6] Similarly, Fourier transform infrared (FT-IR) spectroscopy is used to evaluate the protein and lipid contents of natural rubber simultaneously, as FT-IR is a rapid and non-destructive method.[7] Further characterizations are made according to the application where waste rubber granulates have been used as a filler. Several review articles have focused on the diversified applications of rubbers with their process variables such as oil absorbants,[8] asphalt and cement binders,[9,10] environment friendly polymer composites,[11] and so forth. Determination of processing variables such as temperature, pressure, and bond energies for selective breaking of sulphur crosslinks are carried out with help of the characterization techniques.[12] The properties of a material can be tailored by altering its structure. As surface oxidation of waste rubber crumbs is a popular technique for modifying the properties of the end product. Primarily, the phenomena of oxidation is identified by exploring the functional groups present on the surface with FT-IR.[13] Waste rubber crumbs are often used as fillers in several engineering applications as mentioned earlier and henceforth matrix–interface interaction of filler needs to be recognized with the help of several characterization tools. Thus, evaluation of the properties of rubber or waste rubber products becomes unavoidable due to the practical consequences. A well defined behavioural chart of waste rubber products under several conditions needs to be prepared before its reuse, and characterizing techniques help to predict the performance of a sustainable material.

6.2 What is Rubber?

6.2.1 Natural Rubber

Originally, rubber is obtained in latex form. Rubber is mainly the polymer of isoprene obtained in the sap of a plant, named *Hevea brasiliensis*, a native of the Brazilian forest.

6.2.2 Synthetic Rubber

Chemically modified or chemically synthesized rubber, is categorized as synthetic rubber. Chemical modification by vulcanization allows it to retain its original shape after being stretched. The most primitive synthetic rubber is Buna–S (a styrene–butadiene co-polymer) (Table 6.1).

Table 6.1 Composition of tyre rubber. Reproduced from ref. 36 with permission from Elsevier, Copyright 2001.

Materials		Element			
Name	Content	Name	Signal	Tyre	Plastic
Natural rubber	23	Carbon	C	73.0	65.0
Synthetic rubber	24	Hydrogen	H	6.0	9.5
Oxygen	O	4.0	6.5		
Carbon black	25	Nitrogen	N	1.4	2.5
Sulfur	S	1.3	0.1		
Steel cord	14	Chlorine	Cl	0.07	5.9
Zinc	Zn	1.5			
Weave cord	4	Iron	Fe	13.5	
Ashes	10.5				
Others	10				

6.2.3 Some Specific Elastomers

(i) Styrene Butadiene Rubber: This makes up the major proportion of utilized rubber apart from natural rubber. However, it has poor mechanical properties when compared to natural rubber and thus often needs particle fillers for application in fields requiring appreciable mechanical strength.
(ii) Polychloroprene: This is also known as neoprene, and finds its use because of its similarity in mechanical properties to that of natural rubber.
(iii) Nitrile butyl rubber: Commonly called buna–S, nitrile butyl rubber is available on the market in various grades of viscosity, primarily due to varying proportions of acetonitrile in the elastomer chain. Buna-S characteristically has oil resistant properties, but its weathering properties need to be tailored or improved by the addition or incorporation of various fillers.

6.3 Rubber Testing and Techniques

To describe rubber testing and its techniques, we first need to give a brief introduction of the instruments required in the rubber laboratory. The second half of this section deals with the testing details of the various physical properties of rubber.

6.3.1 Instrumentation

A rubber laboratory requires the following machines for the complete processing of rubber samples, starting with crushing of raw rubber and ending with the curing process.

6.3.1.1 Mills and Internal Milling Machines

Mills are used for the initiation of the rubber mixing process.[14] The raw elastomer, along with the fillers, is introduced into a gap (called the nip) between two inwardly rotating and adjacently placed hollow cylinders. As the rubber is pressed between these rotating cylinders, there is a gradual decrease in its highly viscous nature and it finally becomes reduced to a small paste eligible for further milling processes by the internal milling machines and calendars. Internal milling machines on the other hand, are comprised of the following parts:

(i) Hoppers and ram: Raw samples are introduced into the mixer through these hoppers and are forced down by the ram.
(ii) Mixing chambers: As the name suggests, this is the place where mixing takes place and this requires less time as compared to the mills.

(iii) Cooling channels: These channels are found drilled into the walls of the cooling chamber and water is run through these channels in order to carry out fast reduction of the temperature of the mixing chamber. This cooling is necessary because of the faster rate of operation as compared to the mills.
(iv) Thermocouple: This is found inside the mixing chambers and is required to regulate the temperature of the components.
(v) Rotors: The mixing function is carried out by the rotors and they can be either intermeshing rotors or tangential rotors.
(vi) Hydraulically controlled cylinders: Responsible for controlling the mixing process and are maintained by pressure variations.
(vii) Discharge doors: Required for the discharge of mixed rubber. However, some machines may not have this where the discharging is carried out by mixing in the chamber itself that keeps on rotating.

6.3.1.2 Extruders

Extruders consist of one or two screws that move the material into the mill and mixing machine. The back of the extruders is equipped with hoppers, through which rubber is fed to the extruders.

6.3.1.3 Calendars

This is also an integral part of the entire milling setup. The sample is calendared after it has passed through the mills. Calendaring is a high precision process, that removes any residual bubbles trapped within the already milled sample, or any other roughness related discrepancies of the already milled samples.

6.3.1.4 Curing Equipment

Moulds and autoclave are the primary curing devices in the laboratory. Moulds are hard bound cavities used for the purpose of giving elastomers with a low viscosity value, shape and dimensional stability. In the simplest of the designs, a mould consists of a steel plate with grooves of desired shape and size engraved on it. Listed here are some of the advanced moulding techniques. Injection moulding is a specific technique of moulding where a nozzle-bearing barrel makes the sample-injecting device in this case. The barrel is equipped with rams and screws and is thus designed to be movable enough to deliver the rubber sample, through the nozzle to the tip of a hole situated on a closed mould. The sample is thus introduced into the mould cavity through succeeding holes, and a final cross-linked rubber product is generated at the end. Autoclave curing is the alternative curing process associated with rubber samples. An autoclave is a cylindrical device made up of steel that maintains a constant pressure inside the cylinder, and is used in the rubber laboratory for the process of curing, under specified temperature and pressure conditions. The dimensions can vary, but are generally found to be 30 m in length and having a diameter starting from 3 m.

The heat required for the curing process is supplied inside the cylinder by wet stream that is introduced under pressure, ranging from 0.3–0.7 MPa. For cases requiring higher pressure in the specified temperature range, the wet stream is coupled either with nitrogen, or carbon dioxide streams. The streams may be used individually too, depending on the experimental conditions. However, oxygen streams cannot be used and the cylinder needs to be completely devoid of oxygen, as oxygen can disturb the chemical composition of samples, namely, peroxide cross-linking agents.

6.3.2 Physical Testing

This section describes the testing processes for various physical properties of rubber.

6.3.2.1 Electrical

Generally, rubbers are insulators and they are often used as encapsulating element for cables, wires and various components in electrical appliances. However, coupling of electrical properties with the inherent flexibility of rubbers makes them a fascinating material for suitable electrical applications. In this regard, to make rubbers act as a conducting material, different kinds of fillers such as carbon nanotubes and carbon blacks are incorporated in the rubber matrix.[15] With the use of conductive fillers, conductivity of the elastomers can be enhanced in several orders of magnitude from less than 10^{-11} S m^{-1} to 10^{-2} S m^{-1}.[16] Electrical measurement procedures of rubbers and rubber based materials are similar to the other materials which exhibit results of the same order of magnitude. However, there remains a large possibility of variation in the obtained results with repeatability of the measuring procedures. One of the main reasons is that rubber experiences physical distortion by the measuring electrodes due to their low stiffness. Variability in measurement also comes with the high contact resistance observed at the rubber electrode interface. Variability in the obtained result may also be attributed to the influence of the type of processability, including mechanical stresses. The conductivity of the diene rubbers can be increased with halogens. Similarly, the conductivity of natural rubber can be increased with doping of a Lewis acid such as antimony pentachloride.[17] The term percolation threshold is often used for measurement of electrical properties such as conductivity of a conducing filler incorporated material. The percolation threshold is the maximum limit of filler content in the matrix beyond which no significant change is observed in the electrical properties. For rubbers incorporated with multiwall carbon nanotubes, this percolation threshold can be achieved below 1 phr content of filler.

Often electrical resistance of the material is measured and subsequently the measured resistance is converted into volume resistivity. An altering voltage during the measurement process is needed in order to minimize the background current effect. The governing formula for measuring volume resistivity is provided in the following equation:

$$\rho = \frac{RS}{d}$$

In which "S" is the cross-sectional area perpendicular to the current and "d" is the thickness of the sample between the two electrodes.

Electrical resistivity of thin films varies with strain. Henceforth, an estimation of strain dependence of electrical resistivity can be measured by stretching the test material in a stretching machine. Resistivity considering the effect of strain is measured under the approximation that volume remains constant during the stretching process and it is measured according to the following formula:

$$\rho = \frac{RS_0}{\alpha^2 L_0}$$

In which S_0 represents the initial cross sectional area and L_0 is the jaw length, α is the extension ratio (the length of the sample in the direction of applied strain to the initial length before deformation). Electrical properties depend on the dispersion ability of the fillers in the matrix. The properties can be tailored with alteration in the filler matrix interface.[18]

The ASTM D257-14 standard utilizing guard ring electrodes is widely used for measurement of the electrical characterizations of a material. For the insulating application of cables, the electrical properties of the silicone rubber are carried out at direct current (DC) conductivity. The influence of contact properties of different materials on the silicone elastomer is measured by polarization/depolarization current measurements. Current *versus* time measurements are also carried out in order to ensure the material behaviour over an extended period of time at higher applied field strengths. The results are concerned with establishing the relationship between the repeatability and reliability of its behaviour.[19]

DC conductivity measurements at room temperature are carried out for measurement of the antistatic properties of rubbers. Electrical conductivity measurements of ground tyre rubber (GTR) reinforced composites material exhibits a high degree of disorder which follows sublinear dispersive conductivity defined by:

$$\sigma = \sigma_0 + A\omega^n$$

In which σ_0 is the DC conductivity, and $\omega = 2\pi f$ in which f is the frequency, while A and n (which have values between 0 and 1) are parameters that depend on temperature and materials. At lower frequency conductivity of GTR the reinforced material tends to form a plateau and at the higher frequency conductivity tends to follow a linear curve on the logarithmic scale.[20] Thus, GTR reinforced compounds show the trend of disorder for many polymers and polymeric compounds. Dynamic electrical analysis measures the permittivity of the material with varying frequency. The real and imaginary part of the permittivity is directly proportional to the stored and dissipative energy of the material. The results of complex permittivity help to obtain the results of the electrical modulus. The electrical modulus of a material is used to measure the relaxation analysis in the low frequencies and in the high temperature region by avoiding the conductivity contribution to dielectric losses.

Several resistivities associated with rubbers should be mentioned here. The electrical resistance between opposite faces of a unit cube is defined by the volume resistivity, in which the resistance between opposite sides of a square on the surface is defined as the surface resistivity. Another type of resistance associated with rubber is insulation resistance. This is the resistance measured between any two particular electrodes on or in the rubber and, hence, is a function of both the surface and volume resistivity and of the test piece geometry. Rubbers are also classified according to their resistivity values, which are provided in the subsequent section.

- Insulating rubber: When resistance value exceeds 10^8 ohms.
- Anti-static rubber: Rubber having a resistance value between 10^4 and 10^8 ohms.
- Conducting rubber: Rubber having a resistance value up to 10^4 ohms.

6.3.2.2 Thermal Properties

Basic thermal characterization of rubber and rubber like product involves thermal degradation studies and thermal conductivity studies. Thermal degradation studies involve pyrolysis of waste rubber materials, in particular waste tyre materials. The pyrolysis temperature is determined by any of the conventional thermal analysers or by a combination of several thermal analyser tests, such as TGA and differential thermal analysis (DTA). The tests are performed at inert atmosphere and generally at each degradation temperature the obtained gaseous products are collected and characterized further using gas chromatography (GC) or another analyser. In this way the degradation profile of the rubber or rubber like material is identified. The mechanism of formation of the major decomposed products can be decoded by studying major decomposition products. Pyrolysis of rubber follows a radical mechanism. The decomposition profile varies with the composition of elastomers and for blend rubber like products it varies with the proportion of blending.[21–23] Oil, char and gases are yielded in the pyrolysis process. Fixed-bed (batch), screw kiln, rotary kiln, vacuum and fluidised-bed are the different reactors used in the pyrolysis process. Amongst them, fixed bed is widely used in the pyrolysis analysis of waste rubbers. These reactors are heated in inert atmosphere with an external electrical source. Inert gas acts as carrier gas which condenses the derived gas products downstream. The degradation profile starts at around 350 °C and is carried out in a temperature range of 450–700 °C.[24,25]

The basic characterization principles and objectives of several kinds of thermal analyser remain the same with rubber and rubber derived products as TGA is used for composition identification and analysis of degradation and thermo-oxidation behaviour. Similarly, DSC is used for identifying the changes in thermal transition temperatures and can be used to identify changes in the composition of rubber samples.[26] Devulcanization of the waste rubber product is carried out for successful blending of the product with other polymer matrices. However, the devulcanization process alters the thermal behaviour of rubber products. For instance in microwave devulcanization longer exposure time shifts its glass transition (Tg) (generally measured by DSC, dynamic mechanical analysis (DMA) can also be used) to lower temperatures by modifying the chemical structure of the rubber. Thermo-oxidation behaviour may also shift to a lower temperature limit.[27] For rubbers, DMA is used as an indirect measure for the characterization of rolling resistance and wet skid applications in addition to being used for the measurement of loss and stored energy in the material. The $\tan\delta$ value at 0 °C gives an indirect measure of the wet skid resistance and at 60 °C gives an indirect measure of rolling resistance of the tyre tread compound.[28] In thermal conductivity analysis, heat conductivity, heat resistance, and heat generation in the material are characterized.[29]

6.3.2.3 Permeability

Generally, rubbers either show low transmission rates for the transmission of liquid or gaseous substances, or remain impermeable to these substances. Transmission of these can take place through the rubber sample by the process of absorption and diffusion. Permeability is broadly categorized and tested under two headings, namely, gas permeability and vapour permeability. Constant volume, constant pressure, and carrier gas methods are the process variants used for gas permeability studies. Permeability measurements are useful for applications in which rubber is used as a membrane. Permeability of waste rubber products becomes necessary when they are used in several civil engineering applications, such as GTR filled asphalt pavements.[30] The water permeability depth in the concrete mixtures is characterized when scrap-tyre-rubber is used as a replacement for aggregate and as a filler in concrete.[31,32]

6.3.2.4 Adhesion

Waste rubber products are often used as reinforcement fillers in composite materials. Hence, adhesion means measurement of the bonding strength between the rubber particles and the host matrix. However, the direct measurement of bond strength is impractical for industrial purposes.[33] Certain adhesion tests include adhesion to metal, fabric, cord and so forth. Among them ASTM D4299 is a standard test procedure for the evaluation of adhesion of rubber to rigid metal substrates. The test procedure comprises of several test methods which vary according to sample designation. For instance test method A measures the static adhesion strength of rubber to metal. This method is applicable when the test specimen is a replica of an actual product. Method B determines adhesive strength by a 90° peel strip test, in which the obtained data indicates the adhesive strength across the width of the samples. For conical or cone shape samples test method C of the ASTM D429[34] is used. Similarly, there are several others process variants in ASTM D429. Peel tests and direct tension tests are used for assessment of adhesion properties of rubber to fabrics. ASTM D413-98(2017)[35] is used to measure the adhesion of rubber to flexible substrates (fabric, fiber *etc.*). The standard test procedure predicts the adhesion by the force required to separate a flexible material from the rubber layer.

6.4 Disposal of Waste Rubber: A Serious Threat to Ecology

With the development of the polymer industry, a considerable amount of the waste generated is gradually becoming a matter of concern. As a substantial percentage of the polymer industry is made up of the rubber industry, disposal of rubber waste or discarded rubber has attacted attention. The primary sources of waste rubber include discarded tyres or automobile parts, belts, pipes, and many more household and industrial substances. Rubber, though derived as a latex from biological sources, is not bio-degradable in nature and thus, can be listed among potential ecologically harmful substances. These are therefore not environment-friendly and thereby the need to recycle them arises. Statistics report the annual consumption of rubber to be roughly 15 million tonnes, whereas, the annual output for rubber amounts to as much as 31 million tonnes. Out of these, 10 million tonnes are annually discarded. From the numerous factors demanding the proper recycling of waste and discarded rubber, some of them are discussed as follows:[36]

- **Protection of the environment**: This definitely remains the primary concern behind the entire recycling process. As already mentioned, rubber, and specifically synthetic rubber used in various products, is not bio-degradable in nature and hence is expected to be potential danger to ecology.
- **Conservation of energy**: Petroleum is one of the major sources from which raw rubber is obtained. Thus, if recycled rubber helps in the reduction of the supply of rubber, there is an ultimate curtailment in the requirement of petroleum for the rubber industry. Besides, rubber itself has been reported in the literature to be of high calorific value, amounting to 3.3×10^4 kJ kg^{-1}. Thus, recycling rubber can contribute a significant role in the conservation of energy. In fact, rubber, if not recycled, may account for a net loss of 3×10^4 kJ of energy annually.
- **Use of rubber as a raw material**: As modern lifestyle shows a gradual increase in the consumption of polymeric materials, and the rubber industry is a noticeable segment of these polymeric entities, the huge and constant supply of raw rubber remains a necessity. As mentioned previously, raw rubber largely draws its supply from petroleum, and therein comes the constraint of limited availability of these non-renewable resources and judicious utilization of existing raw materials.

Discarded rubber, if recycled properly, can become a secondary source of the raw input to the rubber industry. Waste rubber, recycled by means of numerous processes as mentioned in the subsequent sections of this chapter, is finding an increased use in various construction fields (Table 6.2). According to the literature, regenerative rubber and powdered rubber are important forms of recycled rubber which are being used to combat the expanding demand for raw material in the rubber industry (see Figure 6.1).

Table 6.2 Statistical information showing the amount of rubber discarded as waste in some parts of the world. Reproduced from ref. 36 with permission from Elsevier, Copyright 2001.[a]

Place	Year	Amount	Disposing						
(10 000 tons)	Thermal-utilizing	Making powder	Regenerated rubber	Renew	Export	Burying	Other		
US	1992	280	23	6	4	3	63	1	
Japan	1992	84	43	12	9	25	8	3	
Germany	1993	55	38	14	1	18	18	2	9
Great Britain	1992	45	9	6	18	67			
EEC	1990	197.5	30	20			50		

[a]EEC, European economic community.

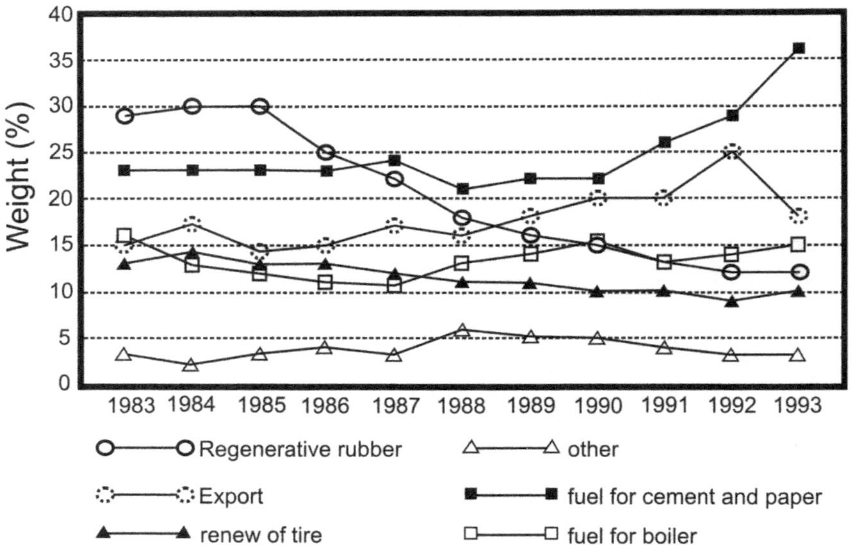

Figure 6.1 Recycling of waste tyres in Japan over the years. Reproduced from ref. 36 with permission from Elsevier, Copyright 2001.

6.5 Possible Explorations of Waste Rubber

Rubber regenerated from scrap tyres or other spare parts are used for the modification of various substances, some of which are listed as below.

6.5.1 Rubber–Rubber Blends

Natural rubber has been extensively exploited in numerous applications. However, blending of more than one type of rubber has also been studied in a number of research articles as a technology to improve various properties such as abrasion resistance, flexibility, and so forth, For example, use of polybutadiene rubber as a blending component of natural rubber has been reported and the subsequent modifications in their properties have been studied.

6.5.2 Concrete Modified by Waste Rubber

Crumb rubber widely finds its use as aggregates in concrete and asphalts. Although this is associated with the lowering of stiffness values of concrete, this hardly affects the strength of the concrete, as avery low volume fraction of rubber is used for incorporation. Numerous studies have presented improvements in the sound and thermal properties of concrete. This is also known to improve the fire properties of high strength concrete. Numerous samples of concrete are prepared, using varying percentages of rubber, depending on the specific needs of the application area.

6.5.3 Asphalt Binders

The ASTM defines asphalt rubber as "a blend of asphalt cement, reclaimed tyre rubber, and certain additives in which the rubber component is at least 15 percent by weight of the total blend and has reacted in the hot asphalt cement sufficiently to cause swelling of the rubber particles."[37]

6.6 Recycling of Rubber

There are various methods of recycling of rubber (devulcanization). They can be broadly categorized into two parts:

(i) Chemical processes: Processes in which crosslinks between polymer chains are broken *via* various chemicals.
(ii) Physical processes: Processes in which crosslinks between polymer chains are broken *via* mechanical, thermo-mechanical, microwaves, and ultrasonic treatments, and so forth.

6.6.1 Thermo-mechanical Recycling of Rubber

The aim of the thermomechanical process of devulcanization is to determine the difference in bond energies of the various bonds present in the rubber polymer chain.[38] Three types of bonds are mainly present in rubber polymer chains, namely C–C, C–S, and S–S bonds. This process aims to break the C–S and S–S bonds. The typical bond energies are shown in Table 6.3.

One of the thermomechanical processes used in the industry is the high shear mixing (HSM) devulcanization process. In this process, as quoted from Brown 2008,[52] "the rubber to be treated is subjected to high levels of shear and extensional stressing that provides an efficient way of selectively breaking

Table 6.3 Bond energy data. Reproduced from ref. 38 with permission from Elsevier, Copyright 2018.

	Bond energies
Bond type	Bond dissociation energy (kJ mol^{-1})
C–C	370
C–S	310
S–S	270

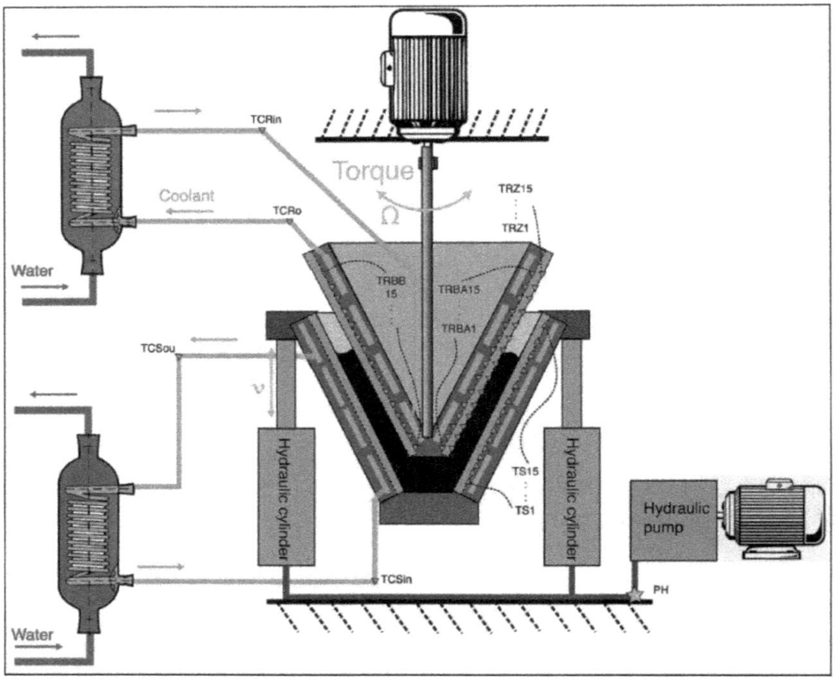

Figure 6.2 Schematic representation showing the different parts of the HSM machine. Reproduced from ref. 38 with permission from Elsevier, Copyright 2018.

cross-linking bonds". This is accomplished by the use of two cones, namely the rotor and stator. There are grooves on the inner surface of the stator and on the outer surface of the rotor. Rubber is filled into the gap between the cones and a high amount of torque is applied, so that a high amount of shear is generated. The rotor rotates in both the clockwise and anti-clockwise direction. Due to this shearing, the rubber becomes devulcanized into smaller particles ultimately resulting in dust formation, as shown in Figure 6.2.

6.6.1.1 Characterization

Four different techniques are followed for the characterization of a HSM product:

(i) Sieving: The HSM product, being heterogeneous in size, is sieved using different sieve sizes so that smaller particles (~100 mm in diameter) become separated from the large particles (~few cm in diameter).
(ii) Swelling: Swelling tests are conducted using cyclohexane as a solvent for determining the crosslink density of the sample. The crosslink density can be calculated using the formulae given by Flory and Rehner and subsequently corrected by Kraus as follows:

$$\nu = \frac{V_{r0} + \chi V_{r0}^{2} + \ln(1 - V_{r0})}{V_{sol}\left(\frac{2}{f}V_{r0} - V_{r0}^{1/3}\right)} \tag{6.1}$$

$$\frac{V_{r0}}{V_{r}} = 1 - \frac{\Phi\left[3c\left(1 - \sqrt[3]{V_{r0}}\right) + V_{r0} - 1\right]}{1 - \Phi} \tag{6.2}$$

$$\Phi = \frac{V_{c}}{V_{T}} = \frac{\frac{m_{c}}{\rho_{c}}}{\frac{m_{c}}{\rho_{c}} + \frac{m_{pol}}{\rho_{pol}}} \tag{6.3}$$

$$V_{r} = \frac{V_{pol}}{V_{G}} = \frac{m_{s} - m_{c}}{(m_{d} - m_{c}) + (m_{s} - m_{c})\left(\frac{\rho_{pol}}{\rho_{sol}}\right)} \tag{6.4}$$

In which, m_s is the swollen mass, m_d is the dry mass, m_c is the mass of carbon black in the dry sample, m_{pol} is the mass of polymer, ρ_c is the carbon black density, ρ_{pol} is the polymer density, ρ_{sol} is the solvent density, c is the correction parameter that depends only on the type of filler and is equal to 1.17, χ is the Flory-Huggins polymer–solvent interaction constant ($f = 0.35$) for the system EPDM-cyclohexane, Φ is the carbon black volume fraction in the dry sample, v is the crosslink density, V_{r0} is the polymer fraction of the swollen mass, V_r is the apparent polymer fraction of the swollen sample, V_{sol} is the molar volume of solvent, V_{pol} is the polymer volume in the swollen sample, V_c is the carbon black volume in the dry sample, V_T is the volume of the dry sample and V_G is the volume of the swollen sample.

(iii) Sol Fraction: Vulcanized rubber can be divided into two fractions, namely the sol fraction (s) and gel fraction (g). The sol fraction consists of free polymer chains not linked to the network and can therefore be separated from the gel phase using a solvent. The gel fraction, on the other hand, is formed by crosslinked polymer chains, non separable from the solvent. The sol and gel fractions can be calculated as follows:

$$s = 1 - g \tag{6.5}$$

Extraction *via* Soxhlet apparatus is the most common method to measure the sol fraction. The extraction can be conducted in two steps:

1. Extraction of polar components such as the free sulphur, accelerators, plasticizers and other low molecular weight substances. This can be done over 24 h in the Soxhlet using acetone as the solvent.
2. Extraction of the sol fraction using toluene as the solvent for 72 h in the Soxhlet.

The mass of each sample was measured before and after extraction and denoted as m_1 and m_2 respectively. Two kinds of sol fractions have been calculated, the sol fraction of vulcanized rubber (s_i) and the sol fraction of devulcanized rubber (s_f).

$$s_f = \frac{m_1(1 - e_N) - m_2}{m_{\text{pol}}} \tag{6.6}$$

$$e_N = \frac{m_{i1} - m_{i2}}{m_{i1}} \tag{6.7}$$

$$s_i = \frac{(2 + \gamma_i) - \sqrt{(\gamma_i^2 + 4\gamma_i)}}{2\gamma_i} \tag{6.8}$$

$$\gamma_i = v_i \frac{\overline{M_n}}{\rho_{\text{pol}}} \tag{6.9}$$

In which, m_{i1} is the mass of the non-treated rubber before extraction, m_{i2} is the mass of the non-treated rubber after extraction, e_N is the mass fraction of non-rubber components that can be extracted by Soxhlet extraction, v_i is the crosslink density of the non-treated rubber, $\overline{M_n}$ is the number average molecular weight of the original polymer chains equalling 150 000 g mol^{-1}, and γ_i is the crosslink index that represents the average number of crosslinks per polymer chain for the non-treated rubber.

6.6.2 Waste Rubber Recycling by Microwave Devulcanization

Devulcanization of waste rubber (GTR) was performed in a system comprising of a conventional microwave oven having a motorized stirring system.[39] The speed of stirring was maintained at 100 rpm using a speed control device. Exposure time was varied between 1–5.5 min. After the process of devulcanization, homogenization of all the samples was performed in a laboratory two-roll mill. Here, the source of waste rubber used was ground waste tyre rubber.

6.6.2.1 *Characterization*

(i) Temperature measurement of devulcanized waste rubber:
Temperature measurement of the treated samples was done *via* a rod thermometer. It was observed that an increase in the exposure time led to an increase in the temperature of the samples.

(ii) Gel content and cross-linking density:
The soluble (sol) and insoluble (gel) fractions of the devulcanized samples were determined by extraction process using toluene as the solvent as per the ASTM 2765-11 standard. The percentages of the sol and gel fractions were calculated according to the equations detailed below.

$$\text{Extract}(\%) = \frac{\text{Weight lost during extraction}}{\text{Weight of original specimen}} \tag{6.10}$$

$$\text{Gel Content} = 100 - \text{Percent Extract} \tag{6.11}$$

After the extraction process, the gel fraction was prepared for measuring the cross-link density as per ASTM D6814-2. The cross-link density was calculated using the equation below.

$$V_e = \frac{-\left[\ln(1-V_r) + V_r + \chi_1 V_r^2\right]}{\left[\dfrac{V_1\left(V_r^{1/3} - V_r\right)}{2}\right]} \tag{6.12}$$

In which, ν_e is the cross-linking density (mol cm^{-3}), V_r is the volume fraction of rubber in the swollen sample, V_1 is the solvent molar volume, and χ is the polymer–solvent interaction parameter.

It was found that, with an increase in the exposure time, the values of the gel fraction and the cross-link density of the samples decreased. This confirmed that devulcanization had taken place (see Figure 6.3).

(iii) Chemical study of waste rubber by ATR-FTIR:

Attenuated total reflectance Fourier transform infrared (ATR-FTIR) analysis was performed on the gel fraction samples, as well as on the samples before extraction, so as to obtain data about the chemical modification that had taken place due to the devulcanization process. The scans ranged from 4000–500 cm^{-1} at a resolution of 4 cm^{-1} over 20 scans at room temperature (see Table 6.4).

(iv) Thermal analysis of waste rubber by TGA:

TGA was conducted to provide information about the thermo-oxidative degradation of the waste rubber after the devulcanization process, as shown in Figure 6.4. The analysis was performed on samples before, as well as after extraction. The first mass loss at about 300 °C refers to the organic additives. The second mass loss at about 300–400 °C refers to NR decomposition. The third mass loss at about 550–800 °C refers to the decomposition of the synthetic rubber. The fourth mass loss, occurring under an oxidizing atmosphere, refers to the carbon black decomposition.

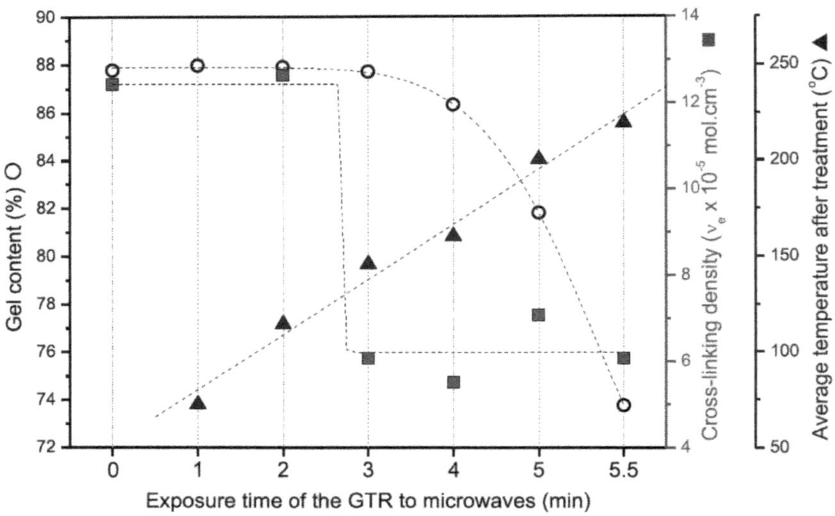

Figure 6.3 Gel content, cross-linking density and average temperature after microwave treatment *versus* exposure time of the GTR to microwaves. Reproduced from ref. 39 with permission from Elsevier, Copyright 2017.

Table 6.4 Peaks observed in the ATR-FTIR spectra.

Peaks (cm^{-1})	Bond
2870	C–H bond of the main chains of NR of GTR
1475–1450	CH_2 bonds, referring to vibrations of stretching and aromatic bending of the SBR
1390–1372	CH_3 bonds, referring to groups of NR
1705	C=O bond, confirming carbon black presence
840	C=C bond presence in NR of GTR
680	C–CH_3 bond presence in SBR of GTR
556, 750 and 1083	Peaks corresponding to S–S, C–S and S–O bonds, respectively

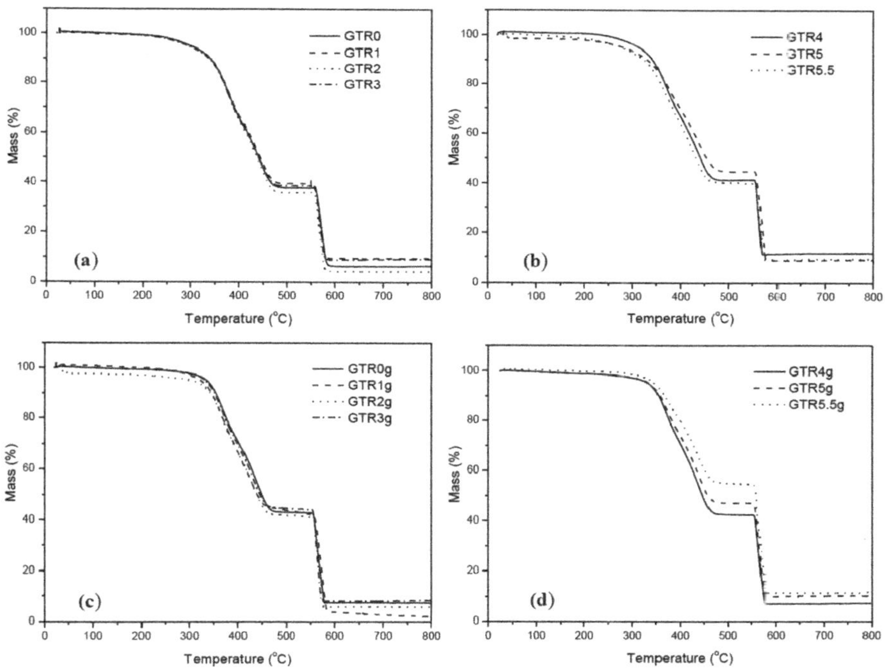

Figure 6.4 TGA curves of the devulcanized GTR (a) and (b), and their gel fractions (c) and (d). Reproduced from ref. 39 with permission from Elsevier, Copyright 2017.

6.6.3 Devulcanization of Natural Rubber by Mechanochemical Means

Devulcanization of natural rubber vulcanizate was performed in an open roll mixing mill by utilising bis(3-triethoxysilyl propyl) tetrasulfide (TESPT), a devulcanizing agent. The extent of devulcanization was monitored by measuring the sol content, gel content, crosslink density, molecular weight between crosslinks and Mooney viscosity of devulcanized rubber.[40]

Characterization:

(i) Evaluation of the extent of devulcanization:

Devulcanized rubber samples were extracted for 48 h using acetone as a solvent. The insoluble (gel) and soluble (sol) fractions of devulcanized rubber were then determined after 72 h Soxhlet extraction using toluene as a solvent followed by drying of the sample in a vacuum oven at 70 °C (see Figures 6.5 and 6.6).

The gel and sol fractions were determined using the following equations:

$$\text{Gel fraction} = 1 - \frac{\text{Weight of dissolved rubber in solvent}}{\text{Weight of the rubber sample after acetone extraction}} \tag{6.13}$$

$$\text{Sol fraction} = 1 - \text{Gel fraction} \tag{6.14}$$

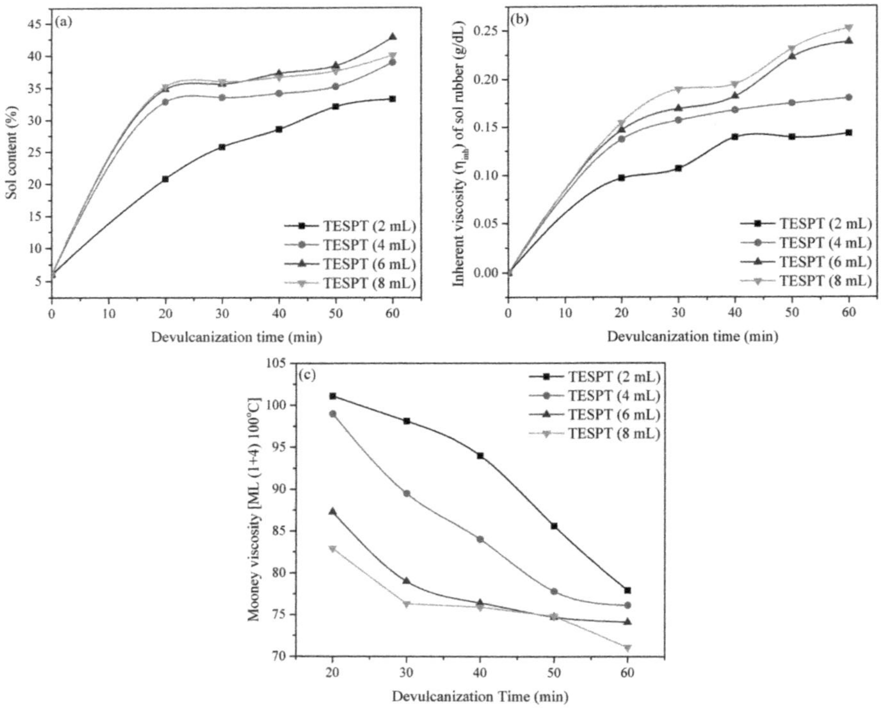

Figure 6.5 Effect of devulcanization time on (a) sol content, (b) inherent viscosity (η_{inh}) of sol rubber and (c) Mooney viscosity of devulcanized rubber. Reproduced from ref. 40 with permission from Elsevier, Copyright 2016.

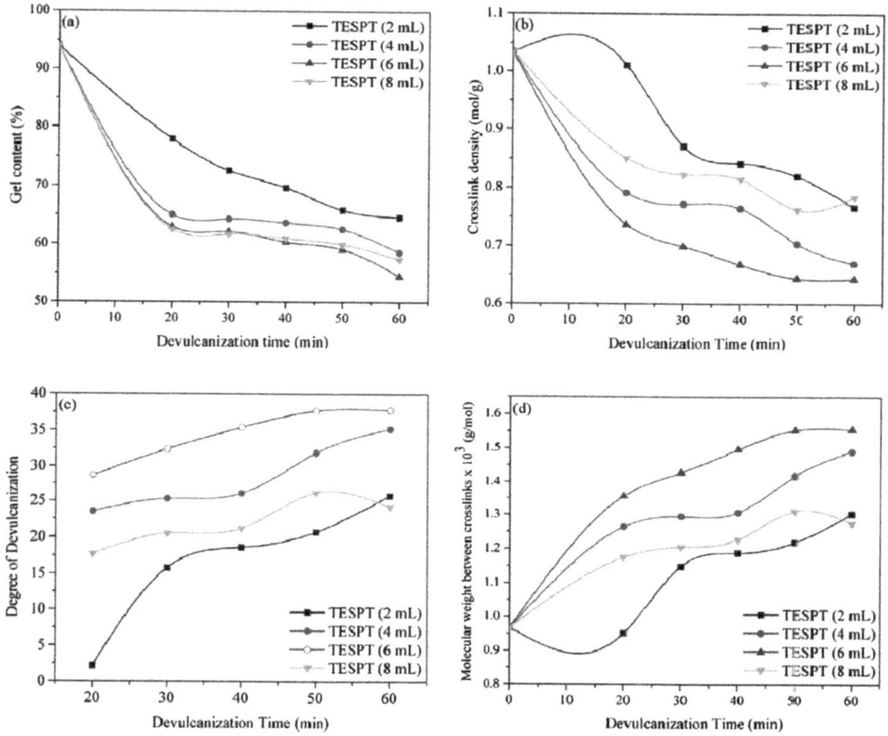

Figure 6.6 Effect of devulcanization time on (a) gel content, (b) crosslink density, (c) degree of devulcanization and (d) molecular weight between crosslinks of devulcanized rubber. Reproduced from ref. 40 with permission from Elsevier, Copyright 2016.

Inherent viscosity of sol rubber was determined *via* the following equation:

$$\eta_{\text{inh}} = \frac{\ln \eta_{\text{r}}}{C} \tag{6.15}$$

In which η_{inh} is the inherent viscosity, η_{r} is the relative viscosity of sol rubber and C is the concentration of sol rubber in (g dL^{-1}).

The Mooney viscosity of devulcanized rubber was measured at ML(1+4) 100 °C as per ASTM D 1646. Equilibrium swelling experiments of devulcanized rubber were carried out in toluene according to ASTM D 471. The swelling value (Q) of devulcanized rubber was calculated as follows:

$$Q = \frac{M_{\text{s}} - M_0}{M_0} \tag{6.16}$$

In which M_s is the swollen weight of the devulcanized rubber and M_0 is the dried weight of devulcanized rubber.

The crosslink density of devulcanized rubber (v_c) was measured using the following equation:

$$v_c = \frac{1}{M_c} = \frac{\ln(1 - V_{rf}) + V_{rf} + \chi V_{rf}}{V_s d_r \left(V_{rf}^{1/3} - \frac{V_{rf}}{2} \right)} \tag{6.17}$$

$$V_{rf} = \frac{\left(\frac{W_2}{d_2} \right)}{\left(\frac{W_1}{d_1} \right) + \left(\frac{W_2}{d_2} \right)} \tag{6.18}$$

In which M_c is the molecular weight between crosslinks of devulcanized rubber, V_s is the molar volume of the solvent equal to 106.2 cm^3 mol^{-1}, V_{rf} is the volume fraction of devulcanized rubber in the swollen specimen, χ is the interaction parameter equal to 0.393, d_1 is the density of solvent, d_2, and d_r is the density of devulcanized rubber, W_1 is the weight fraction of the solvent and W_2 is the weight fraction of devulcanized rubber.

The degree of devulcanization of the devulcanized rubber was calculated as follows:

$$\%\text{Devulcanization} = \left(1 - \frac{\xi}{\xi_0} \right) \times 100 \tag{6.19}$$

In which ξ_0 is the crosslink density of vulcanized rubber and ξ is the crosslink density of devulcanized rubber.

6.7 Characterizing Recycled Rubber Products

Recycled rubber widely finds applications in the modification of products, as mentioned in an earlier section. This section describes the techniques associated with the characterizations of those recycled rubber incorporated substances. These characterizations show how the waste rubber content affects the original properties of the material.

6.7.1 Characterizing Cross-link Density in Rubber–Rubber Blends

Characterizing cross-link density in rubber–rubber blends and composites involves study of the network chain and quantifying the number of cross links within a given volume. This section describes some of the spectroscopic techniques involved in this type of characterization.

6.7.1.1 Swollen State NMR Spectroscopy

With the increase in cross-linking among the monomer molecules, a reduction in sample mobility occurs.[41] This leads to broadening of the NMR peaks and forms the basic principle underlying the characterization techniques involved in determination of the crosslinking density of these blends. The vulcanized blends need to be swollen to an equilibrium position in order to give signals from different environments.

6.7.1.2 Solid State NMR Spectroscopy

As mentioned in an earlier section, analysis of the degree of crosslinking based on the increase in signal sharpness is limited to liquid samples only and rubbers cured by vulcanization need to be analysed by swollen state spectroscopy. Similarly, the existing technology is not suitable for obtaining signals for solid samples. 13C-NMR is used to obtain NMR signals from solid polymeric materials.[42]

6.7.2 Morphological Characterizations for Rubber–Rubber Composites

This section discusses the use of various microscopic techniques in the study of the morphology of rubber–rubber blends. These characterizations provide an insight into the structural properties of many rubber composites or rubber–rubber blends and are useful for comparison studies involving varying compositions of recycled rubber. Each of the techniques has a distinct procedure and is associated with particular advantages and disadvantages.

6.7.2.1 Scanning Electron Microscopy

This is a microscopic technique based on imaging of the surface topography by a fine electron beam scanning the surface of the sample. Specimen preparation in SEM requires the sample to be coated with conductive material such as Au, Au/Pd, as the material to be scanned by the electrons need to electrically conductive in nature.[43,44] SEM is associated with significantly high focal depths, which are responsible for bright three-dimensional pictures, which are used for clarity in the analysis of surface structures. SEM is generally used for fractography studies, in which the fracture surface of the samples are studied. Rubber samples need to be reduced dimensionally so as to fit into the sample holder.

6.7.2.2 Transmission Electron Microscopy

A transmission electron micrograph is the image of the interaction of a fine beam of electrons with the sample when the beam is transmitted through the body of the sample. Transmission electron microscopy (TEM) procures better images than SEM, as TEM images are of a better resolution than the SEM images. However, this microscopy technique requires ultra-thin samples and the quality of the micrograph obtained varies with the thickness of the sample. Therefore, TEM becomes quite a difficult technique, requiring appreciable precision in terms of sample preparation. Cryomicrotomy and the focused ion beam method are the two methods largely followed for sample preparation in TEM.[44] Although a number of examples are known in which TEM images have been used for morphological analysis of rubber–rubber blends, Figure 6.7 shows an example of TEM micrographs taken for the analysis of the tensile fracture surface of a number of rubber–polybutadiene rubber blend samples, prepared with varying blend ratios.[45]

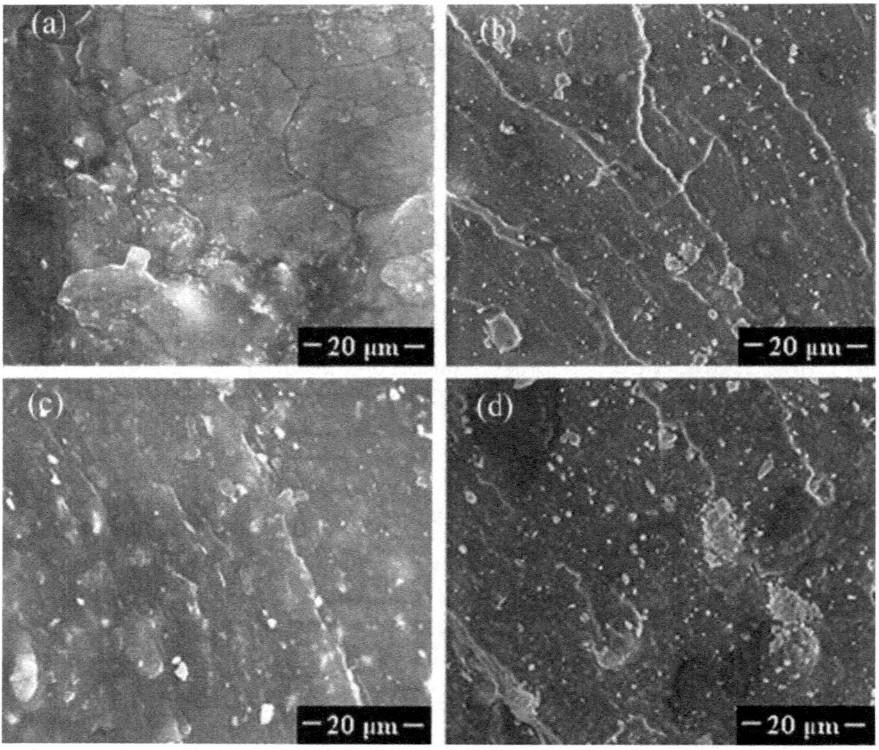

Figure 6.7 TEM images showing the tensile fractured surfaces of samples with varying proportions of reclaimed rubber: (a) NR–PBR (100), (b) NR–PBR/RR (80/20), (c) NR–PBR/RR (60/40) and (d) NR–PBR/RR (40/60) blend systems, where NR stands for natural rubber, PBR stands for polybutadiene rubber and RR stands for reclaimed rubber. Reproduced from ref. 45 with permission from Elsevier, Copyright 2013.

6.7.2.3 *Scanning Probe Microscopy*

The basic principle of this technique involves taking images of the surface, where a probe is mechanically scanned over the sample surface and the image is a result of a study of the probe-surface interaction presented as a function of the position.

- Atomic force microscopy (AFM):

This technique is widely used, and at times preferred over SEM owing to certain advantages. SEM still remains a costly process, requiring the maintenance of a vacuum environment. SEM also demands for a coating to be placed over the sample using a metal or graphite. Moreover, the AFM technique provides 3-D surface images as compared to the 2-D micrographs obtained from SEM.[44]

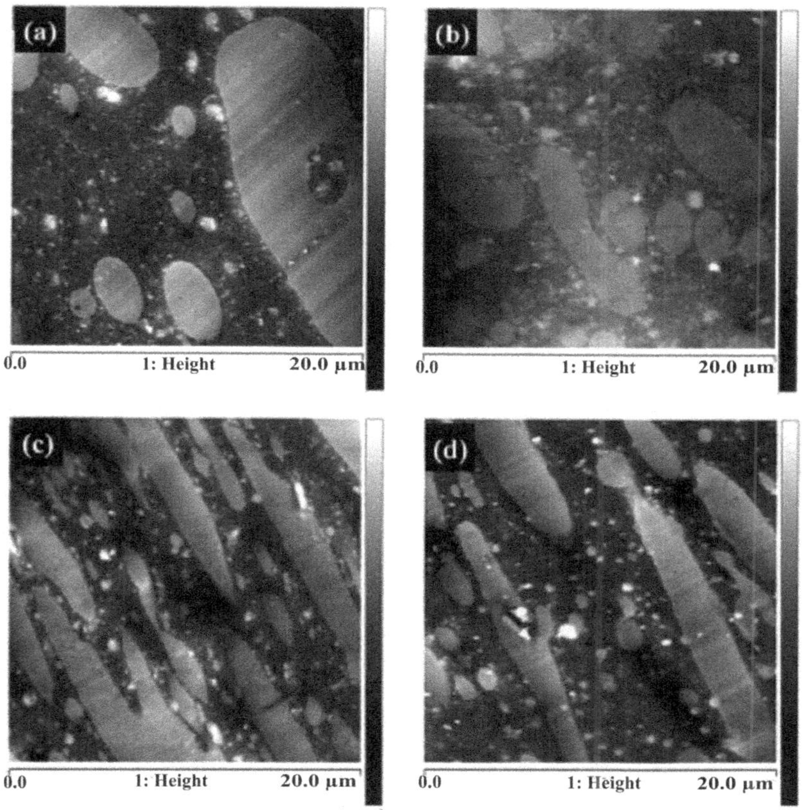

Figure 6.8 AFM height images of natural rubber (NR), natural butadiene rubber (NBR) and epoxidized natural rubber (ENR) composites with varying compositions of epoxidized natural rubber. (a) The NR/NBR (70/30) composite and (b–d) NR/NBR/ENR (70/30/10) composites with the different degrees of epoxidation of ENR: (b) ENR25, (c) ENR40 and (d) ENR50. Reproduced from ref. 46 with permission from Elsevier, Copyright 2016.

As an example, Figure 6.8 shows how AFM has been used to study the morphological characteristics of natural rubber and natural butadiene rubber composites prepared with variable compositions of epoxidised natural rubber, which had been used as a compatibiliser in the composite.[46]

6.7.2.4 *Chemical Staining*

In order to produce alternate regions of dark and light shaded areas in a microscopy image, a process called chemical staining is carried out.[43] This is carried out in order to bring in some chemical difference between the elastomers blended. Most of the time this is done by the incorporation of heavy atoms into the chemical structure of one of the elastomers. For example, osmium tetra-oxide is commonly used as a staining agent, as it reacts with the unsaturated double bonds present in the elastomer. Different elastomers possess different degrees of unsaturation and thus are differentially stained using osmium tetroxide. Staining is associated with a brighter image as staining results in a bulkier backflow of electrons.

6.7.3 Characterizing the Mechanical and Thermal Properties of Devulcanized Rubber/Polypropylene Blends

Ground tyre rubber was devulcanized *via* the mechanical–chemical process. The obtained devulcanized rubber was then blended with previously melted polypropylene in different proportions in a mixer.[47] Devulcanized rubber and melted polypropylene were mixed for 10 min at 190–195 °C to achieve a uniform dispersion of devulcanized rubber in polypropylene. Composites were formed *via* compression moulding at 10 MPa pressure and 195 °C for 5 min. Mechanical and thermal characterizations were performed to study the tensile, elongation at break and the thermal properties of the blends. FTIR spectroscopy was performed to ascertain the bonds present or newly formed in the blends.

- Mechanical analysis

The mechanical properties of the blends depend, to a large extent, on the proportion of devulcanized rubber present in the blends. Mechanical analysis shows that with the increase in the content of devulcanized rubber, the tensile strength of the blend decreases. This trend may be credited to poor distribution of the devulcanized rubber in the polypropylene matrix and also due to incompatibility of viscosity ratio difference between the devulcanized rubber and polypropylene. In addition, the tensile strength increased with gamma irradiation up to 50 kGy, but decreased with a further increase in irradiation dose. This behaviour may be due to the completion of the cross-linking of polypropylene.

In the case of elongation at break, the values increase with an increase in the loading of devulcanized rubber till 75 wt% and thereafter decreases with an increase in devulcanized rubber loading. This may be attributed to the fact that an increase in the proportion of devulcanized rubber particles limits the flow of the blends, weakening the uniformity of the matrix and thereby reducing the elongation at break values. Also, as the radiation dosage increases, the blends get stiffer, reducing their elasticity and thereby reducing the elongation at break values.

- Thermal analysis

Differential Scanning Calorimetry (DSC) was performed to determine the structural and thermal properties of the blends. Two distinct peaks were observed, one at about 161.4 °C, related to te hα-phase and another one at about 143.9 °C, related to the β-phase of polypropylene, as shown in Figure 6.9. It was observed that when adding devulcanized rubber and increasing its proportion thereof, the peak related to the β-phase of polypropylene reduced and eventually disappeared whereas, the peak related to the α-phase of polypropylene reduced in size. These changes in peaks may be suggestive of the role of devulcanized rubber in the improvement of the thermal stability of the blends.

Thermogravimetric Analysis was performed to determine the degradation behaviour of the blends. Thermal degradation of polypropylene mainly occurs due to random splitting along the backbone chain, resulting in formation of long radical fragments. A mass loss occurs around 250–385 °C caused by the formation of volatile products due to random splitting of the chain. It was observed that incorporating devulcanized rubber into polypropylene improved its thermal stability. It was also observed that the thermal

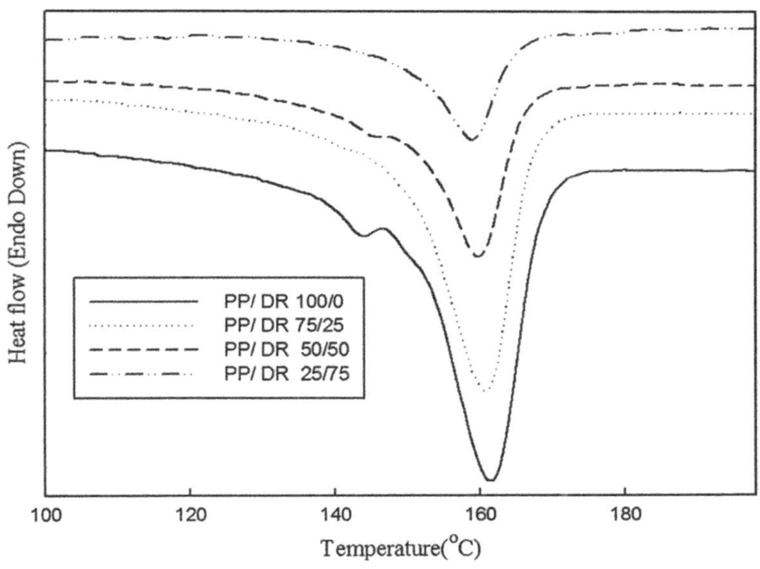

Figure 6.9 DSC thermographs of DR/PP blends at different concentrations. Reproduced from ref. 47 with permission from Elsevier, Copyright 2013.

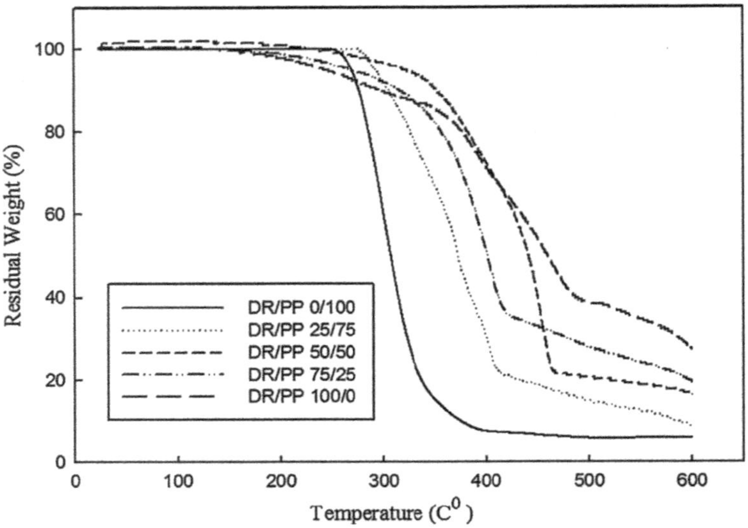

Figure 6.10 TGA thermograms of DR/PP blends of different compositions. Reproduced from ref. 47 with permission from Elsevier, Copyright 2013.

stability of the blends continued to increase up to incorporation of 50 wt% devulcanized rubber into polypropylene. This resulted in a rise in the initial degradation temperature up to 393 °C from 293 °C. A further increase in the proportion of devulcanized rubber reversed the trend and reduced the initial degradation temperature to 368 °C (Figure 6.10).

6.7.4 Concrete Modified by Waste Rubber

Incorporation of waste rubber into the constituent mixture of concrete is associated with certain alterations in the original properties of concrete, which are summarized in this section.

(i) Compressive strength and flexural strength

A gradual increase in the percentage of waste rubber in concrete is associated with a corresponding decrease in both the compressive strength[48,49] and flexural strength of the concrete.[48,50] The various factors justifying the abovementioned decrease in the compressive strengths are as follows:

- Rubber aggregates have a lower density as compared to the other ingredients of concrete.
- Rubberized concrete samples have substitutes that show poor stiffness.
- The stress applied is not uniformly distributed throughout the bulk due to inappropriate bonding at the interface of rubber aggregates and cement pastes.[48] This leads to cracks.

- As rubber is not properly bonded at the interface, rubber modified concretes sometime show heterogeneity in samples. Due to its low specific gravity,[48] rubber tends to form an upper surface layer during vibration. These non-homogeneous concrete samples are of relatively low strength and are prone to cracks.

(ii) Modulus of elasticity

Concrete is a composite material of cement and aggregate materials. Therefore, the modulus of elasticity of the component material is supposed to affect the overall modulus of elasticity of the material. As rubber is characterized by a low value of modulus of elasticity, use of rubber instead of other aggregates with cement lowers the modulus of rigidity of the concrete.

(iii) Tensile strength

Being a soft material, rubber is expected to act as a barrier against crack growth in concrete. This hypothesis could lead to the assumption that incorporation of rubber into concrete would result in an increase of the tensile strength. However, this is not the case and concrete shows a decrease in the tensile strength with an increase in the percentage of rubber incorporated.[48,50] The prime reason for this observation is also attributed to the fact that rubber has extremely weak bonding interactions with cement paste. Crack propagation becomes easier at the interface.

6.8 Rheological Properties of Asphalt Binders Modified with Devulcanized Rubber

Asphalt binder, PG 64-22, was used as base binder and devulcanized rubber was used as one of the modifiers. The binder and devulcanized rubber were blended at 177 °C for 45 min at a 2000 rpm stirring speed.[51] The apparent viscosities of the un-modified, as well as the modified binders, were measured using a rotational viscometer as per AASHTO T316. A dynamic shear rheometer (DSR) was used to measure the rheological properties. The properties were measured as per AASHTO T315.

6.8.1 Apparent Viscosity

As previously reported in the literature, the apparent viscosity should not be more than 3.0 Pa·s at 135 °C. This viscosity is very important for good construction handling. The analysis showed that, for blended binders with an increase in temperature, the apparent viscosity values reduced and at 135 °C, were lower than 3.0 Pa·s. The same trend was observed in the case of base binders and their values were lower than those of the blended binders. It was observed that the apparent viscosity values of the devulcanized rubber blended binder were higher than those observed in other blended binders.

6.8.2 Performance Grade Critical Temperature

Performance grade critical temperature/failure temperature is the temperature at which the G*/sin δ value equals 1.0 kPa for a virgin binder and 2.2 kPa for rolling thin-film oven (RTFO) an aged binder. From the analysis, it was observed that the failure temperature values for virgin base binders were in the range of 64–70 °C. In the case of virgin modified binders the failure temperature values were observed in the range of 76–82 °C. In the case of RTFO aged base binders, the failure temperature values were observed in the range of 68–74 °C. In the case of RTFO aged modified binders, the failure temperature values were observed in the range of 76–82 °C. From this analysis it can be said that the RTFO aged base binders performed better than virgin base binders. Whereas, there was no significant difference in the performance of the virgin modified binders and RTFO aged modified binders.

6.8.3 Rutting Resistance Factor

The rutting resistance of an asphalt pavement is characterized by the G*/sin δ value of the binder. The higher the G*/sin δ value of the binder, the better the rutting resistance of the asphalt pavement that has been paved using the same binder will be. From the analysis, it was observed that, with a rise in temperature, the values of G*/sin δ decreased for all the binders, base and modified both. The G*/sin δ values of base binders were found to be lower than those of the modified binders.

6.8.4 Phase Angle

Phase angle is the time lag occurring between the applied load and the deformation corresponding to it. Also, as phase angle increases so does viscosity of the asphalt binder. From the analysis of phase angles, it was observed that, the base binders had more phase angle values than those of modified binders. Among the modified binders, the ones modified with devulcanized rubber showed slightly smaller phase angles, thereby showing more elasticity at the highest temperatures.

6.9 Conclusion

Some basic characterizations of a material are carried out in order to study the signature properties of the material. Some of the primary properties studied include morphology, chemical structure, chemical reactivity, mechanical characteristics and so forth. When the material under consideration is rubber, a few other studies should be highlighted, including elasticity, stiffness, rheological properties and so on. Proper study of these material properties is required, so that the material can be properly exploited in various fields of application, based on its characteristics.

Previously, studied properties could be, in many instances, further modified according to the needs of the particular application field. These modified materials are again characterized to determine the improved properties.

References

1. C. Chandrasekaran, in *Anticorrosive Rubber Lining: A Practical Guide for Plastics Engineers*, ed. C. Chandrasekaran, William Andrew Applied Science Publishers, Elsevier, 1st edn, 2017, ch. 19, p. 165.
2. D. Dobrota and G. Dobrota, *J. Cleaner Prod.*, 2018, **172**, 3591.
3. B. Adhikari, D. De and S. Maiti, *Prog. Polym. Sci.*, 2000, **25**(7), 909.
4. V. V. Rajan, W. K. Dierkes, R. Joseph and J. W. M. Noordermeer, *Prog. Polym. Sci.*, 2006, **31**(9), 811.
5. M. J. Fernández-Berridi, N. Gonzalez, A. Mugica and C. Bernicot, *Thermochim. Acta*, 2006, **444**(1), 65.
6. S. Datta, J. Antos and R. Stocek, *Polym. Test.*, 2017, **59**, 308.
7. S. Rolere, S. Liengprayoon, L. Vaysse, J. Sainte-Beuve and F. Bonfilsa, *Polym. Test.*, 2015, **43**, 83.
8. B. Wu and M. H. Zhou, *Waste Manage.*, 2009, **29**(1), 355.
9. X. Shu and B. Huang, *Constr. Build. Mater.*, 2014, **67**(B), 217.
10. D. L. Presti, *Constr. Build. Mater.*, 2013, **49**(B), 863.
11. M. Sienkiewicz, H. Janik, K. Borzędowska-Labuda and J. Kucińska-Lipka, *J. Cleaner Prod.*, 2017, **147**, 560.
12. K. Fukumoria, M. Matsushitaa, H. Okamotoa, N. Satoa, Y. Suzukib and K. Takeuchi, *JSAE Rev.*, 2002, **23**(2), 259.
13. L. H. Chou, C. K. Yang, M. T. Lee and C. C. Shu, *Composites, Part B*, 2010, **41**(8), 613.
14. A. Ciesielski, in *An Introduction to Rubber Technology*, ed. A. Ciesielski,Rapra Technology Ltd., UK, 1st edn, 1999,ch. 4, pp. 49–75.
15. R. Brown, in *Physical Testing of Rubber*, ed. R. Brown,Springer ,USA ,4th edn, 2005,ch. 13, p. 259.
16. P. S. Thomas, A. A. Abdullateef, M. A. Al-Harthi, M. A. Atieh, S. K. De and M. Rahaman, *et al.*,*J. Mater. Sci.*, 2012, **47**(7), 3344.
17. M. Najidha, N. S. Saxena, R. Sreeja, C. H. Unnithan and P. Predeep, *Mater. Lett.*, 2005, **59**(27), 3431.
18. L. Bokobza, *eXPRESS Polym. Lett.*, 2012, **6**(3), 213.
19. C. F. Niedik, C. Freye, F. Jenau, D. Haring and G. Schröder, *et al.*, *IEEE Xplore*, 2016, DOI: 10.1109/ICD.2016.7547814.
20. R. Mujal-Rosas, J. Orrit-Prat, X. Ramis-Juan and M. Marin-Genesca, *J. Thermoplast. Compos. Mater.*, 2014, **27**(9), 1209.
21. F. Chen and J. Qian, *Waste Manage.*, 2003, **23**(6), 463.
22. J. Haydary, Z. Koreňová, Ľ. Jelemenský, and J. Markoš, in *Thermophysics*, ed. J. Leja and S. Prváková, Vydavateľstvo STU, Kočovce, 1st issue, 2008, ch. 7, pp. 62–68.

23. E. Grieco, M. Bernardi and G. Baldi, *J. Anal. Appl. Pyrolysis*, 2008, **82**(2), 304.
24. P. T. Williams, *Waste Manage.*, 2013, **33**(8), 1714.
25. J. D. Martínez, N. Puy, R. Murillo, T. García, M. V. Navarro and A. M. Mastral, *Renewable Sustainable Energy Rev.*, 2013, **23**, 179.
26. H. Liang, D. Rodrigue and J. Brisson, *J. Appl. Polym. Sci.*, 2015, **132**(42), 42692.
27. C. H. Scuracchio, D. A. Waki and M. L. C. P. da Silva, *J. Therm. Anal. Calorim.*, 2007, **87**(3), 893.
28. P. Sarkar and A. K. Bhowmick, *ACS Sustainable Chem. Eng.*, 2016, **4**(10), 5462–5474.
29. C. E. Barnett, *Ind. Eng. Chem.*, 1934, **26**(3), 303.
30. C. Chiu, *Resour., Conserv. Recycl.*, 2008, **52**(3), 522.
31. E. Ganjian, M. Khorami and A. A. Maghsoud, *Constr. Build. Mater.*, 2009, **23**(5), 1828.
32. M. Gesog and E. Güneyis, *Constr. Build. Mater.*, 2011, **25**(8), 3319.
33. R. Brown, in *Physical Testing of Rubber*, R. Brown, Springer, USA,4th edn, 2005,ch. 18, p. 364.
34. ASTM D429-14, *Standard Test Methods for Rubber Property—Adhesion to Rigid Substrates*, ASTM International, West Conshohocken, PA, 2014.
35. ASTM D413–98(2017), *Standard Test Methods for Rubber Property—Adhesion to Flexible Substrate*, ASTM International, West Conshohocken, PA, 2017.
36. Y. Fang, M. Zhan and Y. Wang, *Mater. Des.*, 2001, **22**(2), 123.
37. G. Dondi, P. Tataranni, M. Pettinari, C. Sangiorgi, A. Simone and V. Vignali, *Constr. Build. Mater.*, 2014, **68**, 370.
38. R. Diaz, G. Colomines, E. Peuvrel-Disdier and R. Deterre, *J. Mater. Process. Technol.*, 2018, **252**, 454.
39. F. D. B. de Sousa, C. H. Scuracchio, G. H. Hu and S. Hoppe, *Polym. Degrad. Stab.*, 2017, **138**, 169.
40. S. Ghorai, S. Bhunia, M. Roy and D. De, *Polym. Degrad. Stab.*, 2016, **129**, 34.
41. A. J. Tinker, in *Blends of Natural Rubber Novel Techniques for Blending with Speciality Polymers*, ed. A. J. Tinker and K. P. Jones,Chapman & Hall, Great Britain, 1st edn, 1998,ch. 2,pp. 9–16.
42. A. B. Ahmad and A. B. Amu, in *Blends of Natural Rubber Novel Techniques for Blending with Speciality Polymers*, ed. A. J. Tinker and K. P. Jones,Chapman & Hall, Great Britain, 1st edn, 1998,ch. 4, pp. 40–45.
43. P. Cudby, in *Blends of Natural Rubber Novel Techniques for Blending with Speciality Polymers*, in A. J. Tinker and K. P. Jones,Chapman & Hall, Great Britain, 1st edn, 1998,ch. 3,pp. 24–36.
44. H. Miyagawa, in *Rubber Nanocomposites, Preparation, Properties and Applications*, ed. S. Thomas and R. Stephen, John Wiley & Sons (Asia) Pte Ltd., Singapore,1st edn, 2009, ch. 11,pp. 259–280.
45. D. De, P. K. Panda, M. Roy and S. Bhunia, *Mater. Des.*, 2013, **46**, 142.
46. X. Zhao, K. Niu, Y. Xu, Z. Peng, L. Jia, D. Hui and L. Zhang, *Composites, Part B*, 2016, **107**, 106.

47. M. M. Hassan, N. A. Badway, M. Y. Elnaggar and E. A. Hegazy, *J. Ind. Eng. Chem.*, 2013, **19**(4), 1241.
48. A. Sofi, *Ain Shams Eng. J.*, 2017, DOI: 10.1016/j.asej.2017.08.007.
49. M. Bravo and J. de Brito, *J. Cleaner Prod.*, 2012, **25**, 42.
50. E. Ganjian, M. Khorami and A. A. Maghsoudi, *Constr. Build. Mater.*, 2009, **23**(5), 1828.
51. O. Xu, P. R. Rangaraju, S. Wang and F. Xiao, *Constr. Build. Mater.*, 2017, **154**, 841.
52. C. Brown, *US Pat.*, 7461970 B2, 2008.

CHAPTER 7

High Performance Flooring Materials from Recycled Rubber

RAGHVENDRA MISHRA[a], M. K. ASWATHI[a] AND SABU THOMAS*[a,b]

[a]International and Inter University Centre for Nanoscience and Nanotechnology, Mahatma Gandhi University, Kottayam, Kerala, India, 686 560; [b]School of Chemical Sciences, Mahatma Gandhi University, Kottayam, Kerala, India, 686560
*E-mail: raghvendramishra4489@gmail.com

7.1 Introduction

Waste polymeric materials are one of the main growing public concerns for the environment. Polymer waste from various sources is expected to reach about 10.5 million m^2 or £70 million per year. The majority of these types of waste are disintegrated in land fill or burned.

Recycling of polymeric materials is used in industrial and post-consumer material waste handling processes, in this way, the recycling of polymeric materials into high performance materials is an excellent option to reduce the volume of waste polymeric materials and the cost of materials with desirable physical, chemical, thermal and acoustic properties.[1,2] By adopting a

Green Chemistry Series No. 59
Rubber Recycling: Challenges and Developments
Edited by Jin Kuk Kim, Prosenjit Saha, Sabu Thomas, Józef T. Haponiuk and M. K. Aswathi

Published by the Royal Society of Chemistry, www.rsc.org

suitable recycling method, this recycled polymer waste can be converted into viscoelastic materials which can be used to make sound shielding, carpet underlay material and used in many other applications.[3,4]

Rubber is a type of polymeric material, it has been used in various products, however, the majority of the rubber is used in the tyre industry. Rubber based product are made of vulcanized (*i.e.* cross-linked polymer chains) rubber and various reinforcing materials through compounding. Co-polymer styrene-butadiene (SBR) or blends of natural rubber and SBR have generally been used as the most common rubber matrix, with carbon black, textile or steel fibres, extenders and vulcanizing agents being used to improve strength and abrasion resistance, control viscosity, and reduce internal friction and cross-linking of the rubber.[5,6] Due to the lack of a suitable disposal method, this produces environmental hazards. Therefore, there is a need to recycle waste rubber, which could reduce the environmental problem and wasting of energy. The main problem of waste rubber is the thermosetting structure of the vulcanized and compounded rubber, therefore it is difficult to melt and recycle the waste rubber because the chain of the vulcanized rubbers are cross linked by covalent bonds and simple methods are not well suited to breaking these bonds. However, there are some methods that have been used to recycle the crumb rubber such as grinding (ambient grinding, cryogenic grinding) and blending with thermoplastics.[7,8] These methods are desirable for the recycling of waste rubber and for converting it into a suitable product. Therefore, recycled rubbers are being used in automotive parts, and for vibration damping and acoustic insulation with various shapes, good flexibility and stiffness.

It has been reported previously that the acoustic performance of recycled rubber based materials depends on the size of the rubber waste granules and consolidation methods of the rubber waste granular blend.[9] Acoustic shielding materials refer to sound absorbing materials, generally noise can have an adverse impact on technological development. The excellent mechanical and thermal insulation of properties as a function of a binder are another application of recycled rubber, it has been reported that the bound crumb rubber bodies show good absorption, transmission impact sound insulation, mechanical and thermal insulation as a function of the binder loading and chemistry.[10,11] Rubber-based acoustic materials are used to reduce the impact sound in multi floor buildings. Density, dynamic stiffness and compressibility are important parameters for impact sound insulation characteristics of a floating floor.[12]

The mechanical and thermal properties of the concrete, based on the crumb rubber and steel or plastic fibres partially coated with rubber (FCR) have been reported. The main advantage of this study is that steel or plastic FCR is cheaper than crumb rubber, and FCR can enhance the energy absorption of rubberized concretes. In addition, thermal insulation can also be improved by the addition of steel or plastic FCR aggregates in rubberized concretes. Compressive, bending, impact and wear resistance is improved in concrete with steel or plastic FCR aggregates compared with conventional rubberized concrete.[13,14] It has also been reported that the high volume of recycled rubber aggregates can influence the acoustic properties of rubber containing concrete.[15]

Better compatibility and interaction of rubber particles and asphalt binders has shown an enhancement in the properties and performance of crumb rubber in asphalt paving mixtures. The performance and properties of recycled waste rubber in Portland cement concrete depends on the compatibility of the recycled waste rubber in Portland cement concrete and the stress concentration.[16] The operating temperature is another factor, which affects the comprehensive strength of different high strength concretes. A compressive strength of 10%, 15%, and 20% silica fume (SF) loading can decrease the weight of concrete, and improve the compressive strength in a range of operating temperatures below 300 °C. A sharp decrement in compressive strength was found in the range of 300–400 °C, which is due to volume expansion of the concrete matrix and resultant internal cracks. With a 3% mass fraction fine recycled rubber loaded high strength concrete can show better compressive strength with increasing temperature.

Tensile thermal properties of the recycled rubber membrane reinforced with HDPE particles were mentioned under different conditions.[17] In this study, the properties of recycled high-density polyethylene (HDPE) pellets and the rubber matrix membrane are discussed. Crushed polyethylene particles can form channels inside the membranes that offer a reinforcing effect to the structure of the composite membrane material.

Flooring materials are subject to mechanical wear, ageing, soiling and maintenance, and the performance of flooring materials in daily use depends on the variations in the tribological characteristics of flooring materials.[19] In sports rooms, mechanical erosion, routine cleaning processes and material transfer from shoe soles (elastomer and dirt particles) alter the flooring surfaces. The slip resistance of floors was measured in terms of the surface changes from smoothing to roughening through mechanical wear.[20] The walking and mechanical behaviours of flooring materials are quantified by the slip resistance, static and dynamic friction coefficients.[21] The friction varies with various conditions such as whether the flooring is dry, wet, or is subjected to a water/detergent dilution, oil, or a water/oil dilution.[22]

The main aim is to recycle rubber and convert it into high-performance flooring materials, therefore, various methods have been used to recycle the waste rubber, such as retreading, pyrolysis and incineration, grinding, pulverization, reclaiming, microwave and ultrasonic methods. This chapter summarizes the recent advances in the use of waste rubber as a high-performance flooring material.

7.2 Types of Flooring Materials

The objects and occupants within a building are supported by the floor. Cement or lime concrete, bricks, flagstones, marble, glass, ceramic, plastic, mud and murram, wood, cork, linoleum, asphalt, and rubber are the main types of flooring materials.[18] The use of particular flooring materials is based on the specific application. Wood-based flooring materials are a leading origin of formaldehyde emission (FE), a harmful gas released within buildings. Many compounded wood panel goods such as particleboard (PB), plywood (PLW), medium density fibreboard (MDF) and high-density fibreboard (HDF), have been frequently popular, and continue to be applied in the manufacturing of furniture, cabinets, and various building products. These goods are principally bonded with formaldehyde-based resin adhesives such as urea-formaldehyde (UF) resin, melamine-urea-formaldehyde (MUF) resin, phenol-formaldehyde (PF) resins, and so forth. Amongst these, the inclusion of melamine into the UF resin increases the stability and humidity resistance properties of this adhesive. In addition, MUF has drawn consideration as an alternative option to melamine formaldehyde (MF) resins due to the lower cost.[23]

There are three kinds of wood-based flooring materials, laminate wood flooring, engineered flooring, and solid wood flooring. Laminate wood flooring is comprised of HDF as the core material, while engineered flooring is made of PLW, with a thin veneer bonded to the face of the PLW using UF and MF resins as hot press adhesives.[24] Laminate wood flooring is comprised of four principal elements bonded together. A wood content-based core and a wear-resistant, ornamental surface made from a resin-based MF resin and aluminium oxide shackled to a moisture resistant material. A balancing support is tied to the underside of the core, on the top, there is a thin cap sheet of aluminium oxide, which affords protection and stain resistance. In the expansion, polyvinyl acetate (PVAc) was included as a substitute for MF resin in the formaldehyde-based resin system to overcome the FE from the adhesives used between PLW and the fancy veneers.[4] A small amount of laminate flooring may lead to impurity of indoor air with the aid of floor heating.[25,26]

In portions of Europe in which winters are moderate, farm goats are often put in uninsulated houses as insulated homes are important in countries that have cold winter weather. The goats react to the low temperature by decreasing their resting time, though extending the time consumed working and eating. This small time physiological feedback in the cold season may symbolize a raised metabolism. On average temperature days, straw was the least and metal was the most approved flooring, while solid wood and mattresses were most favoured in the cold period.[27]

Recently, remarkable studies have been used to examine the common burning types of gasoline on various flooring materials and the impact on the quantity and flow behaviour of the gasoline on the burning spots. The nature of the burning spot is influenced by the character of the combustion itself, as well as the features of the matrix, while the environment of the fire source, such as the area, the direction of the wind and the burning season, will have an effect on the spots. Once the rubber pad was filled with gasoline, there were remarkable durable spots which showed a colour change. Nevertheless, the pad has not been burned, but there was a small malformation. Although, as with the burnt EVA mosaic floor, the carbonization spots will be held in the matrix, it is possible for the EVA to burn within its edge with a common bulging appearance on the floor material, which was produced if the surface was rough enough. For the two floor materials, there was significant flowing of combustion. When the fire was extinguished using water, the burning spots were noticeably affected by the water, particularly for the one filled with gasoline, as the water altered the flow of the combustion.[28]

Previous and current studies have been used to investigate whether the indoor environment of an institution may influence the health, comfort, and performance of students. Also, considering the number of students and their actions, the windows and outside conditions, and the kind of air-conditioning present, the interior flooring materials are an essential factor that should be taken into account in inside atmosphere conditions. It appears that flooring materials (having the largest surface area) are capable of affecting indoor air, acoustics, lighting and thermal quality.[29]

The fire risk of rubberized flooring, wool carpet flooring, and also various combinations of floor carpet materials have been studied by cone calorimetery at a variety of heat fluxes. Evaluating and analysing the evidence, the fitted equation of product samples, as well as the fire risk indexes, for example, fire growth index (FGI), entire heat discharge, overall heat discharge, and poisonous gas creation rate index revealed that the fire hazard from higher to lower sequence is rubberized carpet, wool floor, blended materials based floor or carpet. Rubber carpet is actually accessible to ignition; the flame retardant activity is actually inadequate. The fire risk of rubber floors is utmost and the wool floor has the lowest risk. The choice of flooring material for various facilities demands attention due to various factors such as the occupant outcomes, safety, aesthetics, durability, maintenance, manufacturer's information, personal experience and costs.[32]

The reflection, absorption and transmission of electromagnetic waves from flooring materials at microwave and radio frequencies are another advanced application of flooring materials. Cut-pile type carpets do not exhibit good reflection properties; if they are placed on a plywood layer, the reflection from this layer is disrupted. The inclusion of an underpad to the carpet causes even poorer reflection, which is expected due to the scattering, although loop-pile carpets display more reflectivity than cut-pile carpets. Hardwood and vinyl materials are primarily employed with just a plywood layer as a supporting material in-house flooring constructions. A very satisfactory reflection is expected from this type of material owning to a shiny surface with a very little roughness. A better reflection from vinyl compared to hardwood can be observed. The inclusion of a plywood supporting layer does not alter the reflectivity of the hardwood and vinyl flooring materials.[30]

Flooring building materials such as granite, ceramic tiles and marble are now employed around the world. Most of the building materials are generally constituted of rock and soil and these few raw materials possess typical radioactive isotopes such as ^{232}Th and ^{238}U decay series and ^{40}K. Gamma emission is generated from naturally available radioisotopes, for example ^{40}K and also the radionuclides from ^{232}Th as well as the ^{238}U series, in addition to their decay outputs (additionally referred to as terrestrial background radiation), which occur at trace levels in all ground formations. In particular, natural condition radioactivity due to gamma radiation depends principally on the geographic and topographical environments and is found various levels in the soil of the various regions. Natural radionuclides are found in air, food, water, soil, bodies and in building materials such as local rocks, stone, sand, gravel, cement, concrete, brick, tile, gypsum, and so forth. The gamma radiation originating from the walls, floors and ceilings, and Radon and Thoron and their progeny are the main origins of radiation hazards. Natural radionuclides in building materials may lead to both external hazards produced by their direct gamma radiation and also internal hazards from radon gas. Marble and granite are more costly than ceramic tiles. Ceramic tiles have been adopted worldwide due to being a beautiful building material with a low price. They are composed of a compound of earth minerals that have been finished into a shape and ignited at high temperature. The body of ceramic tiles is a blend of various raw materials including clays, quartz materials, and feldspar, and due to the presence of zircon in the finish, ceramic tiles can produce natural radioactivity that is considerably greater than the common building materials.[31,32]

The cleanability of materials can be considered as a main consideration in flooring materials. Cleaning costs are expected to describe nearly one-third of the managing expenses of buildings. The cleanability of plastic materials is enhanced by various coatings or processes. The transparent layer or erosion layer of low-plasticized polyvinyl chloride (PVC) on vinyl flooring materials enhances the cleanability of spots of dirt. Acrylics or styrene acrylic floor polymers, usually referred to as PVC floor covers, promote maintenance. These polymers possess excellent water resistance and security to block marking, scuffing and dust accumulation. Cleanability of PVC flooring material can be enhanced by the inclusion of a thin coating of polyurethane on the top of the PVC flooring material surface.[33,34] The principal kinds of polyvinyl floorings are vinyl tiles or sheets, and vinyl containing tiles that are classified as heterogeneous or homogenous vinyl floorings. In addition, polymers such as acrylic, epoxy, polyurethane, polyester or vinyl ester resins have been used to fabricate the most common floors. The static dissipative, impact-sound insulating, electrostatic conductive properties, and foam backed or even advanced slip retardant surfaces, are the main advantages of polymer flooring materials, the main types of polymer flooring materials are mentioned in Table 7.2.

Composites containing epoxy loaded with metallic materials such as iron, brass, and copper can be used as flooring materials for reducing the electric static charge generation. In addition, low thermal transmittance, wet floors, and recycling of materials are the primary concerns for developing new materials. Therefore, there is a growing need to overcome the risk of falling and to improve walking confidence inside buildings and other mentioned environments. Rubber carpet is suitable for accessible flooring materials expected to improve convenience, by adding a cushioning effect to the knees during walking or roaming. Accidents resulting from falling and slipping are linked to the materials present in footwear, floor, impurities, the environment, and the geometrical shape of the rubber carpet.[36,37]

Reclaimed rubber is utilized over fresh rubber in flooring due to the special features and its durability. Rubber floorings are usually applied in health clubs, townhalls, apartment hallways, institutions and schools, and working areas, as well as fire and police force headquarters. The excellent mechanism for walking on rubber mats laid over cement is owing to the higher effective delivery of forces from the foot to the elastomer, dispersing the forces into deformation energy within the materials.[38] Reclaimed rubber from automotive tyres is a suitable alternative for flooring materials, however, rubber tyres contain heavy metals, which makes the quality of rubber originating from waste tyres highly uncertain. In addition, rubber tyres contain various chemical constituents, which can contaminate water resources if rubber disposal is not carried out in an appropriate manner, initially burning of tyre was used for disposal of tyres, this procedure is not well suited to the current climate situation, therefore recycling of rubber tyres can fulfil current demands with inherent properties. For these reasons the study of recycled rubber as asphalt have been carried out, these studies show that using recycled shredded rubber in normal road asphalt can offer a higher frictional response, higher expected lifetime, greater durability, and lower maintenance costs compared to a normal road.[39]

Recycled rubber floorings can be made in the form of floor tiles with relatively high flexibility, long service life, impact absorption, and sound absorption properties. Recycled rubber is also added to virgin rubber mixtures, which leads to a reduction in the price of the rubber mixtures and the use of new virgin elastomers. If a rubber mixture could be produced that including greater than 50% scrap rubber and showed good physical properties and was inexpensive, rubber goods production could be improved and it would be an efficient application of waste rubber. However, compounding of waste rubber greater than 50% with virgin natural rubber has been observed to be quite challenging and the physical characteristics also decrease considerably.

Leather is a common natural polymer comprised of a three-dimensional cross linked structure of collagen fibres. When waste leather fibres are added to elastomers, the leather could work as a short fibre reinforcement for the matrix, which contributes essential fibrous characteristics from the leather, and the fibrous nature is preserved during processing. If the processing of elastomers is performed at comparatively high temperatures, it is difficult to control the fibrous nature of leather. In this case, leather waste is present in the form of particulates in elastomeric formulations. The leather also affects the vulcanization characteristics and vulcanized features of the rubber, this is due to many reactive functional groups in leather, which enhance the mechanical properties and vulcanization characteristics. The leather

Table 7.1 Polymer based flooring materials.[35]

Polymer	Additives	Flooring materials
Poly(vinyl chloride) as a binder	Plasticizers, stabilisers, fillers, pigments, impact modifiers	Vinyl flooring consisting of colour, flexibility, hardness and sheen
Polyolefin	Calcium carbonate-filled polyolefin plastomers	Polyolefin floor
Plastics	Various nanoparticles such as titanium dioxide (TiO_2) particles	Self cleaning plastic floor

particles are neutralized to remove its acidic nature and the enhancement of the vulcanization characteristics decrease the swelling of the vulcanized rubber in organic and aqueous media.[40]

When comparing the static and dynamic coefficients of friction, the static coefficient of friction is always greater than the dynamic coefficient of friction. It was reported that rubbers and plastics exhibit a continuous change in coefficient of friction with respect to velocity. This condition is serious once the coefficient of friction has drop down with respect to the velocity. Temperature and impurities can change the coefficient of friction, and flooring materials should be regularly cleaned to reduce the impurities.[41] Floor properties mentioned above, such as the surface coefficient of friction (COF), abrasiveness and softness are the main parameters to that must be considered for solving the problems of slip and fall accidents.[42] (Table 7.1)

7.3 Recycled Rubber as Flooring Materials

An active conversation on the opportunities of waste rubber recycling has been found as a preferred point through the decades. Although the issue of rubber scrap is still increasing, rubber recycling began at various points due to rubber technology. Several procedures have been created to make this material re-usable throughout the years. Cured rubber has been applied as a particulate in virgin rubber commodities, such as carpet or flooring materials. The procedure adopted for grinding the rubber and separation of impurities or pollutant has advanced as a result of the rising quality demands in terms of homogeneity and cleanliness of rubber powders and grinds. The subsequent action in the promoting of this material is a collapse of the vulcanization network, and transformation of this material into its primary unvulcanized state, this has not been possible so far. Several recycling methods have been explained, but the reported methods do not fulfil the quality required for recycled rubber. However, recycled rubber is extensively being utilized in rubber mixing creating high-quality tyre mixtures, the leading influence being the positive impact on the method performance of the material.

Table 7.2 Layer components of recycled and commercial underlays and their physical properties at a mixing temperature range of 20–30 °C.[110] Reproduced from ref. 110 with permission of Elsevier, Copyright 2011.

Sample	GTR grain size (mm)	Underlay components	Density (kg m^{-3})	Thickness (mm)	Porosity Ω
M1	<2.0	Ground tyre rubber (50%)/vermiculite (20%)/EPS (20%)	429.5	10.9	0.62
M2	<3.5	Ground tyre rubber (90%)	430.9	12.0	0.60
M3	0.5–1.0	Ground tyre rubber (85%)	570.0	10.1	0.36
M4	1.0–3.0	Ground tyre rubber (85%)	558.5	9.9	0.54
M5	1.0–3.0	Ground tyre rubber (93%)	544.7	9.8	0.59
M6	0.5–1.0	Ground tyre rubber (90%)	594.6	9.9	0.27
M7	<0.5	Ground tyre rubber (90%)	398.0	10.2	0.31
M8	<2.0	Ground tyre rubber (40%)/cement mortar (60%)	284.7	10.0	0.23
Impactodan 10	—	Reticulated polyethylene	21.8	10.0	—
Texsilen plus	—	Expanded closed-cell polyethylene	29.0	5.0	—
Trosilen AL-	—	Expanded polyethylene with aluminium film	48.3	3.0	—
Inasonic 120/20	—	Expanded polyurethane foam	152.6	20.0	—

The importance of the reuse of waste rubber in safeguarding the climate and saving energy is of considerable interest in the current technological competitive market. Rubbers are similar to thermosetting materials in their handling and crosslinked structure, and as a result, rubbers cannot be softened or reformed in shape by heating again. A sustainable process was revealed for reusing rubber scrap materials (*e.g.* waste tyres) to produce precious fuels in an oxidative decoupling process.[43] This particular method involves the breakdown of rubber through selective oxidation decoupling of C–C, C–S, and S–S bonds by using water as a solvent at, or near its supercritical condition.

Adkins developed a technique for preparing used tyres for the recuperation of oil, steel, vinyl chloride and carbon. The procedure involved compounding of the crushed car tire with a lot of isocyanides, polyurethane, latex and soybean oil, and the compound was then baked at 700 °F for 10 min to produce the above-mentioned items. The inclusion of soybean oil to the compound allows a more secure and more conventional process.[44]

Recovering of scrap rubber is, hence, a fascinating approach to deal with the rubber disposal issue. Recovering of scrap rubber items, such as utilized car tires and tubes, shock absorbers, common carrier lines and so forth, involves the conversion of a three dimensionally interlinked, insoluble and infusible solid thermoset polymer (vulcanized rubber) network to a two-dimensional, delicate, plastic, low modulus, processable and vulcanizable thermoplastic item, which can recreate a number of the properties of virgin rubber. Recycling and reuse of rubber from utilized and scrap rubber items can consequently give some valuable oil assets, and in addition provide a solution for the scrap/squandered rubber disposal issues.[45–47]

Recycling procedures for the scrap rubber can be comprehensively categorized into two types: physical recycling methods and chemical recycling methods. In a physical recycling method scrap/squandered rubber items are recycled with the assistance of external energy. Accordingly, in a physical recycling method, the three-dimensional system of the crosslinked rubber network is fragmented in the vicinity of various energy sources. Due to the breaking of the network structure the polymer rubber chain is changed into short molecular weight segments, so it can be effectively miscible with the virgin rubber as a result of mixing. Consequently, recycled rubber achieved by the physical recycling procedure can be utilized as a non-strengthening filler. Yet, in the procedure, a particular level of energy is utilized which is adequate to fragment just the crosslinked bonds, allowing a decent amount of recycled rubber will be received which will be plastic in character. The mechanical, thermo-mechanical, cryo-mechanical, microwave, and ultrasonic methods are the main physical methods used for reclaiming the waste rubber.[48–54]

Most of the reused rubber based companies utilize a chemical reclaiming agent for the production of recycled rubbers, organic disulfides or mercaptans agents are particularly employed with mechanical processing at an elevated temperature. In view of these chemicals, many procedures have been established. Aside from these, a couple of inorganic mixtures have been attempted to be used as recycling agents. Moreover, biological assault of natural rubber latex is remarkably simple. Researchers have attempted numerous possibilities to deal with elastomeric materials as a hotspot for microbial attack. Clearly, nature can deal with its own waste issues, yet when the man noticeably changed the natural rubber polymer into a specific material using sulfur and various different additives and ingredients, biological assault has become reduced.

A fascinating approach reported the use of a chemolithiotrope bacterium in aqueous suspension for assaulting the surface of powdered rubber for the devulcanization of waste rubber by keeping the broken scrap rubber in a bacterial suspension of chemolithotropic microorganisms and performing air transfer until the natural sulfur or sulfuric corrosive fumes withdrew. This is a fascinating procedure which produced recycled rubber and sulfur in a convincing way with the help of bacteria.[55,56]

There are various other methods for recycling of rubber, which will be discussed in the next section. School, college, university sports grounds and various institutions have progressively set a program of sporting and related activities in their outline requirements for students. Consecutive sports programs generally play a vital role in the quality, status, and identity of various schools and colleges. This attention on games means that institutions regularly host several regions, covering exercise and fitness centres, high-impact move and force development, game tracks and indoor game courts. The quality of a facility is broadly described by the open areas, tracks, courts, performance spaces and play surfaces, and the type of floor.

Most sports flooring materials are divided into various group such as hardwood, carpet, synthetic grass, and flexible flooring. Excellent grade rubber tiles contribute a long lasting surface. Introduction of ground flooring is necessary for fitness applications, particularly when considering the density. Health club equipment may appear reasonable stationary, but it may create damage and cut into a ground floor surface, particularly when equipment is relocated repeatedly for cleaning or for arranging in a different pattern. A recycled rubber based floor surface is an individual ground floor option that stand up quite satisfactory under these circumstances. Rubber is the leading flooring material amongst flexible floor substitute alternatives, which is further comprised of (polyvinyl chlorate) and polyurethane due to the flexibility, sturdiness, slip resistance, and moisture resistance of rubber. It is hard to beat the flexibility, sturdiness, slip resistance and moisture resistance of rubber, making it the principle flooring option for weight rooms, locker rooms, home, educational areas, and various gyms. Rubber can stand up to magnificent inflictions, it is uninfluenced by a dropping unfastened weight or the periodically bump of exercise movements and machines.[56–60] In terms of a ground sheet item and interlocking tiles, rubber gives the maximum design versatility. Polyurethane is commonly is poured onto rubber web as a substitute.

A rubber background offers appropriate pressure mitigation to lessen the prevalence of bone and knee accidents due to impact. It is hard to reduce the use of wood in basketball arenas, but due to the lower maintenance and higher safety that can be achieved vinyl combined with a recycled rubber in the background is used to provide the appearance of timber with flexibility. Rubber is a good shock absorber, and single damaged tiles can be replaced easily without removing the complete rubber floor. Recycled rubber maintains a high coefficient of friction when the floor is wet; it provides one of the genuine wet-floor surfacing choices.

Reclaimed rubber flooring can be used to address areas that need to be quieter by diminishing sound transmission. The application of recycled rubber decreases the environmental issues and hazards. The production of building conditions which give excellent interior air quality improves the health, performance and life-cycle of building occupants. Rubber surfacing presents an approximately unlimited variety of fascinating colours and floor designs.[61–69] Rubber ground flooring is available in diverse forms which includes rubber flooring rolls, rubber flooring tiles, and fashionable rubber mats. Rubber floor overlaying can save people from physical damage coming from the existing flooring. Timber, tile, and laminate styles of floors present a greater risk of injury. Overlaying is prepared to appear in the shape of heavy furniture, workout gadgets, or maybe for a regular circulation of heavy foot fall. By including rubber floors in a residence, a prevailing substitute floor may be required as a safeguard against chips and cracks to a floor that may be very costly to renew. In addition, rubber floors frequently appear with patterned surfaces, to offer a much more resilient and reliable flooring alternative. The probabilities of uncomfortable and embarrassing slip-and-fall injuries taking place are decreased whilst a rubber ground shield is in the area. Rubber surfaces additionally assist in the countering of moisture, even though a sign is frequently left out signifying the chance of accident, water can be dangerous and cause excessive damage. Firstly, surfaces such as timber and tiles are very unsafe when wet, and secondly, if sufficient water is left over a period of time the surface can become more slippery due to microbial by-products. Rubber floors avoid this as rubber has an inherent potential to withstand water. It prevents the accumulation of microbial products like mould, making the area more secure.[70]

As an added advantage, rubber basement floors can also assist to maintain the warmth. Basements and bunkers are usually cold places due to the atmosphere in their underground vicinity. The present cement or timber floors of a cellar do not shield or provide resistance to winter and cold temperatures, whereas rubber flooring offers a notable means of insulation and to keep a basement warm. Recently various types of flooring materials have been used such as recycled rubber based basement flooring, rubber based eco-friendly flooring, recycled rubber based garage covering floor, gym recycled rubber flooring, kennel recycled rubber flooring, recycled rubber based modular floor tiles, recycled rubber based play ground surfaces, recycled rubber based outdoor flooring and so forth.[71–74]

7.4 Recycling and Processing of Scrap Rubber

The terms used for the recycling of rubber are recycling, reclaiming, devulcanization, desulphorisation, depolymerisation, incineration, surface modification, and back to feedstock. Back to feedstock is applied to make a cured rubber into a low molecular weight chemical compound, surface modification is employed to produce powdery and viscous materials, and incineration is an energy based process, which is performed by the energy release from burned rubber, depolymerisation is used as a reverse polymerisation to obtain the monomers of the polymers, it is completed by the breaking of the C–C bonds. The sulphur content is removed from scrap rubber by desulphurisation. Devulcanization, is employed for breaking the cross linked sulphur bond in rubber. Reclaimed rubber can be described as "the outcome goods" on account of the treatment of vulcanized scrap rubber, tires, tubes and diverse waste rubber things by the use of heat, mechanical power and chemical reagents, wherein regeneration of a rubber mixture to its authentic plastic state is produced, as a consequence allowing the goods to be processed, compounded and vulcanized. Finally, recycling is defined as the utilization of scrap rubber in useful products. Rubber is harvested from natural or synthetic sources.

Rubber is constituted of natural or synthetic sources. Green rubber is collected from the milky white fluid called latex, determined in lots of plants;

synthetic rubbers are constituted of unsaturated hydrocarbons. Long before Columbus reached the Americas, the local South American people had been utilizing rubber to make a plenty of waterproof merchandise. The people of Spain attempted to reproduce this merchandise, such as shoes, coat, caps, but it wasn't until the 18th century that European experts and manufacturers started to apply rubber accurately for a business reason. The British scientist and chemical expert Charles Macintosh built a factory in Glasgow for the production of the water-resistant products and rainproof clothes in 1823. A primary step forward took place in the mid nineteenth century with the addition of the procedure of vulcanization. This procedure offered increased maturity, elasticity, and stability of rubber to the variation in temperature. It additionally provides a rubber barrier to gases and resistance to heat, electricity, chemical movement and abrasion. Vulcanised rubber also revealed the frictional properties preferred for pneumatic tyre utilities. The predominant applications of vulcanised rubber are for car tyres and conveyor belts, impact absorbers and anti-vibration mountings, pipes and hoses. It also serves a few specific purposes which include electricity transmission belting, water lubricated bearings, pipes for managing of abrasive sludges and so forth.

Natural rubber is harvested from rubber yielding vegetation, extensively from the tree *Hevea brasiliensis* which emanates from the south of the United States. These days, the greater percentage of all the natural rubber is produced from these trees in the rubber farms of Indonesia, Malay Peninsula and Sri Lanka, The general term for this kind of rubber is para rubber. There are several artificial rubbers. Those are made in a comparable manner to plastics, through a chemical procedure referred to as polymerisation. Those synthetic rubbers are Buna rubbers, neoprene, and butyl rubber. Artificial rubbers have generally been produced with precise properties for particular uses. The artificial rubbers usually adopted for tyre manufacture are styrene-butadiene rubber and butadiene rubber (both individuals of the buna own family). Butyl rubber, due to the fact it is gasoline-impermeable, is typically employed for internal tubes. The present practice of rubber manufacture includes a complicated series of methods inclusive of mastication, mixing, shaping, moulding and vulcanisation. Various ingredients are incorporated throughout the rubber compounding technique to give favoured traits to the finished product.

Fillers help to reinforce or strengthen rubber. Carbon black is an anti-abrasive and is typically used in tyre manufacturing. Pigments encompass zinc oxide, lithopone, and some natural dyes. Softeners are essential as the mixture is simply too stiff for correct incorporation of the numerous additives. The moulding of the compound is performed once the preferred mixing (compounding) has been completed and vulcanisation is often executed on the moulded product.

Recycling of rubber is a difficult process. However, there are various purposes to reclaimed or recycled rubber, such as the fact that recycled rubber can be half the price of that of natural or artificial rubber. Additionally, recycled rubber has a few features that are more favourable than the ones of virgin rubber, meaning rubber from reclaim requires much less energy for the general manufacturing method compared to virgin rubber. Reclaiming the rubber is a nice approach to reduce the unwanted rubber stock that is regularly difficult. There is an immense ability for recycling and reclamation of rubber in developing countries.[75–78]

7.4.1 Mechanical Reclaiming Process

In the mechanical reclaiming method, scrap rubber is fed into an open-roll mixing mill and milling is performed at excessive temperatures, as shown in Figure 7.1. In this procedure, a drastic molecular weight decline occurs due to mechanical shearing at excessive temperatures. It is a physical method for the reclaiming of vulcanized rubber and refining of the reclaimed rubber. The operation may be facilitated with the aid of mixing an appropriate quantity of vulcanized rubber through supplemental heating. Blends of fresh rubber with reclaim rubber showed that an increase in the reclaim rubber content improved the cure rate but the scorch time, curing time and reversion resistance decreased. As the percentage of reclaim rubber within the blends increases the modulus, abrasion loss, compression set and hardness boom at the same time as the tensile energy, elongation at damage, tear strength, resilience and flex resistance decrease. The result shows that a change in the percentage of reclaim rubber changes the crosslink density.[79,80]

7.4.2 Thermo-mechanical Reclaiming Process

This process entails the thermo-mechanical deterioration of the rubber vulcanizate network. The vulcanizate is swollen in an appropriate solvent and then brought to a mill to produce a granular powder with a few μm diameter. This powdered rubber is vulcanized with curing constituent components. The products received therefore exhibit dependent properties compared to the ones of the main vulcanizates.[81]

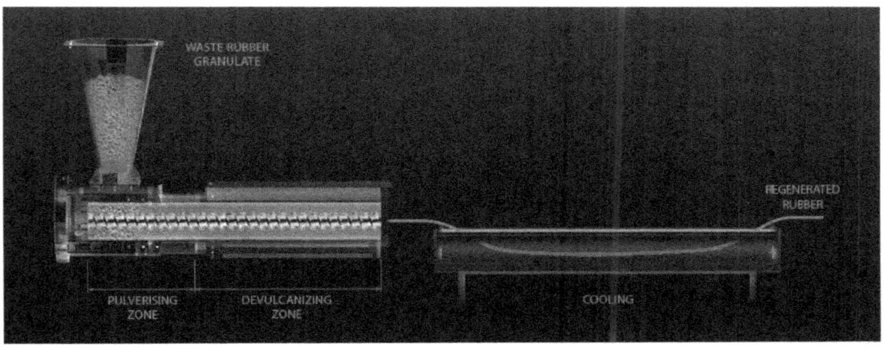

Figure 7.1 Mechanical reclaiming processes through twin screw extruder.

7.4.3 Cryomechanical Reclaiming Process

This reclaiming technique entails placing small portions of vulcanized rubber into liquid nitrogen which are placed into a ball mill to produce the fine powder. The particle dimension of the powder is altered through the immersion time in the liquid nitrogen and monitoring the grinding chamber of the mill.[82]

7.4.4 Wet or Solution Grinding

In this method, the particle size of rubber is reduced through grinding of scrap rubber in a liquid medium. The method proceeds by putting rough ground rubber crumbs into a liquid medium, typically water, and grinding between two grinding wheels, and the particle size depends on the grinding time.[83]

7.4.5 Microwave Method

In the microwave technique, carbon–carbon bonds are cleaved by using microwave energy at a specified frequency and energy. This method has several advantages such as being economical and ecological without depolymerization, and is an effective method.[84]

7.4.6 Ultrasonic Method

Ultrasonic waves have been applied for the devulcanization of crosslinked rubber. In this method solid rubber scrap, including tyres, is immersed in a liquid and then turned into a place with a supply of ultrasonic power whereby the majority of the rubber is correctly disintegrated and dissolved into the liquid. In this method, the ultrasonic irradiation is within the range of about 20 kHz and at an intensity of higher than 100 W.[85–87]

7.4.7 Chemical Reclaiming Processes

An incredible range of the reclaim rubber producers use a chemical reclaiming agent for the production of reclaim rubbers. The disulfides or mercaptans are the common organic reagent, which are particularly used throughout the mechanical process at elevated temperature. Diphenyl disulfide, dibenzyl disulfide, diamyl disulfide, bis (alkoxy aryl) disulfides, butyl mercaptan and thiophenols, xylene thiols and other mercaptans, phenol sulphides and disulphide reclaiming agents have been used for this process.[88–91]

There are some other methods, which can be used for reclaiming of waste rubber such as reclaiming by inorganic compounds,[92,93] reclaiming by chemical degradation,[94] pyrolysis of waste rubber,[95,96] the de-link process,[97] and reclaiming processes using renewable resource materials.[98] However, the aging features of the vulcanizates comprising reclaim rubber should be given specific consideration as reclaimed rubber itself is a tainted material.

Reclaim rubber obtained by the use of renewable resource materials was examined for its aging performance. A renewable resource material reclaims rubber–fresh rubber compound exhibited enhanced aging features which have been measured after various times aging at fixed temperature in an air aging oven.[99–101] Polymers being used in various application through macro, micro and nanotechnology, and the recycling of any polymer are useful ways to save the money, energy, and reduce the problem arising from waste.[102–105]

7.5 High Performance Flooring Applications of Recycled Rubber

In recent years, a lower percentage of waste tyres are intended for landfill. The recycled tyre rubber is being turned into new tyres, oil, used in civil engineering application and goods, agricultural, sports surfaces and flooring. The advantages of the application of rubber modified flooring are being utilized, and the inclusion of tyres into floors has been increased, rubberised asphalt is shown in Figure 7.2.[106]

Recently, research has provided the simplest approach to the solution of the difficulties facing highway authorities and planners. Tyres can be used in remarkable applications, such as for artificial rock, rubber dams, erosion control, parks, and highway boundaries. Scrap rubber may be included

Figure 7.2 Scheme of rubberised asphalt production through the "wet process".[106] Reproduced from ref. 106 with permission from Elsevier, Copyright 2013.

into rubber films and moulded commodities which include floor mats, car guards, and carpet underlays, crumb rubber can also be blended with plastics to offer ground mats, adhesives, rubber playground surfaces, roads, special athletic surfaces, and refuse cans.

Noise is a problem for lots domestic regions and influences every day visitors. As the source of noises are rarely ever visible and hard to quantify, noise abatement has received a lower precedence than specific commercial or climate issues. The symbol applied to estimate sound ranges is a logarithmic ratio of acoustic pressure in decibels (dB). Reused rubber tyre goods serving in the role of a sound damping component in wall sound barriers is one of the profitable rubber waste material applications. Concrete block walls are widely used in developing country, adding partially shredded rubber pieces during building construction will improve the noise shielding capabilities of the wall.[107] The order of the equivalent number of vehicles and the comparable traffic level of noise, throughout two periods of time have been reviewed, after installation of the highway and at a 8 month period after construction, as shown in Figure 7.3.[108]

Reusing waste tyre rubber by means of blending it into concrete materials has become a fascinated approach to prevent accumulation of waste tyres. In this view, the consequence of the volume constitution of waste rubber, as well as treatment techniques, indicated an influence on the performances of concrete. Many experts have mentioned that the mechanical strength of waste rubber concrete has dropped, while durability is enhanced with an increase of the waste rubber amount. A 20% replacement of fine aggregate and a 5% replacement of the overall mix presented specific properties for possible applications. Similarly, the consequences indicated that the modifiers had an excellent effect on the mechanical and durability of scrap

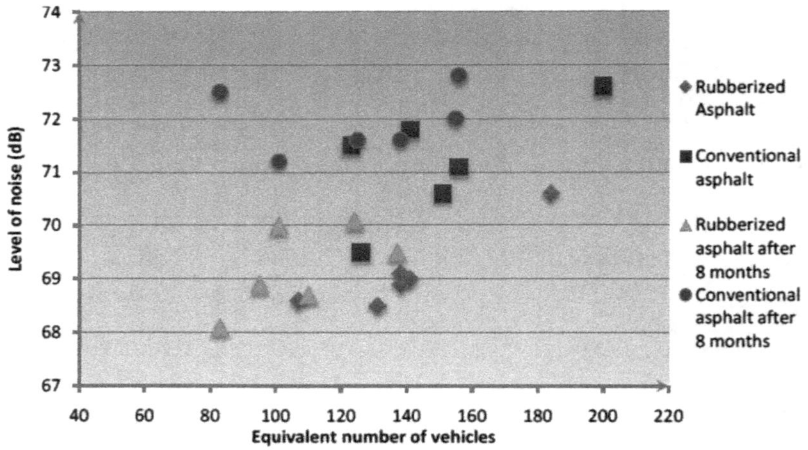

Figure 7.3 Combination of level of noise and the equivalent number of vehicles.[108] Reproduced from ref. 108 under the terms of the CC BY 4.0 licence, https://creativecommons.org/licenses/by/4.0/.

rubber concrete. This averted the drawback of waste rubber concrete having a lower strength and offers an advantage for the construction of modified scrap rubber concrete.[109] The objective of the various studies was to experimentally examine the acoustical characteristics of modern underlays produced from granulated rubber from automotive tyres. In one of the studies, researchers compared the properties of recycled underlay samples with the physical properties of recycled and commercial underlays, the formulation is shown in Table 7.3. The resonance frequencies, and consequently the

Table 7.3 Dynamic stiffness, impact sound improvement index, layer components and thickness of recycled and existing commercial underlays.[110–114] Reproduced from ref. 110 with permission of Elsevier, Copyright 2011.[a]

Sample	Dynamic stiffness σ′ ($MN\ m^{-3}$)	Impact sound improvement index ΔL_w(dB)	Underlay components	Thickness (mm)
M1	48.5	23	GTR (50%)/vermiculite (20%)/EPS (20%)	10.9
M2	14.3	31	GTR (90%)	12.0
M3	29.9	26	GTR (85%)	10.1
M4	33.4	25	GTR (85%)	9.9
M5	48.5	23	GTR (93%)	9.8
M6	31.3	25	GTR (90%)	9.9
M7	16.0	30	GTR (90%)	10.2
M8	49.2	22	GTR (40%)/Cement mortar (60%)	10.0
Texsilen plus 5 (*TEXSA*)	47.4	23	Expanded closed-cell polyethylene	5.0
Trosilen AL-3	124.4	16	Expanded polyethylene with aluminium film	3.0
Inasonic 120/20 (*INASEL*)	24.3	27	Expanded polyurethane foam	20.0
ImpactoDAN (*DANOSA*)	31.3	25	Reticulated polyethylene	10.0
ConforDAN (*DANOSA*)	100.0	18	Reticulated closed-cell polyethylene	3.0
ChovAIMPACT plus (*ChovA*)	24.0	24	Polyethylene closed-cell	9.0
ChovAIMPACT AR 10 (*ChovA*)	7.9	20	Not reticulated closed-cell polyethylene	10.0
ChovAIMPACT AR 5 (*ChovA*)	32.0	20	Not reticulated closed-cell polyethylene	5.0
Panel PST (*ISOVER*)	17.0	22	Rockwool	22.0
Ultracoustic floor TP (*Knauf*)	27.0	—	Mineral wool agglomerated with resin	20.0
Ultracoustic floor TPT 01 (*Knauf*)	10.0	—	Mineral wool	20.0

[a]GTR, ground tyre rubber.

dynamic stiffness, are related between the commercial underlays tested in the laboratory for this study and the new resilient layers, as mentioned in Table 7.3.[110–114]

7.6 Advantages and Disadvantages of Rubber Flooring

7.6.1 Advantages

The main advantages of rubber flooring compared to other conventional flooring materials are durability, low maintenance, water resistance, fire and burn resistance, sound properties, colour choices, textures, resilience, anti-static, anti-fatigue, environmentally friendliness, non-slip surface and so forth.[115]

7.6.1.1 Durability

The most highlighted benefit of rubber flooring, durability, is due to the fact that it is strong, tough, and resilient against a variety of conditions. The reused rubber floor coverings are swiftly achieving popularity in the role of a floor overlay in a number of interior as well as external locations. This is because of its durability, strength, and ease of maintenance found in hard tile materials, with a soft, yielding feel that is comfortable to stand on, and safe to fall on. Due to its extreme durability and resilience, rubbery material is difficult to damage. Depending on the type of tile used, and the environment it is installed in, a properly cared for rubber floor should be able to last twenty years, if not longer.

7.6.1.2 Water Resistance

Rubber floor coverings are nonporous and as a result, there is no need to bother about problems from plain liquid spills. This indicates that it is a preferred floor for wet conditions, for instance washrooms, kitchen areas, and also laundry rooms. Nevertheless, in the event that a rubber floor is placed at grade or below grade environment, a vapour shield may need to be taken into account to avoid moisture from seeping up from underneath. A number of rubber floors tiles are additionally available, and which are designed to deal with moist conditions.

7.6.1.3 Fire and Burn Resistance

This particular type of flooring material is retardant to burns from cigarette butts as well as other compact heat resources, however, a hot enough flame will probably burn it. Additionally, it is non-poisonous, which enables it to not discharge dangerous smoke into the air in the event of the fire. If adhesives were employed in the course of installation this may lead to generation of toxic chemicals.

7.6.1.4 Sound/Acoustic Properties

According to the thickness of the floor tiles, rubber is able to work as an effective noise barrier between floor surfaces in a building. The genuine flexibility may also bring down noise from walking as well as rolling carts by approximately 18 decibels. The noise absorbency of rubber floors helps to prevent the problem of noise. Rubber absorbs sounds instead of reflecting them. In extremely populated locations, wherein people are resting or working throughout the day, minimizing the sounds of children at play is essential. The sound of noisy children playing cannot be entirely annihilated, however it is dropped considerably.

7.6.1.5 Colour Choices

There are lots of colour tones as well as designs in the marketplace for selecting rubber floor tiles. These may have a polished surface or various other styles. Marbling can be copied as well as random speck designs on the rubber floors, which are generally found in natural stone materials, for example, neon tones are used to produce loud as well as brash flooring patterns.

7.6.1.6 Textures

If rubber floors are being formulated they can be placed in a mould in order to provide a number of textures which include dimples, rivets, studs, and waffles. This can be great for improving the traction of the floor, making it more slip resistant in wet areas, especially in commercial and high traffic situations.

7.6.1.7 Low Maintenance

Maintenance of rubber flooring is very easy compared to other conventional flooring materials. Polishing the rubber flooring surface with a water soluble wax emulsion will also make it more resistant to damage and discoloration. It requires no maintenance, and it naturally resists stains, water, dirt and mildew. The process of maintaining it over time is simply a matter of keeping the surface clean of dirt and debris by sweeping, vacuuming, or wet mopping with mild soap and water.

7.6.1.8 Softness

Despite the fact that it has commercial grade durability, rubber flooring is actually quite soft to the touch. This is can be important for helping to relieve stress fatigue associated with standing or walking for long periods of time, making this a popular flooring choice in retail locations. The soft yielding nature of rubber also makes it an appropriate floor for playrooms, bedrooms, and other living areas. This effect can be increased by purchasing tiles with fabric, cork, or foam-rubber backing.

7.6.2 Disadvantages of Rubber Tile Floorings

There are several disadvantages of rubber tile floorings,[116] these are detailed below.

7.6.2.1 Expensive

The most important problem with rubber flooring is the fact that it may be very costly as per requirements. Manufacturers may be able to find low end materials for as little as 2.00 USD per square foot, but this will limit the thickness, quality, and style choices. More expensive options can range as high as 12–15 USD per square foot, making this material comparable to stone or woodblock flooring.

7.6.2.2 Slippage

Smooth, untextured rubber tiles can become rather slippery when water is present. This effect is even worse if the material has been polished.

7.6.2.3 Above Grade

Despite the fact that rubber is resistant to water damage it is actually prone to seepage problems, such as loosening of adhesive, and also curling of floor tiles. A water vapour barrier can be employed at particular grade applications. The particularly manufactured waterproof rubber flooring materials must be used for below grade installations.

7.6.2.4 Staining

While rubber flooring is resistant to most staining agents, there are a few products that can discolour its surface. Detergent and other abrasive cleaning liquids can be especially dangerous to rubber floors. Consumers should avoid detergents and harsh cleaning agents when caring for a rubber floor as they can discolour the surface of the material. Grease can also have a drastic negative effect if it is not wiped up immediately.

7.7 Conclusions

The primary features of rubber flooring are its durability, genuine resiliency as well as minimal maintenance cost. Rubber floors are also very reliable. All of these features have allowed rubber floors to become a suitable alternative for all kinds of flooring applications. Healthcare, institution, workplaces and store installations are the major place for rubber flooring applications. The permanence and long life-cycle of rubber carpets are attractive to those people who look for environmentally friendly building flooring, minimizing the requirements for consistent elimination, as well as disposal of rubber. Recycled tyre flooring is usually made of rubber floor tiles, or sometimes into large rolls of flooring. When rubber flooring is installed in a space it can be a beautiful design element, available in virtually any colour or style. There are speckled rubber floor tiles manufactured by mixing of chips of various types of rubber together, in addition to these there are also certain appealing and convincing faux wood, granite, as well as marble, which can indeed contribute character to a room. Distinct textures can be imprinted into the rubber for a more contemporary appearance. Many of these possibilities are needed to create fascinating recycled tyre flooring, which could be one of the most adaptable materials in the marketplace in the near future.

References

1. I. M. Rushforth, K. V. Horoshenkov, M. Miraftab and M. J. Swift, *Appl. Acoust.*, 2005, **66**, 731.
2. R. Maderuelo-Sanz, A. V. Nadal-Gisbert, J. E. Crespo-Amorós and F. Parres-García, *Appl. Acoust.*, 2012, **73**, 402.
3. M. Miraftab, I. Rushforth and K. Horoshenkov, *Autex Res. J.*, 2005, **5**, 96.
4. M. Miraftab, I. Rushforth, K. Horoshenkov and M. Swift, *Institutional Repository*, University of Bolton, 2004.
5. M. Myhre and D. A. MacKillop, *Rubber Chem. Technol.*, 2002, **75**, 429.
6. Y. Y. Su and R. M. Shemenski, *Int. J. Adhes. Adhes.*, 2001, **21**, 481.
7. M. Pittolo and R. P. Burford, *Rubber Chem. Technol.*, 1985, **58**, 97.
8. M. Tonigold, J. Hitzbleck, S. Bahnmuller, G. Langstein and D. Volkmer, *Dalton Trans.*, 2009, 1363.
9. K. V. Horoshenkov and M. J. Swift, *Appl. Acoust.*, 2001, **62**, 665.
10. H. Benkreira, A. Khan and K. V. Horoshenkov, *Chem. Eng. Sci.*, 2011, **66**, 4157.
11. M. Sobral, A. J. B. Samagaio, J. M. F. Ferreira and J. A. Labrincha, *J. Mater. Process. Technol.*, 2003, **142**, 427.
12. R. Maderuelo-Sanz, M. Martín-Castizo and R. Vílchez-Gómez, *Appl. Acoust.*, 2011, **72**, 823.
13. K. M. Raghvendra and L. Sravanthi, *Mod. Chem. Appl.*, 2017, **5**, 1.
14. N. F. Medina, D. F. Medina, F. Hernández-Olivares and M. A. Navacerrada, *Constr. Build. Mater.*, 2017, **144**, 563.
15. N. Flores Medina, D. Flores-Medina and F. Hernández-Olivares, *Constr. Build. Mater.*, 2016, **129**, 25.
16. X. Shu and B. Huang, *Constr. Build. Mater.*, 2014, **67**, 217.
17. J. Norambuena-Contreras, E. Silva-Robles, I. Gonzalez-Torre and Y. Saravia-Montero, *J. Cleaner Prod.*, 2017, **145**, 85.

18. https://theconstructor.org/building/types-of-flooring-materials-uses-building/16992 (last accessed Jul 2017).
19. I. J. Kim, H. Hsiao and P. Simeonov, *Appl. Ergon.*, 2013, **44**, 58.
20. W. R. Chang, I. J. Kim, D. P. Manning and Y. Bunterngchit, *Ergonomics*, 2001, **44**, 1200.
21. K. W. Li, R. Yu and X. L. Han, *Appl. Ergon.*, 2007, **38**, 259.
22. Y. M. El-Sherbiny, A. M. Samy and W. Y. Ali, *KGK Rubberpoint*, 2012, **65**(3), 37.
23. O. Yong-Sung, *For. Prod. J.*, 1999, **49**(11/12), 31.
24. S. Kim and H.-J. Kim, *Indoor Air*, 2005, **15**, 317.
25. R. Wiglusz, E. Sitko, G. Nikel, I. Jarnuszkiewicz and B. Igielska, *Build. Environ.*, 2002, **37**, 41.
26. M. Z. M. Salem, M. Böhm, J. Srba and J. Beránková, *Build. Environ.*, 2012, **49**, 86.
27. K. E. Bøe, I. L. Andersen, L. Buisson, E. Simensen and W. K. Jeksrud, *Appl. Anim. Behav. Sci.*, 2007, **108**, 45.
28. J. Jin, B. Chen and J. Wang, *Procedia Eng.*, 2016, **135**, 112.
29. P. M. Bluyssen, *Cogent Psychol.*, 2016, **3**, 1268774.
30. J. Ahmadi-Shokouh, S. Noghanian, E. Hossain, M. Ostadrahimi and J. Dietrich, in *GLOBECOM 2009-2009 IEEE Global Telecommunications Conference*, IEEE, 2009, p. 1.
31. G. Senthilkumar, Y. Raghu, S. Sivakumar, A. Chandrasekaran, D. Prem Anand and R. Ravisankar, *J. Radiat. Res. Appl. Sci.*, 2014, **7**, 116.
32. E. Pesonen-Leinonen, R. Kuisma, I. Redsven, A. M. Sjöberg and M. Hautala, *Tenside, Surfactants, Deterg.*, 2005, **42**, 148.
33. S. Turri, M. Scicchitano, G. Simeone and C. Tonelli, *Prog. Org. Coat.*, 1997, **32**, 205.
34. T. Temtchenko, S. Turri, S. Novelli and M. Delucchi, *Prog. Org. Coat.*, 2001, **43**, 75.
35. E. Pesonen-Leinonen, Dissertation, University of Helsinki, 2005.
36. O. Hernandez-Mendo, M. A. G. von Keyserlingk, D. M. Veira and D. M. Weary, *J. Dairy Sci.*, 2007, **90**, 1209.
37. J. Rushen, D. Haley and A. M. de Passillé, *J. Dairy Sci.*, 2007, **90**, 3647.
38. Y. M. El-Sherbiny, M. A. Ramadan and W. Y. Ali, *EGTRIB J.*, 2014, **11**(2), 39.
39. https://www.um.edu.mt/library/oar//handle/123456789/16830 (last accessed 2014).
40. K. Ravichandran and N. Natchimuthu, *Polímeros*, 2005, **15**(2), 102.
41. D. I. James, *Ergonomics*, 1983, **26**(1), 83.
42. H. von Wachenfelt, S. Pinzke and C. Nilsson, *Biosyst. Eng.*, 2009, **104**(4), 534.
43. S. Lee, F. O. Azzam and B. S. Kocher, US Pat. 5516952 A, 1996.
44. L. Adkins, US Pat. 5618852 A, 1997.
45. J. C. Elgin and E. F. Sverdrup, US Pat. 2653349 A, 1953.
46. B. Adhikari, D. De and S. Maiti, *Prog. Polym. Sci.*, 2000, **25**(7), 909.

47. W. C. Warner, *Rubber Chem. Technol.*, 1994, **67**(3), 559.
48. A. A. Phadke, A. K. Bhattacharya, S. K. Chakraborty and S. K. De, *Rubber Chem. Technol.*, 1983, **56**(4), 726.
49. A. I. Isayev, J. Chen and A. Tukachinsky, *Rubber Chem. Technol*, 1995, **68**(2), 267.
50. A. R. Tripathy, J. E. Morin, D. E. Williams, S. J. Eyles and R. J. Farris, *Macromolecules*, 2002, **35**(12), 4616.
51. A. Zanchet, L. N. Carli, M. Giovanela, J. S. Crespo, C. H. Scuracchio and R. C. R. Nunes, *J. Elastomers Plast.*, 2009, **41**(6), 497.
52. D. De, A. Das, D. De, B. Dey, S. C. Debnath and B. C. Roy, *Eur. Polym. J.*, 2006, **42**(4), 917.
53. D. A. Jones, T. P. Lelyveld, S. D. Mavrofidis, S. W. Kingman and N. J. Miles, *Resour. Conserv. Recycl.*, 2002, **34**(2), 75.
54. A. I. Isayev, S. P. Yushanov and J. Chen, *J. Appl. Polym. Sci.*, 1996, **59**(5), 803.
55. E. Abraham, B. Cherian and P. Elbi, 2. *Recent Advances in the Recycling of Rubber Waste*, 2011, vol. 661, pp. 47–100.
56. J. A. Beckman, G. Crane, E. L. Kay and J. R. Laman, *Rubber Chem. Technol.*, 1974, **47**(3), 597.
57. J. L. Zuckerman, C. R. Eckardt and R. V. Bertotti, US Pat. 3937861 A, 1976.
58. W. A. Jank, US Pat. 5749787 A, 1998.
59. R. R. Stowell, C. A. Gooch and S. Inglis, Performance of tunnel ventilation for freestall dairy facilities as compared to natural ventilation with supplemental cooling fans, in *Livestock Environment VI, Proceedings of the 6th International Symposium 2001*, American Society of Agricultural and Biological Engineers, 2001, p. 29.
60. C. Bosco, R. Saggini and A. Viru, *Ergonomics*, 1997, **40**(6), 670.
61. R. L. Propst, US Pat. 5333423 A, 1994.
62. R. Yokubison, T. D. Mohr, T. Haney and D. F. Smith, US Pat. 7900416 B1, 2011.
63. W. G. Corson, US Pat. 2512310 A, 1950.
64. A. Suzuki, R. C. Kerr and A. K. Streeton, US Pat. 6589631 B1, 2003.
65. R. M. Krijnen, E. M. de Boer, H. J. Adèr, D. S. Osinga and D. P. Bruynzeel, *J. Occup. Environ. Med.*, 1997, **39**(9), 889.
66. R. C. Roga, US Pat. 4822669 A, 1989.
67. N. A. Ahmad, M. M. Tap, A. Syahrom and J. M. Rohani, The Relationship Between Coefficient of Friction (COF) with Floor Slipperiness and Roughness, in *Quantitative and Qualitative Factors that Lead to Slip and Fall Incidents*, Springer, Singapore, 2017, pp. 21.
68. G. Maddaloni, N. Caterino and A. Occhiuzzi, *Struct. Control Health Monit.*, 2017, **24**(5), e1906.
69. D. R. Barlow, US Pat. 9631375 B1, 2017.
70. http://www.rubbercal.com/rubber-flooring/(last accessed May 2017).
71. M. S. Al-Homoud, *Build. Environ.*, 2005, **40**(3), 353.
72. A. Konya, *Design Primer for Hot Climates*, 1982, vol. 3.

73. A. Jones, *Soc. Sci. Med.*, 1998, **47**(6), 755.
74. R. W. Day, *Pract. Period. Struct. Des. Constr.*, 1996, **1**(4), 104.
75. A. R. Kemp and W. G. Straitiff, *Ind. Eng. Chem.*, 1944, **36**(8), 707.
76. R. B. Bennett and G. E. P. Smith, *Rubber Chem. Technol.*, 1955, **28**(1), 308.
77. V. V. Rajan, W. K. Dierkes, R. Joseph and J. W. M. Noordermeer, *Prog. Polym. Sci.*, 2006, **31**(9), 811.
78. F. D. B. de Sousa, in *Elastomers*, InTech, 2017, pp. 209–229.
79. G. L. Tao, Y. P. Xia, H. R. Gen, J. Yang and W. Wang, *Key Eng. Mater.*, 2017, **732**, 43.
80. B. Maxwell, US Pat. 4146508 A, 1979.
81. M. Myhre and D. A. MacKillop, *Rubber Chem. Technol.*, 2002, **75**(3), 429.
82. A. A. Phadke and S. K. De, *Conserv. Recycl.*, 1986, **9**(3), 271.
83. C. Tzoganakis, US Pat. 7189762 B2, 2007.
84. M. Sienkiewicz, H. Janik, K. Borzędowska-Labuda and J. Kucińska-Lipka, *J. Cleaner Prod.*, 2017, **147**, 560.
85. F. D. B. de Sousa, C. H. Scuracchio and C. De Engenharia, *Mater. Res.*, 2015, **18**(4), 791.
86. F. D. B. de Sousa, C. H. Scuracchio, G. H. Hu and S. Hoppe, *Polym. Degrad. Stab.*, 2017, **138**, 169.
87. J. Shi, H. Zou, L. Ding, X. Li, K. Jiang, T. Chen, X. Zhang, L. Zhang and D. Ren, *Polym. Degrad. Stab.*, 2014, **99**(1), 166.
88. J. C. Elgin and E. F. Sverdrup, US Pat. 2653348 A, 1953.
89. D. De, S. Maiti and B. Adhikari, *J. Appl. Polym. Sci.*, 1999, **73**(14), 2951.
90. E. F. Sverdrup, US Pat. 2494593 A, 1950.
91. F. J. Webb, W. S. Cook, H. E. Albert and G. E. P. Smith, *Ind. Eng. Chem.*, 1954, **46**(8), 1711.
92. R. D. Myers, P. Nicholson, J. B. Macleod, M. E. Moir, US Pat. 5602186 A, 1997.
93. S. Yamashita, N. Kawabata, S. Sagan and K. Hayashi, *J. Appl. Polym. Sci.*, 1977, **21**(8), 2201.
94. N. Kawabata, B. Okuyama and S. Yamashita, *J. Appl. Polym. Sci.*, 1981, **26**(4), 1417.
95. H. Huang, L. Tang and C. Z. Wu, *Environ. Sci. Technol.*, 2003, **37**(19), 4463.
96. A. Quek and R. Balasubramanian, *J. Anal. Appl. Pyrolysis*, 2013, **101**, 1.
97. U. Ishiaku, C. Chong and H. Ismail, *Polym. Test.*, 1999, **18**(8), 621.
98. A. A. Yehia, M. N. Ismail, Y. A. Hefny, E. M. Abdel-Bary and M. A. Mull, *J. Elastomers Plast.*, 2004, **36**(2), 109.
99. N. Rattanasom, A. Poonsuk and T. Makmoon, *Polym. Test.*, 2005, **24**(6), 728.
100. X. Guo, D. Xiang, G. Duan and P. Mou, *Waste Manage.*, 2010, **30**(1), 4.
101. T. D. Sreeja and S. K. N. Kutty, *Polym. Plast. Technol. Eng.*, 2000, **39**(3), 501.
102. R. K. Mishra, H. J. Maria, K. Joseph and S. Thomas, in *Micro and Nano Fibrillar Composites (MFCs and NFCs) from Polymer Blends*, Elsevier, 2017, p. 1.

103. V. K. Abitha, V. R. Remya, A. V. Rane, S. Jadhav and R. Kumar Mishra, *J. Mater. Sci. Eng. Adv. Technol.*, 2016, **13**(1), 13.
104. R. K. Mishra, J. Abraham, N. Kalarikkal, K. Jayanarayanan, K. Joseph and S. Thomas, in *Polyurethane Polymers*, Elsevier, 2017, p. 203.
105. R. K. Mishra, J. Abraham, K. Joseph, K. Jayanarayanan, N. Kalarikkal and S. Thomas, in *Polyurethane Polymers*, Elsevier, 2017, p. 365.
106. D. L. Presti, Recycled tyre rubber modified bitumens for road asphalt mixtures: a literature review, *Constr. Build. Mater.*, 2013, **49**, 863–881.
107. N. Mokhtar, *Int. J. ChemTech Res.*, 2016, **9**(12), 14.
108. F. Kehagia and S. Mavridou, *Open J. Civ. Eng.*, 2014, **4**(3), 198.
109. K. Gabryś, W. Sas, E. Soból and A. Głuchowski, *Appl. Sci.*, 2017, **7**(7), 741.
110. R. Maderuelo-Sanz, M. Martín-Castizo and R. Vílchez-Gómez, The performance of resilient layers made from recycled rubber fluff for impact noise reduction, *Appl. Acoust*, 2011, **72**(11), 823–828.
111. http://portal.danosa.com/danosa/ (last accessed May 2017).
112. http://www.knaufinsulation.es/files/ki_es/upload/documents/ (last accessed May 2017).
113. http://www.chova.com/esp/productos.asp (last accessed May 2017).
114. http://www.isover.net/productos/ing/ingles.htm (last accessed may 2017).
115. http://flooring-professionals.com/flooring-resources/benefits-of-rubber-flooring/ (last accessed May 2017).
116. https://www.thespruce.com/durability-of-rubber-flooring-tiles-1314691 (last accessed May 2017).

CHAPTER 8

Recycling of Individual Waste Rubbers

S. SAIWARI[a,b], W. K. DIERKES*[a] AND J. W. M. NOORDERMEER[a]

[a]Faculty of Engineering Technology, Department of Elastomer Technology and Engineering, University of Twente, 7500AE Enschede, The Netherlands; [b]Department of Rubber Technology and Polymer Science, Faculty of Science and Technology, Prince of Songkla University, Pattani campus, Pattani 94000, Thailand
*E-mail: W.K.Dierkes@utwente.nl

8.1 Introduction

Recycling of rubber worldwide is of growing importance due to increasing raw material costs, diminishing resources and growing awareness of environmental issues and sustainability. A variety of different recycling processes for rubber have been investigated and developed throughout the years. Several in-depth reviews discussing the state-of-the-art of rubber recycling were published in 1974, 2002 and 2012.[1–3] The main recycling methods hierarchically classified in the order of environmental and economic preference are shown in Figure 8.1.[4]

The lowest level is burning of vulcanized rubber for energy recovery, and on the second level resides back-to-feedstock methods, such as pyrolysis of used rubber to recover gas, oil and chemicals. The third method on the next

Green Chemistry Series No. 59
Rubber Recycling: Challenges and Developments
Edited by Jin Kuk Kim, Prosenjit Saha, Sabu Thomas, Józef T. Haponiuk and M. K. Aswathi

Published by the Royal Society of Chemistry, www.rsc.org

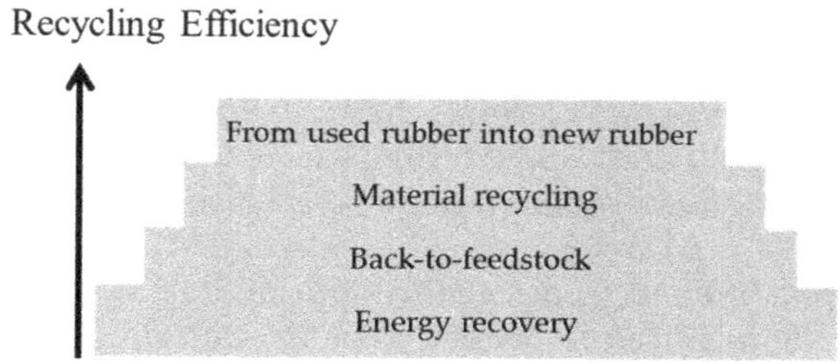

Figure 8.1 Rubber recycling ladder.

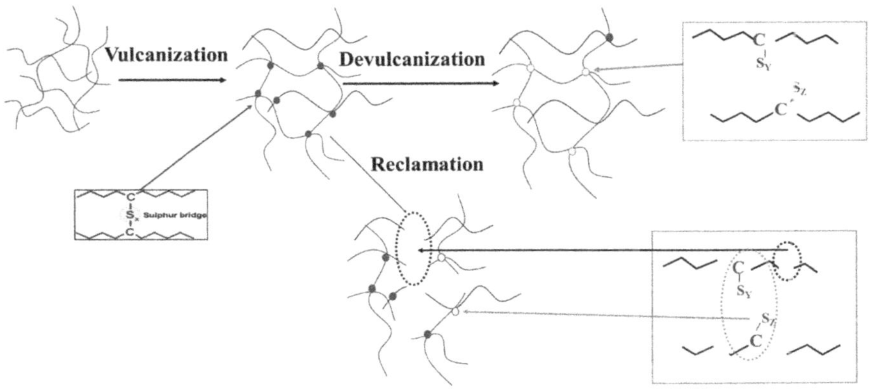

Figure 8.2 Simplified scheme of the 2 reactions occurring during rubber recycling processes: reclamation and de-vulcanization.

level is material recycling by transforming used rubber into products with inferior quality compared to the original material or using recovered rubber for the production of new rubber products. Finally, the most efficient rubber recycling route, or the highest level of the recycling ladder, is transforming used rubber into products of characteristics equal to that of the original materials, for example used tire rubber back into tires.

In the rubber recycling process, two types of rubber network breakdown simultaneously occur: reclamation and de-vulcanization, as shown in Figure 8.2.

Reclaiming and de-vulcanization are often referred to as similar processes. In spite of the fact that they are similar in the procedure, they are fundamentally different in the degree of rubber network breakdown and the molecular structure of the polymeric material. In other words, the difference between "reclaimed" and "de-vulcanized" rubber lies in different ratios of crosslink and polymer chain scission.

De-vulcanization, the most ideal way of rubber recycling, is the process that aims to reverse vulcanization as far as possible without damaging the polymer. In sulfur vulcanization, formation of a rubber network by both, carbon–sulfur bonds (C–S) and sulfur–sulfur bonds (S–S), takes place, therefore only these bonds should be broken during de-vulcanization. De-vulcanization is the process of cleaving the monosulfidic (C–S–C), disulfidic (C–S–S–C) and polysulfidic (C–S_x–C) crosslinks of vulcanized rubber. An efficient de-vulcanization is needed in order to achieve a high-quality recycled rubber.

Reclamation is different from de-vulcanization due to the scission of the carbon–carbon bonds of the polymeric chains. Reclaiming is a process in which vulcanized rubber is converted into a state that it can be mixed, processed, and vulcanized again by using conventional processes. Transforming the cured rubber into a re-processable material is done by breaking the links between and partly within the polymer chains. The general problem of the current reclaiming processes of rubber is the fact that in addition to the sulfur-crosslinks, the main polymer chains are also broken, and this influences the properties and reduces the quality of the recycled material. This technology is sometimes incorrectly referred to as de-vulcanization. For reclaimed rubber, relatively poor mechanical properties are frequently reported, originating from structural changes to the molecular structure of the polymer that occur during the reclaiming process. Extensive polymer scission and a partial re-combination result in highly branched chain segments that differ greatly from the virgin rubber.

In view of these arguments, the conversion of used rubber into a re-processable and re-usable form by the currently used physical/chemical processes should be called reclamation rather than de-vulcanization. The broken polymer chains generated in the reclaiming processes influence the properties and reduce the quality of the recycled material. If the rubber is de-vulcanized, ideally only sulfur crosslinks are broken while the polymer chains remain intact. Consequently, the de-vulcanizate resembles the original material in structure and quality. An improvement of the properties of recycled rubber by developing a more selective breakdown process or "*efficient de-vulcanization*" is an important issue for rubber recycling technology and a global challenge.

8.2 Theoretical Background

8.2.1 Agents for Selective Scission of Sulfur Crosslinks

Recycled rubbers mostly exhibit poor mechanical properties because of main chain scission of the polymer. Schnecko reviewed elastomer recycling and reported the development of chemical probes for de-vulcanization of crosslinked rubber.[5] These chemical probes selectively cleave carbon–sulfur and sulfur–sulfur bonds, but they do not cleave carbon–carbon bonds. The agents for selective scission of sulfur crosslinks are thiol-amine reagents, methyliodide, triphenylphosphine, lithium aluminium hydride, hydroxyl ions, Raney nickel and sodium dibutyl phosphate. Most agents for selective scission of sulfur crosslinks act through nucleophilic displacement reactions. Different

Table 8.1 Properties of different chemical probes.[6]

Name of the chemical probes	Mechanism
Triphenylphosphine	Polysulfides transformed into monosulfides and to a lesser extent to disulfidic links
Sodium di-*n*-butyl phosphite	Di- and polysulfide crosslinks are broken
Propane-thiol/piperidine	Polysulfidic links are broken
Hexane-1-thiol	Poly- and disulfidic links are broken
Dithiothreitol	Disulfide bonds split into two thiol groups
Lithium aluminium hydride	Disulfide and polysulfide bonds are broken
Phenyl lithium in benzene	Polysulfide and disulfide bonds are broken
Methyl iodide	Monosulfidic links are broken

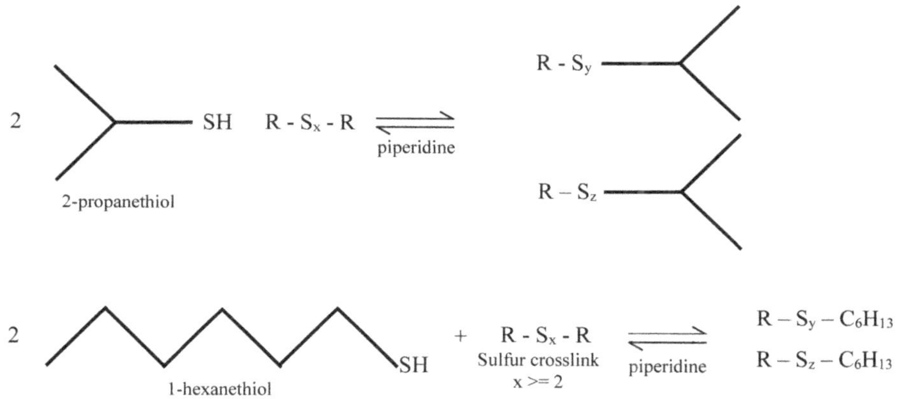

Figure 8.3 Schematic representation of the reaction of thiol-amine reagents with sulfur crosslinks.[7]

chemical probes are presented in Table 8.1. However, the major disadvantages of chemical probes are the toxicity of the additives and the reaction conditions that make them difficult to be applied on an industrial scale. The reaction of thiol-amine reagents with sulfur crosslinks is schematically depicted in Figure 8.3.

The thiol-amine combination forms a complex, possibly a piperidium propane-2-thiolate ion pair, in which the sulfur atom has an enhanced nucleophilic character and is capable of cleaving organic trisulfides and higher polysulfides. Disulfides react at a rate which is slower by a factor of one thousand. The polysulfide cleavage is faster due to p_π-d_π delocalization of the displaced σ-electron pair of RSS^- as outlined in Figure 8.4.

Other reaction mechanisms of chemical probes for sulfur crosslink scission, triphenylphosphine, lithium aluminum hydride, and methyliodide, are shown in eqn (8.1)–(8.4).

Reaction of triphenylphosphine with a polysulfidic crosslink:

$$RSS_xSR' + Ph_3P :\rightarrow xPh_3PS + RSSR' \quad (8.1)$$

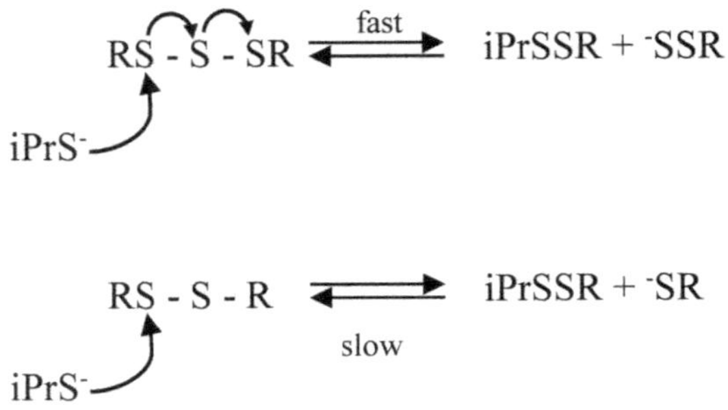

Figure 8.4 Cleavage of poly- and disulfidic crosslinks and p_π-d_π delocalization ($IPrS^-$ = nucleophilic thiol-amine associate).

Reaction of lithium aluminum hydride with polysulfides:

$$RSS_XSR' \xrightarrow{LAH, H^+} RSH + xH_2S + HSR' \tag{8.2}$$

Reaction of lithium aluminum hydride with disulfidic bonds:

$$RSSR' \xrightarrow{LAH, H^+} RSH + HSR' \tag{8.3}$$

Reaction of methyliodide with monosulfides:

$$R_2S + CH_3I \rightarrow R_2^+SCH_3I^- \tag{8.4}$$

8.2.2 Radical Scavengers

Various chemicals that were used in rubber recycling acted as radical stabilizing agents. Disulfides, thiols, phenols and phenolic compounds were added in order to scavenge radicals formed during the reclaiming process. A mechanism that is frequently proposed for the reaction of radical scavenger reclaiming agents with sulfur vulcanizates is the opening of crosslinks or the scission of polymer chains by heat and shearing forces, and the reaction of fragments with disulfide based radicals, which prevent recombination. In 1980, Okamoto used readily available chemicals, thiols and disulfides, on a two roll-mill to breakdown the vulcanized rubber network.[8] He suggested that when a vulcanized rubber compound is plasticized by using mechanical shear forces, for example milling, free radicals are produced that can cause the formation of main chain radicals. Chemicals such as thiols and disulfides react with these radicals, thus preventing their recombination. Main chain scission occurs and crosslinks are opened, leading to a reduction in viscosity.

Disulfides are well-known reclaiming agents, especially for natural rubber. As discussed in the rubber recycling review from 2002, De and co-workers prepared reclaimed rubber using a two roll mill and two different reclaiming agents.[2] One agent was a diallyl disulfide (DADS), and the other one was a renewable resource material (RRM), a vegetable product which is eco-friendly, and contained a disulfide as the major component. They suggested that these reclaiming agents form radicals with increasing temperature by the action of shear. These radicals then combine with the polymer main chain radicals preventing their recombination. Studies of mechanochemical recycling processes of natural rubber and scrap rubber from tires indicated that DADS played an important role in the processes.[9,10] As a result, the mechanical properties, for example, the tensile strength, modulus and tear strength, of revulcanized DADS-containing rubber were higher than those of revulcanized rubber which was obtained without DADS.

Recently, the efficiency of various disulfides as recycling agents for natural rubber (NR) and ethylene propylene rubber (EPDM) vulcanizates was reported.[11,12] While de-vulcanization was observed on sulfur-cured NR at 200 °C, a decrease in crosslink density of 90% was found when EPDM sulfur vulcanizates were heated to 275 °C with diphenyldisulfide (DPDS) in a closed mold for 2 hours.[13] At the same time, EPDM cured with peroxide showed a decrease in crosslink density of about 40% under the same conditions, showing that under these conditions a significant amount of carbon–carbon bonds are broken. The effect of DPDS content on a continuous de-vulcanization of EPDM using a co-rotating twin screw extruder were studied.[12] The re-vulcanized EPDM showed an increase of elongation at break values with an increase of DPDS concentration in the reclaiming process. Diphenyldisulfide was found to be an effective reclaiming agent for natural rubber by Rajan *et al.*[14] They observed that the reclaiming agent helped in preventing the broken rubber chains to recombine. The ratio of main-chain scission to crosslink scission of natural rubber depends on both the concentration of the reclaiming agent and the temperature of the process. The radical moieties of the chains can quickly recombine, unless a diphenyldisulfide molecule is present to act as a radical scavenger. In addition to this, the diffusion speed of the disulfide into the polymer matrix increases at high temperatures, enhancing the chance of combination with a rubber radical. A simplified reaction scheme proposed for the reclamation of natural rubber with diphenyldisulfide is shown in Figure 8.5. The radicals formed by scission of the disulfide are capable of hydrogen abstraction or addition to the double bonds in natural rubber.[15] Hydrogen abstraction is relatively easy because the allylic hydrogens are activated by the double bonds. The benzene-sulfide radical, therefore, abstracts the allylic hydrogen from the natural rubber vulcanizate to form benzenethiol and a natural rubber vulcanizate radical. The polymer radical can now undergo main-chain scission and/or crosslink scission.

However, one of the most important shortcomings of reclaiming with disulfide chemicals is the unpleasant odor during the reclaiming process and of the final reclaim. Tetramethyl thiuramdisulfide (TMTD) was introduced as an alternative by De *et al.*[16–18] TMTD is a multi-functional molecule and acts as a reclaiming agent during the reclaiming process and as a curing

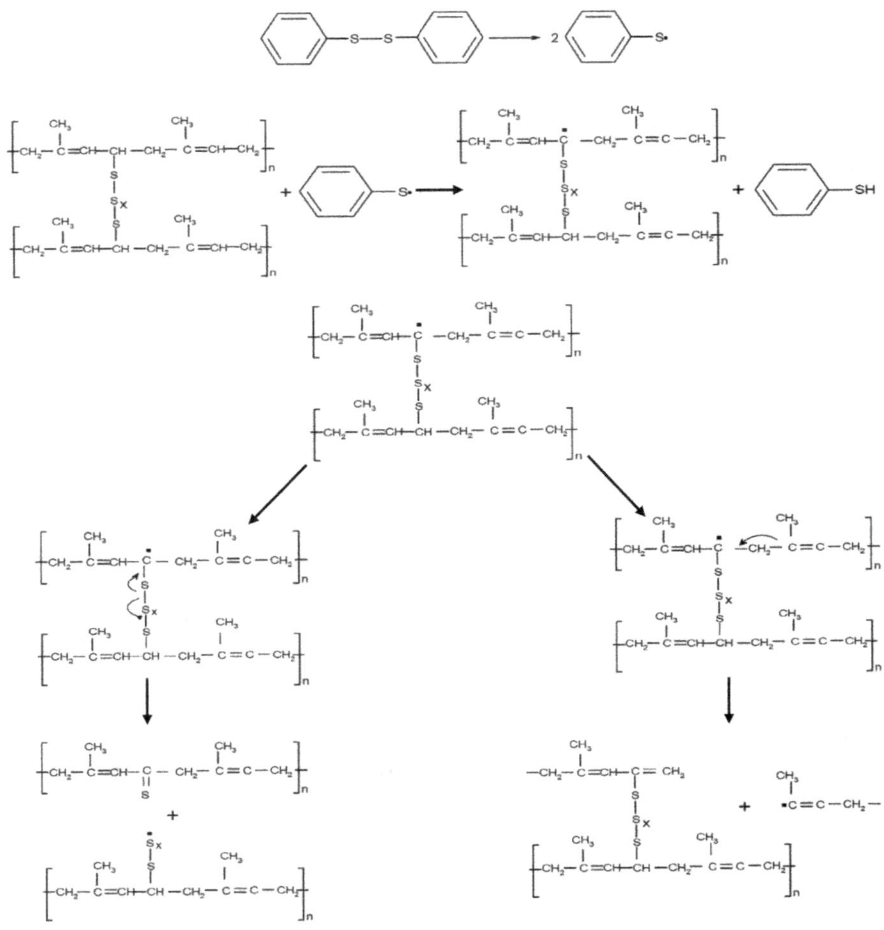

Figure 8.5 Simplified reaction scheme proposed for the reclamation of natural rubber by diphenyldisulfide.[14]

agent during revulcanization of the reclaimed rubber. Another advantage of this reclaiming agent is the reduced smell during the reclamation process and of the final reclaims. Utilization of a suitable amount of TMTD caused a decrease in the gel fraction and crosslink density. A reaction scheme proposed for the reclamation of natural rubber with TMTD is shown in Figure 8.6, mechanical shearing forces break both the polymer chains and crosslinks simultaneously with breaking of TMTD to form thiocarbamate radicals as shown in Figure 8.6, Scheme 1. Generally, the aliphatic disulfides have bond strengths of the order of 293 kJ mol^{-1} in the central S–S bond, which is too high to create radicals at moderate temperatures. In thiuram disulfides, potential resonance stabilization of the radicals should

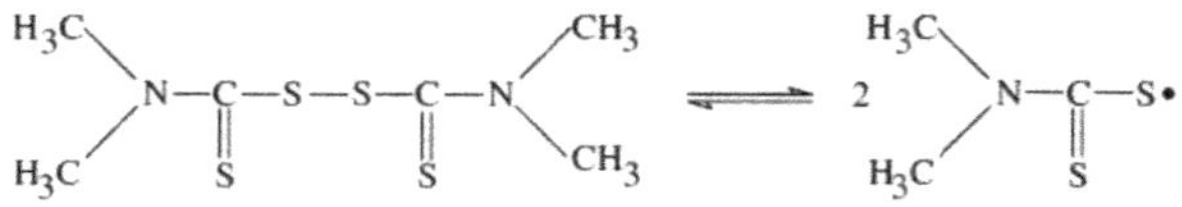

Scheme 1.

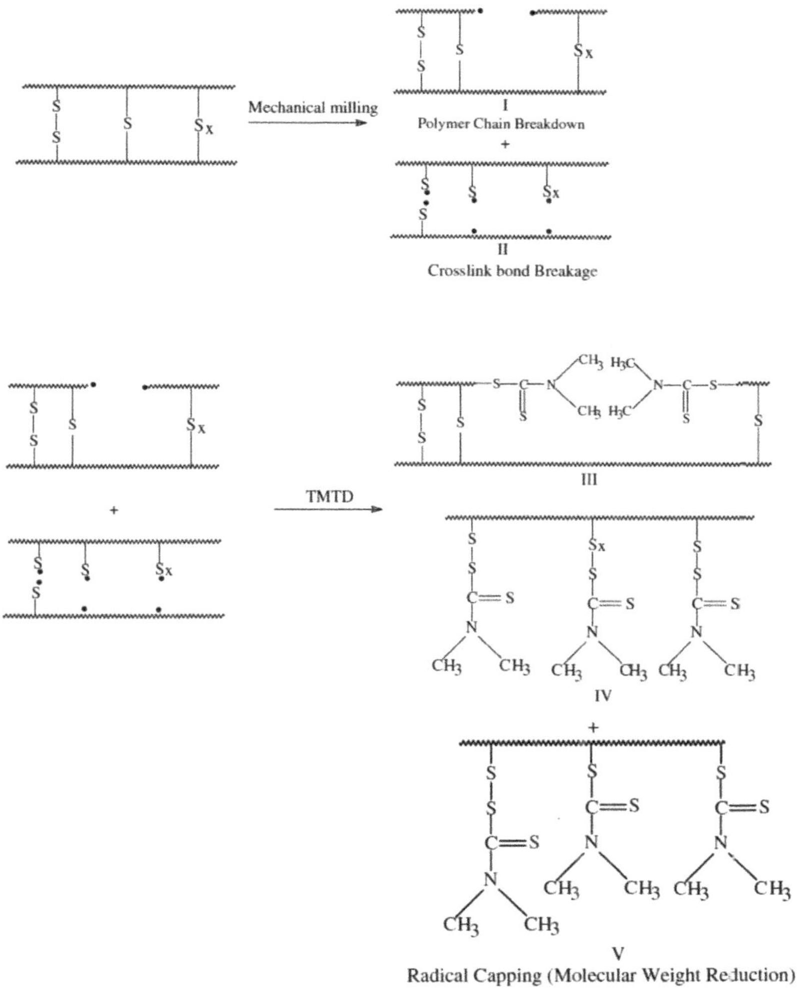

Scheme 2.

Figure 8.6 Mechanical shearing forces breaking TMTD to form thiocarbamate (Scheme 1) and reaction scheme proposed for the rubber reclamation by TMTD (Scheme 2).[16]

appreciably weaken the central bond and facilitate the formation of the thiocarbamate radical.[16] This radical may combine with a broken polymer radical as shown in Figure 8.6. When the reclaimed rubber, prepared using TMTD as the reclaiming agent is mixed with virgin SBR, the tensile strength increases by about 19% and 115% for 20% and 60% of reclaim respectively. Moreover, aging characteristics of the reclaimed rubber containing vulcanizates are superior compared to that of the control formulation, which does not contain any reclaim rubber. The thermal stability of the vulcanizate increases with increasing reclaimed rubber loading. However, due to high crosslink densities, vulcanizates containing rubber reclaim are vulnerable under mechanical stress.[17]

8.2.3 Model for Analysis of De-vulcanization Efficiency

De-vulcanization efficiency can be analyzed by characterization of the creation or destruction of polymer networks. In rubber recycling processes, the rubber network is broken causing main-chain scission and crosslink breakage. There are two methods to investigate whether main-chain scission or crosslink breakage is the dominant process during de-vulcanization: The first method is based on the amount of soluble polymeric material generated during de-vulcanization, so-called sol–gel analysis. The second method is related to stress relaxation of the rubber network, which is based on the idea that the stress at any state of degradation is proportional to the network chain density.

8.2.3.1 Sol-Gel Analysis

Sol-gel analysis, developed by Charlesby in his study on the relationship between crosslinking, main-chain scission and gel fraction in irradiated polymers, is based on the amount of soluble polymeric material.[19,20] Horikx assumed that these relations were also applicable to determine the sol fraction of a rubber sample undergoing network degradation by aging.[21] Gordon and Dobson developed another relation between the sol fraction and the average fraction of monomers that are part of a crosslink, using Good's theory of stochastic processes.[22–24]

8.2.3.1.1 Network Formation by Crosslinking[7]. According to Charlesby polymers can be classified into two groups, according to their response when exposed to ionizing radiation:[19,20]

1. Crosslinking of the polymer chains, leading first to an increase in molecular weight and, at sufficient doses, to the formation of an insoluble network;
2. Scission of the molecular chains, resulting in a decrease in average molecular weight.

Both processes occur simultaneously in many polymers, and the classification depends upon which of the two is dominant.

On the basis of a small probability, p, of breaking a bond, relative to the probability, q, of forming the bond, Charlesby derived a series of equations for the sol- and gel-fractions during crosslinking, which are generally applicable, irrespective of the molecular weight distribution of the original polymers. Most relevant in the present context are the following quantities and relations:

$$\gamma = \frac{q\overline{M}_n}{w} \tag{8.5}$$

In which γ is the crosslinking index, the average number of crosslinks per original polymer molecule, $\overline{M_n}$ is the number average molecular weight of the original polymer and w is the molecular weight of a monomer in the polymer chain, so that $\overline{M_n}/w$ is the number of monomers in a "number average" original polymer molecule, before crosslinking.

The crosslinking coefficient δ, equal to the average number of crosslinks per "weight average" polymer molecule, is related to $\overline{M_w}$ by:

$$\delta = \frac{q\overline{M}_w}{w} \tag{8.6}$$

In which $\overline{M_w}$ is the weight average molecular size of the original polymer.

The gel- and sol fractions during crosslinking are related to q:

$$g = 1 - s = 1 - \frac{\sum xN_x e^{(-qxg)}}{\sum xN_x} \tag{8.7}$$

in which g is the gel-fraction, s is the sol fraction, x is the number of monomers per primary polymer molecule, N_x is the number of polymer molecules with x monomer units and ΣxN_x is the total number of monomers included in all polymers within the sample.

Eqn (8.3) can be further detailed, if the molecular size (or weight) distribution for the primary polymer molecules is defined. For a mono-disperse distribution, all primary molecules are of the same size and the sol- and gel-fractions can be calculated as follows:

$$g = 1 - e^{(-qxg)} = 1 - e^{(-\gamma g)} \tag{8.8}$$

$$s = e^{(-\gamma g)} = e^{-\gamma(1-s)} \tag{8.9}$$

This is in accordance with the fact that gel formation occurs only if $\gamma > 1$.

For a Poisson distribution, Charlesby defines the molecular weight distribution as:

$$N_x = Np(1-p)^{(x-1)} \approx \frac{Npe^{-px}}{1-p} \approx Npe^{-px} \quad \text{for } p \ll 1 \tag{8.10}$$

In which N is the total number of polymer molecules in the distribution.

On the basis of this distribution the following relationship for the sol fraction s and gel-fraction g is derived:

$$s = 1 - g = \frac{1}{(1 + \gamma - \gamma s)^2} \tag{8.11}$$

An implicit equation for s reads:

$$s = \frac{(2+\gamma) - \sqrt{\gamma^2 + 4\gamma}}{2\gamma} \tag{8.12}$$

There are also other relationships derived for this Poisson distribution, which are needed later in this chapter. The average initial molecular weight $\overline{M}_{gel}$ of the molecules incorporated in the gel, is given by:

$$\overline{M}_{gel} = \overline{M}_n \frac{(1 + \gamma - \gamma s)}{\gamma} \tag{8.13}$$

The crosslinking index of the gelfraction, γ_{gel}, is equal to:

$$\gamma_{gel} = (1+s)(1+\gamma-\gamma s) \quad \text{for } \gamma > 0.5 \tag{8.14}$$

Figure 8.7 shows the sol fraction as a function of γ for the mono-disperse and Poisson distribution. It clearly shows that the sol fraction starts to decrease and gel fraction to increase at $\gamma > 1.0$ for the mono-disperse distribution, as originally derived by Flory.[25] However, for the Poisson distribution the sol fraction already starts to decrease at $\gamma > 0.5$. For this, distribution

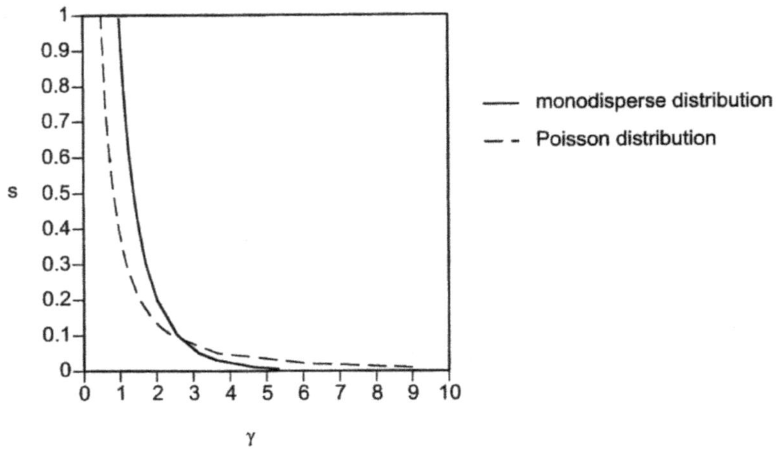

Figure 8.7 Sol fraction *versus* cross linking index according to Charlesby.[20]

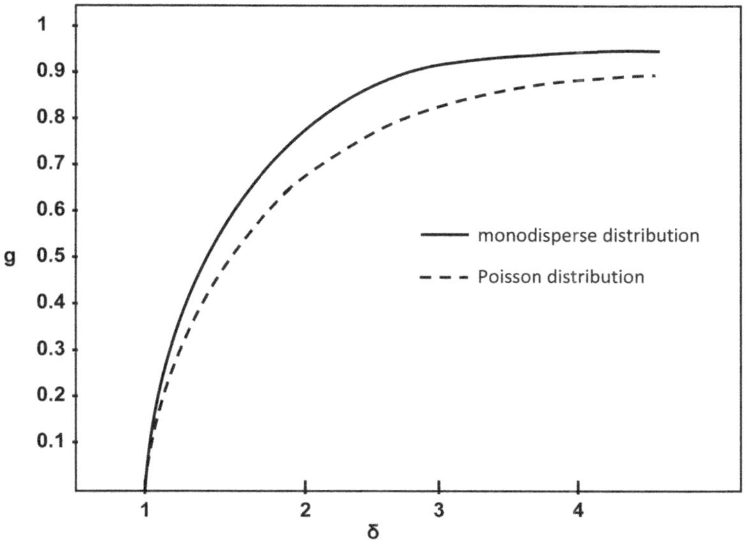

Figure 8.8 Gel fraction *versus* crosslinking coefficient according to Charlesby.[19]

generally holds that Mw:Mn = 2:1. Therefore, both distributions can be merged to start gel formation at a crosslinking coefficient = 1, as shown in Figure 8.8. It stresses the fact that the larger molecules in the distributions are the first to be caught in the crosslink network.

8.2.3.1.2 Network Breakdown (Horikx). Assuming that network breakdown leads to a Poisson distribution of molecular fragments, Horikx used Charlesby's theory to develop a method in which a vulcanized rubber network is broken down again *via* two routes: main-chain scission and selective crosslink breakage.[21]

As a measure of crosslink density, ν_0 is defined as the number of elastically effective polymer network chains per cm^3 including the sol fraction before the network breakdown. In this manuscript, it is referred to as the crosslink density. Inversely, ½ν_0 equals the number of crosslinks per cm^3 of a network for tetra-functional crosslinks. It can be deduced that:

$$\gamma = \frac{\nu_0}{N} \tag{8.15}$$

In which N is the number of primary molecules per cm^3 of polymer.

It follows from eqn (8.9), that the number of primary molecules per gram of network is given by:

$$N_{gel} = \frac{\gamma N}{(1+\gamma-\gamma s)} = \frac{\nu_0}{(1+\gamma-\gamma s)} \tag{8.16}$$

By combining eqn (8.14)–(8.16), it follows that the number of chains in the network, $\nu_{0,\,\mathrm{gel}}$, is given by:

$$\nu_{0,\mathrm{gel}} = \gamma_{\mathrm{gel}} N_{\mathrm{gel}} = \nu_0 (1+s) \tag{8.17}$$

Horikx assumed that the number of network chains, ν_0, is given by the chemically crosslinked chains only. Using eqn (8.11) and (8.17), the following relation is obtained:

$$N_{\mathrm{gel}} = \nu_0 \sqrt{s} \tag{8.18}$$

According to Flory, the number of elastically effective polymer chains in the network, ν_e, as determined by equilibrium swelling measurements, requires a correction for ν_0 to account for dangling chains ends:[25]

$$\nu_e = \nu_0 - 2N \tag{8.19}$$

For tetra-functional crosslinks, it is this ν_e, which is determined *via* swelling using the well-known Flory–Rehner equation:[26]

$$\nu_e = \frac{v_r + \chi v_r^2 + \ln(1 - v_r)}{V_s\left(0.5 v_r - v_r^{1/3}\right)} \tag{8.20}$$

$$v_r = \frac{m_r}{m_r + m_s \dfrac{\rho_r}{\rho_s}} \tag{8.21}$$

In which v_r is the polymer volume fraction of the swollen sample, V_s is the solvent molar volume, m_r is the mass of the rubber network, m_s is the mass of solvent in the sample at equilibrium swelling, ρ_r is the density of the rubber, ρ_s is the density of the solvent and χ is the Flory–Huggins polymer–solvent interaction parameter.

The number of elastically effective chains per unit volume of network is then given by combining eqn (8.16)–(8.18):

$$\nu_{\mathrm{gel}} = \nu_{0,\mathrm{gel}} - 2N_{\mathrm{gel}} = \nu_0 (1+s) - 2\nu_0 \sqrt{s} \tag{8.22}$$

$$\nu_{gel} = \nu_0 \left(1 - \sqrt{s}\right)^2 \tag{8.23}$$

During main-chain scission, the number of crosslinks, ν_0, remains constant and soluble matter is created by randomly detaching small molecular fragments from the network. By virtue of eqn (8.15), γ decreases as much as

N increases. If eqn (8.23) is formulated before (i) and after (f) the network scission and the ratio of both is taken as the relative decrease in the network chain density, it can be expressed as a function of the soluble fractions before and after scission of the sample:

$$1-\frac{\nu_{\mathrm{f}}}{\nu_{\mathrm{i}}}=1-\left[\frac{\left(1-s_{\mathrm{f}}^{1/2}\right)^{2}}{\left(1-s_{\mathrm{i}}^{1/2}\right)^{2}}\right] \tag{8.24}$$

In which s_i is the soluble fraction of the untreated vulcanizate, s_f is the soluble fraction of the fractured network, ν_i is the crosslink density of the untreated vulcanizate and ν_f is the crosslink density of the fractured network.

To determine the soluble fraction in the case of selective crosslink scission, N is taken constant and γ decreases. In this case, ν_0 remains in the equation. The relative decrease in crosslink density is now dependent on the crosslink index γ:

$$\nu_{\mathrm{e}}=\nu_{0,\mathrm{gel}}-2N_{\mathrm{gel}}=\nu_{0}\left(1-\sqrt{s}\right)^{2}=\gamma N\left(1-\sqrt{s}\right)^{2} \tag{8.25}$$

The relation between the effective number of network chains before and after the treatment is now given by:

$$1-\frac{\nu_{\mathrm{f}}}{\nu_{\mathrm{i}}}=1-\left[\frac{\gamma_{\mathrm{f}}\left(1-s_{\mathrm{f}}^{1/2}\right)^{2}}{\gamma_{\mathrm{i}}\left(1-s_{\mathrm{i}}^{1/2}\right)^{2}}\right] \tag{8.26}$$

In which the parameters γ_f and γ_i are the average number of crosslinks per chain in the remaining insoluble network after and before crosslink scission, respectively.

Horikx used the measurement of the network chain density in the gel-fraction by equilibrium swelling together with the value measured of $\overline{M_n}$ measured by GPC to calculate γ_i:

$$\gamma_{\mathrm{i}}=\frac{\nu_{\mathrm{i}}\overline{M}_{n}}{\rho} \tag{8.27}$$

Another way to determine γ_i is by measuring the residual sol content immediately after vulcanization and then making use of eqn (8.12). The experimental data are in better agreement with values calculated with eqn (8.26) if γ_i is calculated according to eqn (8.11), than according to eqn (8.27).[7]

Figure 8.9 gives a graphical representation of eqn (8.24) and (8.26). The curves in the figure correspond to the situation where only main chains are broken (solid curve) and where only crosslinks are broken (dashed curve).

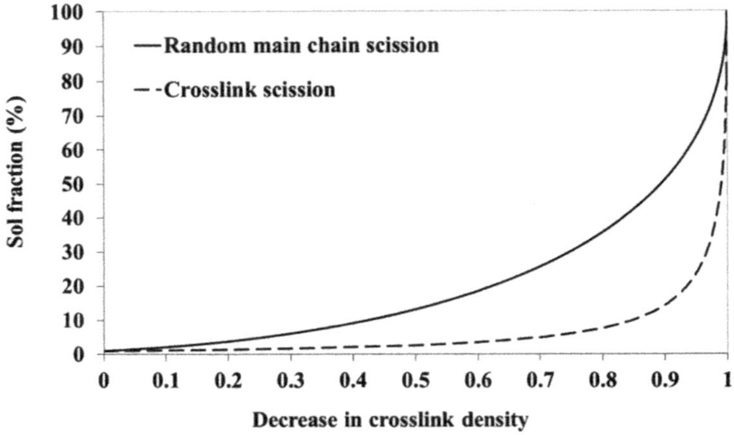

Figure 8.9 Random main-chain scission and crosslink scission curves in a Horikx plot.

In the case of crosslink scission only, almost no sol is generated until most of the crosslinks are broken; only then the long chains can be removed from the network. In the case of main-chain scission, sol is produced at a much earlier stage, because random scission of the polymers in the network results in small loose chains fragments, which can easily be removed.

8.2.3.1.3 Intermediate Stages Between Crosslink Breakage and Main-chain Scission (Yashin and Isayev). The disadvantage of the Horikx treatment is that it describes only the two extremes: pure crosslink breakage or pure main-chain scission. For intermediate cases, Yashin and Isayev used the model of Dobson and Gordon to describe the sol–gel formation of vulcanized rubbers.[27] On the basis of simulations of branched polymers, Dobson and Gordon developed a model for a rubber network.[22–24] This network is represented by a tree, starting from a primary chain or first generation, called the stem, to which chains of the second generation are attached, called the branches, and further branches of the third generation and so forth. Crosslinks connect pairs of chains of succeeding generations. Dobson and Gordon also developed a relation between the sol fraction and the average fraction of monomers forming crosslinks similar to that of Charlesby. For randomly crosslinked mono-disperse primary chains they arrived at exactly the same eqn (8.9). For polymer chains with a Flory or Poisson molecular weight distribution, they arrived at sol fractions as given in the following equation:

$$s = \frac{2p + (1+p)\left[q(1-p) - \left(q^2\{1-p\}^2 - 4q\{1-p\}^2 + 4q\{1-p\}\right)^{1/2}\right]}{2q(1-p)^2} \tag{8.28}$$

This equation deviates from the original equation of Gordon in the sense that the Charlesby formalism for p is used, contrary to the commonly employed p for the extent of polymerization reaction as employed by Gordon.[22]

Yashin and Isayev postulate that under main-chain scission conditions, a Poisson distribution of chain fragments is generated, and for the parameters p and q they propose an exponential decay as given by:

$$1 - p(t) = (1 - p_0)^{e^{-k_p t}} \tag{8.29}$$

$$q(t) = q_0 e^{-k_q t} \tag{8.30}$$

In which k_p and k_q are the scission rate constants for main-chain scission and crosslink breakage, respectively. In the same way as Horikx did, Yashin and Isayev normalized ν_e to the network chain density for the still fully vulcanized sample and did the same for the sol fraction. By simulating network degradation in time and adjusting the ratio of the scission constants, the same type of curves can be constructed as those of Horikx. When k_p/k_q approaches infinity, practically only polymer chains are broken and when k_p/k_q approaches 0, only crosslinks are broken. By taking various ratios between k_p and k_q, intermediate curves can be constructed.

8.2.3.2 Chemorheology

The description of stress relaxation due to chemical effects, so called "chemorheology", was developed to discern between random scission and crosslink scission during ageing processes.[28] Due to the close similarity of degradation by ageing and the controlled network breakage/de-vulcanization, this method could equally well be applied. This theory is based on continuous stress relaxation or intermittent stress–strain measurements of rubber networks during ageing. It is based on the idea that the stress at any state of degradation is proportional to the network chain density:

$$\sigma = \nu_e kT(\lambda - \lambda^{-2}) \tag{8.31}$$

in which σ is the tensile stress (based on the original cross sectional area, the 'engineering stress'), ν_e is the instantaneous effective network chain density, λ is the extension ratio of the stretched to the unstretched length, k is the Boltzmann constant and T is the absolute temperature.

Directly after applying deformation to a network, the retractive force usually declines due to physical rearrangements of the polymer chains and slippage of entanglements. After some time these physical effects come to an end. The further decline of the retractive force is then due to chemical degradation of the network. With exclusion of the physical effects, the ratio of the stress at a certain time to the stress at a normalization time, the moment that all physical effects are over, will thus be the same as the ratio between the network chain density at the same measurement time and the network chain density at the normalization time:

$$\sigma_f/\sigma_i = \nu_f/\nu_i \tag{8.32}$$

In which σ_f is the stress after ageing or de-vulcanization treatment, σ_i is the stress before ageing or de-vulcanization treatment and ν_f and ν_i are the same as in eqn (8.20) and (8.22). Tobolsky showed that the number of bond cleavages, $n(t)$, that have occurred at time t per cm^3 of vulcanizate is:[29]

$$n(t) = \nu_i \ln\left(\sigma_t/\sigma_i\right) \tag{8.33}$$

In which σ_t is the tensile stress at time t.

Tobolsky and coworkers,[29] Ono[30] and Bjork *et al.*[31] indicated that eqn (8.29) can be used to discern between crosslink scission and random scission when rubber compounds are compared that are differing in network chain density. In the case of crosslink scission, $n(t)$ is proportional to ν_i. Normalized curves for continuous stress relaxation *versus* time will then coincide for different network chain densities.

In the case of random scission, $n(t)$ is independent of ν_i, since crosslinks are quite sparse relative to chain bonds. When the initial network chain density is low, breaking a chain will be more likely and detrimental. Therefore, in the case of random scission, the normalized stress relaxation will be greater, the lower the initial crosslink density. From a kinetic point of view, the rate of bond cleavage is constant, because elastically inactive network chains can also be broken in the case of random scission. This leads to a Maxwell type of stress decay in time: the plot of relative stress against logarithmic time should be a straight line. This straight relationship has been difficult to verify for rubbers though, because many chain scission reactions are catalyzed by compounding ingredients, in which case the rate of bond cleavage is not constant.

Decay of stress at constant extensions measures the scission reaction alone. When the sample is kept in a relaxed condition at the elevated temperature and the stress required to attain a certain extension is measured intermittently, the net effect of crosslinking and scission reactions is measured. When the sample is maintained in a stretched condition at elevated temperatures, it can develop a permanent set because of crosslinking. Any crosslinks formed in the stretched state, however, do not contribute to the stress in the stretched state and therefore do not disturb the measurement of the stress–relaxation due to scission reactions.

Mullins and Thomas adapted the description of Tobolsky for the amount of crosslinks removed from the network.[32] On that basis, Mullins and Turner evaluated the probabilities p and q of breaking respectively, creating a crosslink, by combining measurements of stress–relaxation with equilibrium swelling for the determination of N, the number of primary chains.[33]

On the same basis, Ono replaced the ratio σ_f/σ_i by the ratio of the network chain densities.[30] Using these, he calculated the σ from the sol fractions of the undegraded and the degraded vulcanizate at time t. For the degradation of peroxide cured NR networks he showed, that the ratio of stresses and that of the terms containing the soluble fractions agreed excellently with each other. He also showed that the degradation did not obey a Maxwell type of decay in time, but that the stress decreased sharply near the gel point (or the sol fraction increases sharply). This was attributed to the fact that in the later stage of degradation not all the network chains are elastically active anymore. Tobolsky assumed that all the network chains that have not been broken are still elastically effective.

However, this only applies for the initial stage of degradation. It was found that sulfur-cured NR indeed showed this type of degradation behavior, but peroxide-cured NR clearly showed an accelerated decay. When it is assumed that peroxide vulcanizates react through main chain scission, then at a later stage of degradation there will be a lot of clusters of network chains of a small enough size to become soluble and therefore belong to the sol.[30] This makes the assumption invalid.

8.3 De-vulcanization of SBR

Styrene butadiene rubber (SBR) is the main component in whole passenger car tire rubbers and, at the same time, the most critical one in terms of de-vulcanization. De-vulcanization of styrene-butadiene and butadiene based polymers follows a unique pathway due to the specific chemical structure of the polymer.[34] In degradation, polymer breakdown occurs, but at the same time recombination of chain segments takes place. In the present study, special attention was devoted to the investigation of the de-vulcanization efficiency under different conditions: thermal and thermo-chemical de-vulcanization of sulfur-vulcanised SBR. Moreover, an additional positive effect was found when stabilizers were added: they suppress the reverse reaction by radicals or other reactive species and interrupt the oxidation cycle which may result from the presence of oxygen during de-vulcanization.[35]

8.3.1 Thermal De-vulcanization of SBR

In this first part of the investigation, the rubber was subjected to thermal treatment in an internal mixer under three different conditions: in presence of air at all stages (TT); with quenching of the de-vulcanizate in liquid nitrogen after de-vulcanisation (TL); and under nitrogen atmosphere during de-vulcanization as well as with quenching in liquid nitrogen (TN).[35,36] Figure 8.10 shows the experimentally determined sol fractions of the de-vulcanized

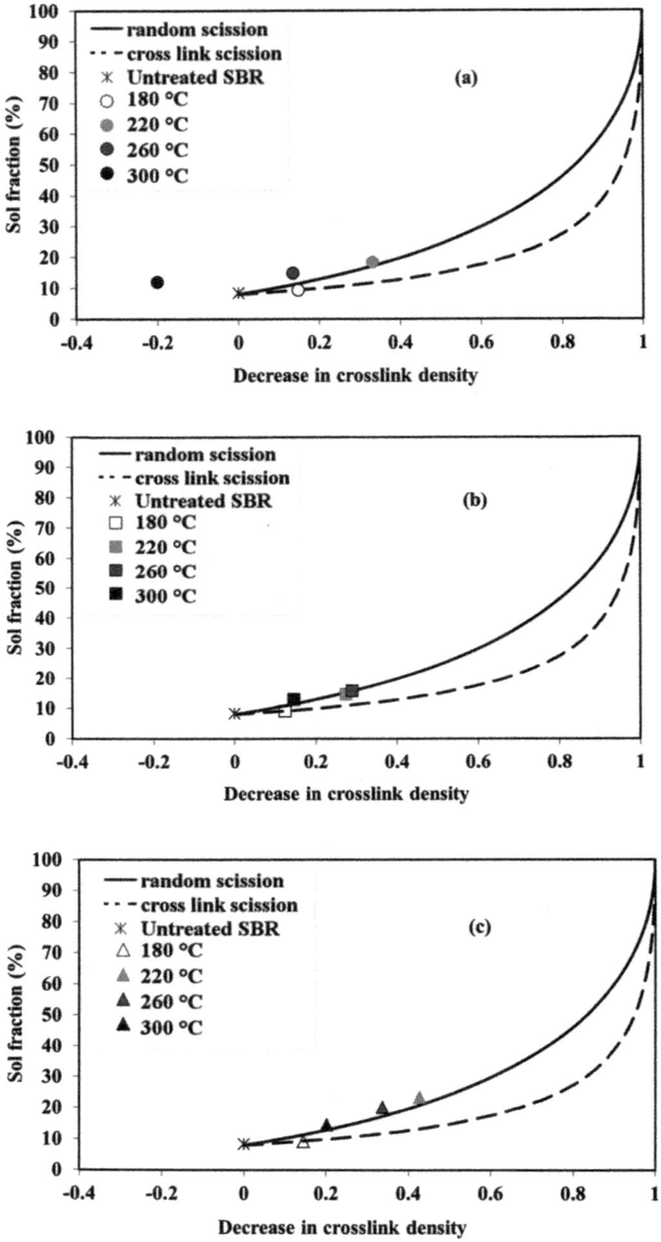

Figure 8.10 Sol fraction generated during de-vulcanization *versus* the relative decrease in crosslink density of de-vulcanized SBR. (a) TT; (b) TL and; (c) TN.

SBR samples at various de-vulcanization temperatures as a function of the relative decrease in crosslink density, the Horikx plot. For the thermally treated material, TT in Figure 8.10a, an increase of the de-vulcanization temperature to 220 °C results in a shift of the data point to the upper right hand side of the graph, which indicates an increase of the sol fraction and decrease of crosslink density. Nevertheless, a further increase of de-vulcanization temperature to 260 °C results in a back turn of the experimental data point to the left, which is the reverse of the expected decrease of crosslink density. This reversion phenomenon is even more pronounced for de-vulcanization at 300 °C; for this temperature the data point is even found at the left hand side of the value for untreated SBR. The detrimental effect of the presence of oxygen in the de-vulcanization process causes inefficient de-vulcanization, in which the crosslink density of the de-vulcanized rubber is even higher than that of the untreated vulcanized starting material.

Suppression of the inefficiency of the de-vulcanization is observed when oxygen is excluded; the results for the TL and TN samples are shown in Figure 8.10b and c, respectively. These illustrate that the experimental data for treatment at 180 °C are situated at more or less the same position as the data for TT. An increase of the de-vulcanization temperature to 220 °C exhibits the same trend as found for TT, in that the percentage of soluble polymer is increased and the crosslink density is decreased. For TL, the values after treatment at 260 °C are more or less at the same position as the values for the materials de-vulcanized at 220 °C, however, the reversion phenomena still occur when the de-vulcanization temperature is further raised above 260 °C. For TN, the data points first move to the right hand side for a treatment temperature of 220 °C, but then turn back to the left hand side for de-vulcanization temperatures of 260 °C and 300 °C.

The detrimental effect of the presence of oxygen in the de-vulcanization is smallest with a nitrogen-blanket present during de-vulcanization and quenching in liquid nitrogen. The reversion phenomenon is less progressive in this case as the crosslink densities observed are still lower than the crosslink densities of de-vulcanized rubber treated at 180 °C and untreated rubber. Thus, it can be concluded that working under exclusion of oxygen during and after de-vulcanization is a requirement for efficient de-vulcanization of SBR, and a temperature of 220 °C is the optimum for de-vulcanization. Above 220 °C reversion phenomena appear, whatever conditions are used. Based on these results, it is to be concluded that the de-vulcanization as obtained under the present circumstances is primarily *via* the mechanism of main-chain scission. Theoretically, the scission of C–S and S–S bonds, which have lower bond strengths than C–C bonds, should be the priority mechanism in network scission. However, in actual practice the data show that main-chain scission still mainly occurs. This is due to uncontrolled scission of the rubber network in the thermal process. As there is a great majority of C–C bonds present in crosslinked SBR compared to C–S and S–S bonds, the final overall probability of cleavage of the polymer backbone is still higher.

8.3.2 Thermo-chemical De-vulcanization of SBR

As thermal de-vulcanization turned out to be non-selective, the effect of a radical scavenger was studied. In this series of experiments, three disulfides were chosen as de-vulcanization aids, differing in chemical structure: an aliphatic one, dibutyldisulfide (DBDS); an aromatic one, diphenyldisulfide (DPDS); and an aliphatic one with an amine groups, di(2-aminophenyl) disulfide (APDS).[35,36]

The experimentally determined sol fractions of SBR de-vulcanized with DPDS at various temperatures as a function of the relative decrease in crosslink density are shown in Figure 8.11a. An increase of the de-vulcanization temperature up to 220 °C results in a shift of the data points to the right hand side of the graph, which indicates an increase in sol fraction and a decrease of crosslink density. Nevertheless, a further increase of the de-vulcanization temperature above 220 °C results in a back-turn of the experimental data points to the left, which is the reverse of the expected decrease of crosslink density. This indicates inefficient de-vulcanization, in which the crosslink density of the de-vulcanized rubber is increased rather than decreased with increasing treatment temperature.

Figure 8.11b and c show the experimentally determined sol fractions as a function of the relative decrease in crosslink density at various de-vulcanization temperatures in presence of DBDS and APDS, respectively. In case of DBDS, an increase of the de-vulcanization temperature up to 220 °C shows the same trend as found for DPDS as a de-vulcanization aid, that is a shift of the data points to the upper right hand side of the graph, while a further increase of de-vulcanization temperature to 260 °C results in a back-turn of the experimental data points to the left. However, in case of APDS de-vulcanized SBR, the back-turn shifting occurs readily at a lower temperature of 180 °C. The reversion phenomenon is most pronounced for de-vulcanization up to 300 °C; for this temperature the data point even moved to the left hand side.

The sol fractions as a function of the relative decrease in crosslink density of SBR de-vulcanizates using all three types of de-vulcanization aids are gathered in Figure 8.12. It can clearly be seen that the reversion phenomenon occurs in all cases, but is more or less pronounced depending on the type of de-vulcanization aid. Each disulfide is effective up to a different particular threshold temperature. Above this threshold temperature, de-vulcanization becomes inefficient. Degradation is more likely to occur at higher de-vulcanization temperatures, which is attributed to an excessive and uncontrolled generation of reactive radicals as mentioned before.

The efficiency for de-vulcanization of these three disulfides is related to their thermal reactivity. DBDS, a representative of aliphatic disulfides, was found to be an effective de-vulcanization aid only at low de-vulcanization temperatures up to 220 °C. This is attributed to its thermal stability and a boiling temperature of 188 °C. One of the aromatic disulfides, DPDS, with a

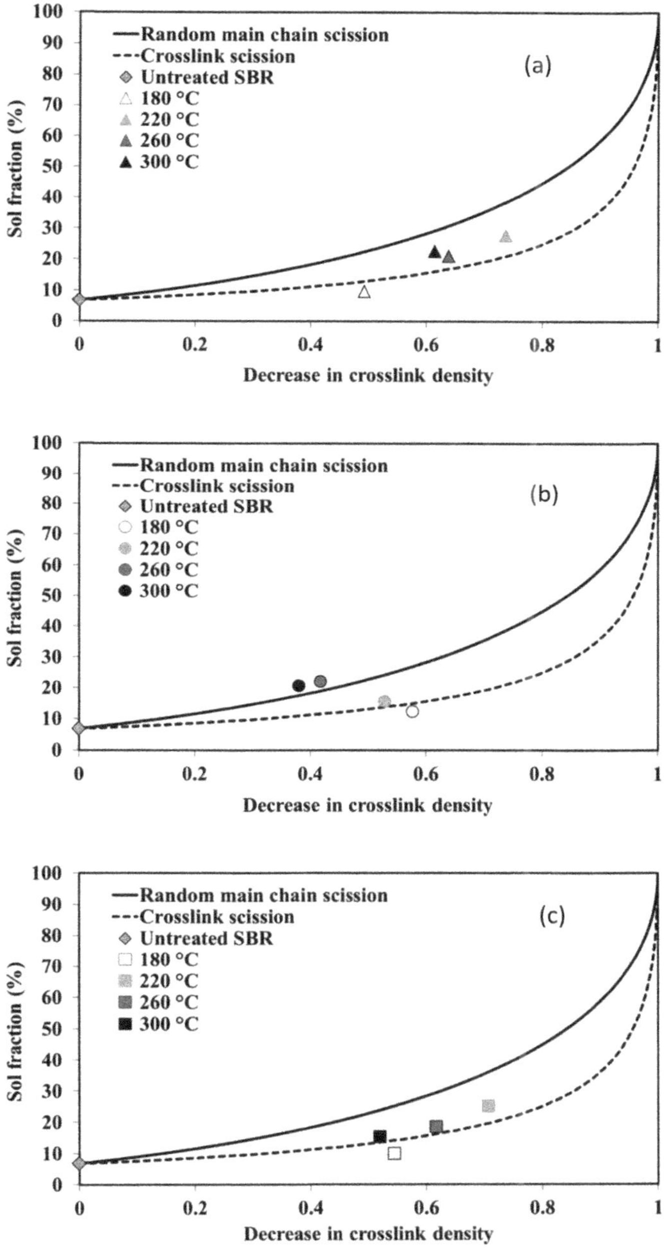

Figure 8.11 Sol fraction generated during de-vulcanization *versus* the relative decrease in crosslink density of de-vulcanized SBR (a) DPDS, (b) DBDS and (c) APDS as de-vulcanization aids.

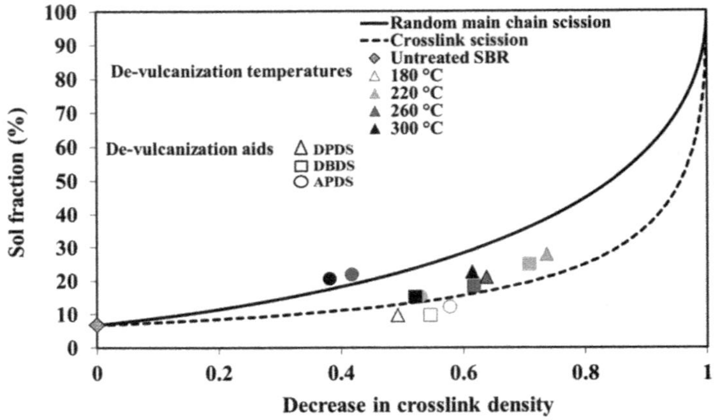

Figure 8.12 Sol fraction generated during de-vulcanization *versus* the relative decrease in crosslink density of de-vulcanized SBR using 3 different types of de-vulcanization aid: DPDS, DBDS and APDS.

Table 8.2 Solubility parameters of SBR and disulfides.

Chemical	Solubility parameter (J cm^{-3})$^{1/2}$
SBR 1723	17.3
Diphenyldisulfide	22.5
Dibutyldisulfide	17.9
Di (2-aminophenyl) disulfide	26.0

higher boiling point than the previous one, was more effective in de-vulcanization at higher temperatures. APDS is an aromatic disulfide which was chosen to study the effect of a combination of a disulfide and an amine group. However, it turned out to be the least effective de-vulcanization aid, even though it has the highest boiling point of the three de-vulcanization aids. This is attributed to the presence of the amine group which reacts according to a nucleophilic mechanism[37] and has a positive effect on the de-vulcanization efficiency only at low treatment temperatures.[38]

Another factor that can play a role in the efficiency of the de-vulcanization aids is the compatibility between rubber and the de-vulcanization aids, which can be estimated by their respective solubility parameters. The calculated values of the solubility parameters of SBR and the three disulfides are shown in Table 8.2. A larger difference in the solubility parameter results in less compatibility of SBR with the de-vulcanization aids, which may affect the efficiency of de-vulcanization. The largest difference is found for the system APDS/SBR, which corresponds with the observation that APDS is the least effective de-vulcanization aid. DBDS and SBR have the closest solubility parameters, but here the low boiling point of DBDS prevails. DPDS is found to be the most effective de-vulcanization aid due to the high boiling point in spite of the difference in the solubility parameter compared to SBR.

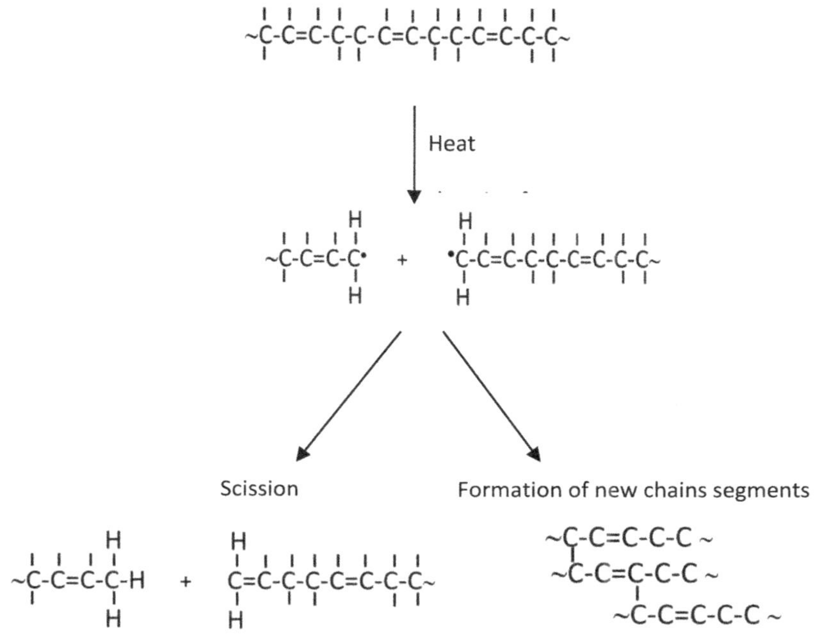

Figure 8.13 Simplified reaction scheme proposed for the degradation of polybutadiene and butadiene based polymers.

Figure 8.13 is a schematic depiction of the degradation of polybutadiene and butadiene based polymers. Basically, two reactions can occur during degradation of polybutadiene, chain scission and formation of active chain segments. Chain scission occurs when breaking of the C–C bonds in the polymer chains is accompanied by hydrogen transfer, and results in polymers with a lower molecular weight.[39] Another reaction, the formation of radically active chain fragments, apparently takes place from three possible reactions as shown in Figures 8.14 and 8.15. Reaction (A) in Figure 8.14 is known as a hydrogen transfer reaction. It is initiated by breaking of the C–C bonds in the polymer chains due to physical forces: heat, light, and mechanical forces. It is accompanied by hydrogen transfer which results in polymer chain scission. However, re-formation of unstable radicals to polymers could possibly take place when scission of C–C bonds in the chain is not accompanied by hydrogen transfer.[39] It is reported that for degradation in an air atmosphere, with the main active component oxygen, hydrogen transfer is less. Therefore, the active radicals tend to recombine to a new polymer.

Reaction (B) in Figure 8.14 is the chemical structure transformation reaction. It is associated with thermal isomerization: *cis–trans* isomerization, cyclization and volatilization of small fragments.[40–42] These chain rearrangements are undeniably enhanced with increasing temperature and occur at the initial decomposition step of SBR.

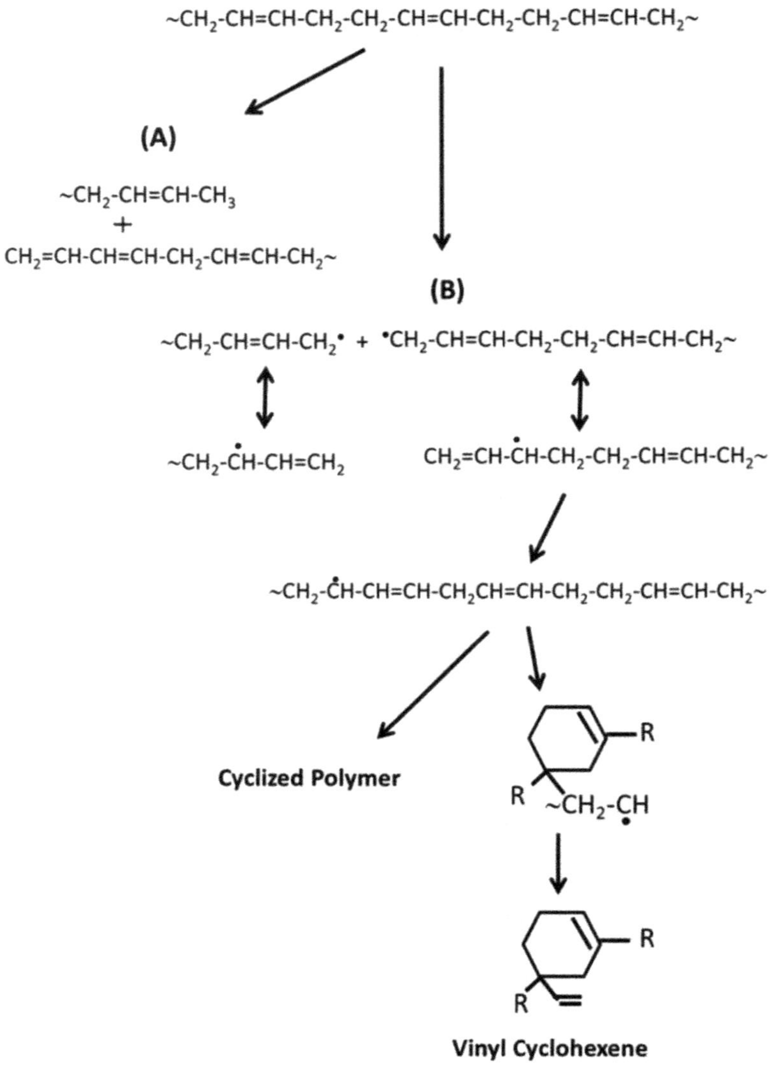

Figure 8.14 Reformation of active bond fragments in the specific degradation mechanism of polybutadiene and butadiene based polymers.

Reaction (C) in Figure 8.15 involves generation of active groups derived from the presence of oxygen. The initial predominant attack of aerial oxygen (O_2) takes place on the double bonds. It continues with the generation of hydroperoxide radicals which then decompose to give carbonyl (>C=O) functionalities.[43] These reactions typically take place at high temperatures. Accordingly, when carbonyl functionalities are generated, the possibility of creating a rubber soluble fraction is less: the bond energy of C=O (724 kJ mol^{-1}) is considerably higher than that of C–C (410 kJ mol^{-1}) or carbon hydrogen bonds, CH (335 kJ mol^{-1}).

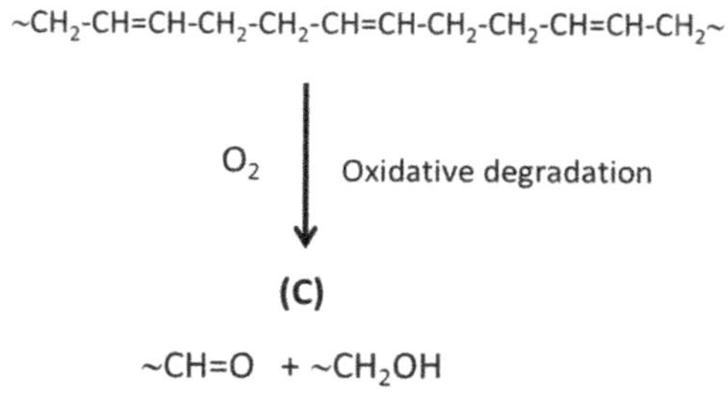

Figure 8.15 Simplified oxidative degradation mechanism of butadiene based polymers.

Table 8.3 Chemical names and structures of thermal stabilizers.

Chemical name	Commercial name	Type	Chemical structure
Pentaerythritol tetrakis (3-(3,5-di-*tert*-butyl-4-hydroxyphenyl) propionate)	Irganox 1010	Hindered phenolic	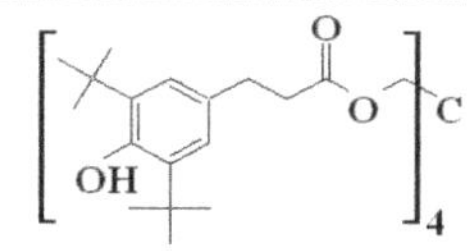
Octadecyl-3-(3,5-di-*tert*-butyl-4-hydroxyphenyl) propionate	Irganox 1076	Hindered phenolic	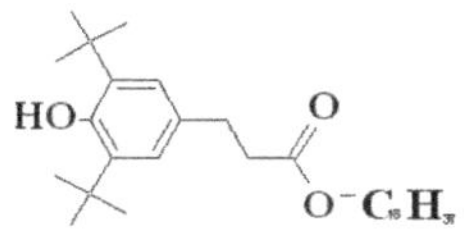
Tris(2,4-ditert-butyl-phenyl)phosphite	Irgafos 168	Phosphite	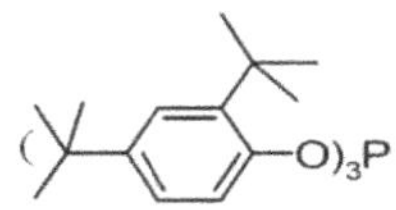

8.3.3 Chemical De-vulcanization of SBR with the Aid of Stabilizers

In this series of experiments, stabilizers as specified in Table 8.3 were additionally added to the de-vulcanization aid, DPDS.[35,36] The experimentally determined sol fractions of DPDS de-vulcanized SBR at various de-vulcanization temperatures as a function of the relative decrease in crosslink density are shown in Figure 8.16. An increase of the de-vulcanization temperature to 220 °C results in a shift of the data points to the right hand side of the graph, which indicates a small increase in the sol fraction and large decrease of crosslink density. Nevertheless, a further increase of de-vulcanization temperature to 260 °C results in a back turn of the experimental data points to the left, which is the reverse of the expected decrease of crosslink density. This reversion phenomenon is even more pronounced for de-vulcanization up to 300 °C; for this temperature the data point is even found at the utmost left hand side. This indicates inefficient de-vulcanization, in which the crosslink density of the de-vulcanized rubber is increased rather than decreased with increasing treatment temperature.

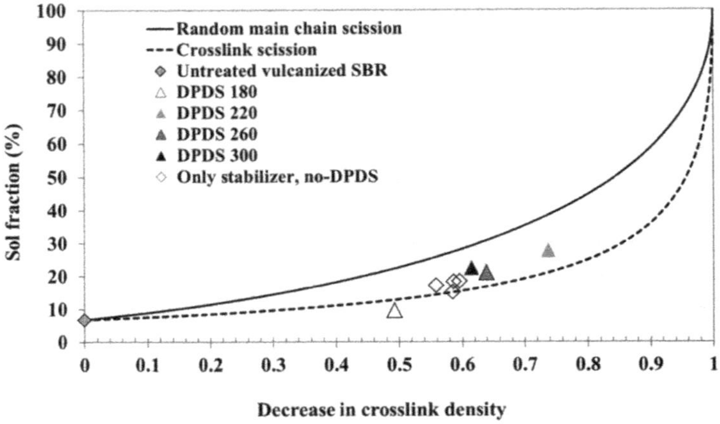

Figure 8.16 Sol fraction generated during de-vulcanization *versus* the relative decrease in crosslink density of de-vulcanized SBR:DPDS *versus* stabilizer only.

A first attempt to improve the de-vulcanization efficiency at high de-vulcanization temperatures above 220 °C was made by replacing DPDS with the oxidation stabilizers. However, it can be seen from Figures 8.16 and 8.17 that there is no significant improvement with switching to the stabilizers. Replacement of DPDS by the oxidation stabilizers shows lower soluble fractions and higher crosslink densities than obtained by using only DPDS. This indicates a more inefficient de-vulcanization, as can be seen in Figure 8.16, where the data points are shifted farther to the left hand side of the graph.

A more efficient de-vulcanization is observed for combinations of DPDS (3% w/w) and the oxidation stabilizers (1% w/w) as de-vulcanization aids. The positive effect of using the combination of these two chemical species is clearly seen in Figure 8.17. The experimental data for treatment above 220 °C are situated at even slightly higher positions than the data for the de-vulcanizate treated at 220 °C with only DPDS. The reversion phenomenon is noticeably absent in this case as there is no move of the data points to the left hand side with increase of de-vulcanization temperature up to 300 °C. Therefore, the utilization of the developed synergism of de-vulcanization aid with oxidation stabilizers results in a more efficient and controlled de-vulcanization.

DPDS is reported to be an effective de-vulcanization aid. A mechanism that is frequently proposed for the reaction of radical scavenger de-vulcanization agents such as DPDS with sulfur vulcanizates is the opening of crosslinks or the scission of polymer chains by heat and shearing forces, and the reaction of fragments with disulfide based radicals, which prevent recombination. A simplified reaction scheme proposed for the rubber de-vulcanization with DPDS is given in Figure 8.18. Based on the present results, it is to be concluded that DPDS is an effective de-vulcanization aid until a temperature up to 220 °C. Above this threshold temperature, inefficient de-vulcanization

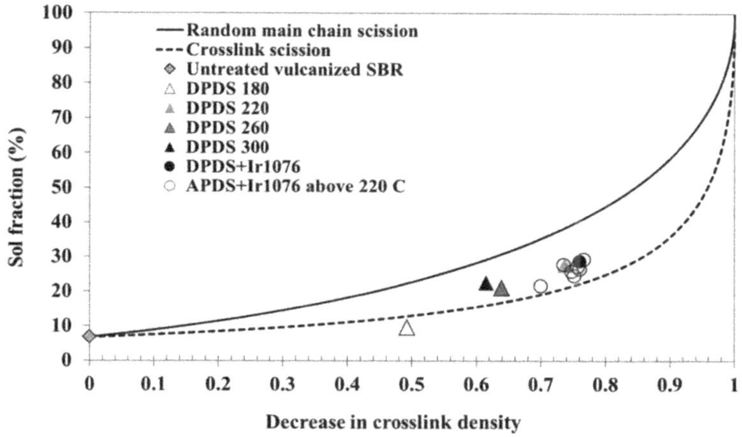

Figure 8.17 Sol fraction generated during de-vulcanization *versus* the relative decrease in crosslink density of de-vulcanized SBR using a combination of DPDS and a stabilizer.

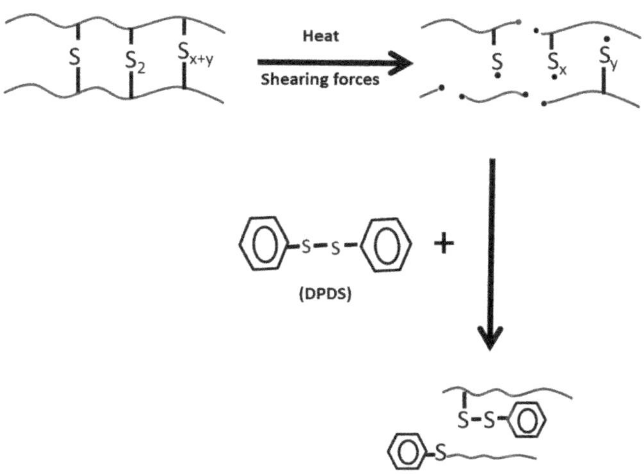

Figure 8.18 Simplified reaction scheme proposed for rubber de-vulcanization using diphenyldisulfide as de-vulcanization aid.

happens. This may be attributed to an excessive and uncontrolled generation of reactive radicals so that degradation is more likely to occur at the higher de-vulcanization temperature.

In general terms, polymer degradation is initiated by breaking C–H and C–C bonds along the polymer chains. Consequently, unstable carbon radicals (R$^{\bullet}$) are formed. These radicals readily react with oxygen (if present) to form peroxy radicals (ROO$^{\bullet}$) which subsequently abstract a hydrogen atom from other polymer molecules to form hydroperoxides (ROOH) and propagate

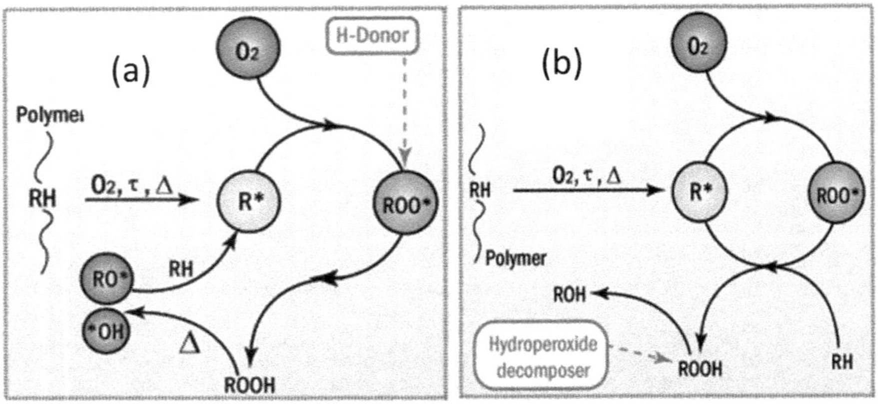

Figure 8.19 Preventing degradation by an oxidation stabilizer:[44] (a) Hindered phenolic stabilizer; (b) phosphite compound.

in an oxidative degradation cycle.[45] Once oxidation starts, it sets off a chain reaction which accelerates degradation unless an oxidation stabilizer is present to interrupt the cycle. The principle mechanism of degradation and the effect of an oxidation stabilizer is shown in Figure 8.19.[46]

Oxidation stabilizers are capable of scavenging or destroying these chain propagation species: alkoxy radicals (RO•), hydroxy radicals (HO•) and peroxy radicals (ROO•). Therefore, the oxidative degradation cycle can be prevented either by using an oxidation stabilizer or by totally eliminating oxygen in the de-vulcanization process. Two types of oxidation stabilizers were used in this present work: hindered phenolic and phosphite compounds. Hindered phenolic stabilizers act as hydrogen donors. The stabilizer reacts with peroxy radicals to form hydroperoxides and prevents the abstraction of the hydrogen from polymer molecules. Phosphite compounds function as hydroperoxide decomposers, the stabilizer prevents the split of hydroperoxide into extremely active radicals, alkoxy and hydroxyl.[47–49]

Efficient de-vulcanization is observed with using a combination of DPDS and oxidation stabilizers as de-vulcanization aids. A combination of two different functional chemicals frequently gives better protection against polymer degradation than would be expected from each compound alone, commonly called synergism. A decrease in crosslink density without creating more sol fraction is found for the combination of an oxidation stabilizer and DPDS. Figure 8.20 emphasizes the importance of the developed "synergistic de-vulcanization aid".

In Figure 8.20, reaction (A), the initial step is the opening of crosslinks or the scission of polymer chains by heat and shearing forces. Unstable radicals are formed. Generally, this happens to a small extent at low temperatures, but is of greater importance with increasing temperature.[40] Subsequently, these

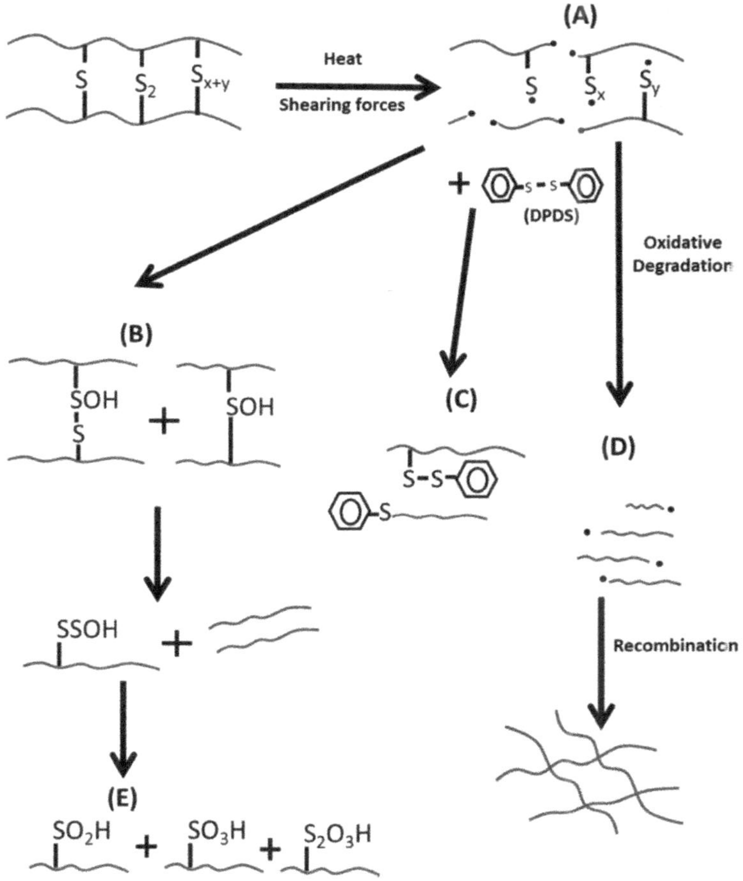

Figure 8.20 Simplified reaction scheme proposed for thermo-chemical de-vulcanization. For reaction pathways (A) to (E) see text.

unstable radicals further react in secondary reaction steps in the presence of oxygen, Figure 8.20 reactions (B)–(D). These three reactions can jointly occur depending on the de-vulcanization conditions.[39,50,51]

In Figure 8.20, Reaction (B), as the hydroperoxide is formed in subsequent oxidative aging of the vulcanizate, a variety of oxidized organic sulfur groups will result from reaction of the various mono-, di-, as well as polysulfide groups with hydroperoxides. The sulfenic acids (RSOH) and thiosulfoxylic acids (RSSOH) formed from the oxidized sulfur network, are expected to react with hydroperoxides to form sulfinic (RSO_2H), sulfonic (RSO_3H) and thiosulfuric (RS_2O_3H) acids. All of these acids can contribute to the destruction of hydroperoxides to non-radical products and thus function as preventive antioxidants.[51]

Figure 8.20, Reaction (C) shows the reaction of active radical fragments with disulfide based radicals, which prevent recombination of rubber molecules. In addition to this, the diffusion speed of the disulfide into the polymer matrix increases at high temperatures, enhancing the chance of combination of one radical with another rubber radical.[52]

In Figure 8.20, Reaction (D) the main-chain scission is caused by degradation. Higher treatment temperatures further activate the developing reactions, the probability of the generated reactive groups to undergo oxidative degradation increases. The number of radical products formed during the high treatment temperature, that may participate in chain transfer, is increased. This results in severe main chain scission of the rubber network and leads to a high degree of network recombination as shown in Figure 8.20. This reaction needs an oxidation stabilizer to prevent the uncontrolled breakage of the rubber network.

At low temperatures, the occurrence of reactions (B) and (C) is predominant; therefore DPDS functions satisfactorily under these circumstances. However, uncontrolled generation of reactive groups occurs at higher temperatures, which leads to the predominant occurrence of reaction (D). The stabilizers prevent this to a large extent.

8.4 De-vulcanization of BR

Within this study, butadiene rubber (BR) was de-vulcanized and investigated concerning its tendency for crosslink *versus* main-chain scission, using the thermo-chemical de-vulcanization process.[35,36] The sol fraction and crosslink density of BR de-vulcanizates (D-BR) as a function of de-vulcanization temperature are shown in Figure 8.21. A de-vulcanization temperature of 220 °C results in a significant increase in sol fraction and a decrease in crosslink density; the values of the material de-vulcanized at 260 °C show values in-between. The sol fraction of BR de-vulcanizates as a function of the relative decrease in crosslink density according to Horikx is shown in Figure 8.22.[21] At a de-vulcanization temperature of 220 °C, all experimental data are situated above the line of main-chain scission with about a 60% decrease in crosslink density compared to the untreated BR; the data points of BR de-vulcanizates at a de-vulcanization temperature of 260 °C are closer to the line of main-chain scission but at a much lower relative decrease in crosslink density. This means that a de-vulcanization temperature of 220 °C is more effective than a temperature of 260 °C, as the percentage of soluble polymer is higher and the final crosslink density is lower. However, a considerable amount of main chain scission still occurs in both cases, as the sol content is rather high.

The differences in the de-vulcanization behaviour at the two different de-vulcanization temperatures may be attributed to the complexity of the degradation mechanism of the polybutadiene based polymer.[34,53–56]

Degradation of polybutadiene and other butadiene based polymers follows a unique pathway due to the specific chemical structure of the polymer.[29]

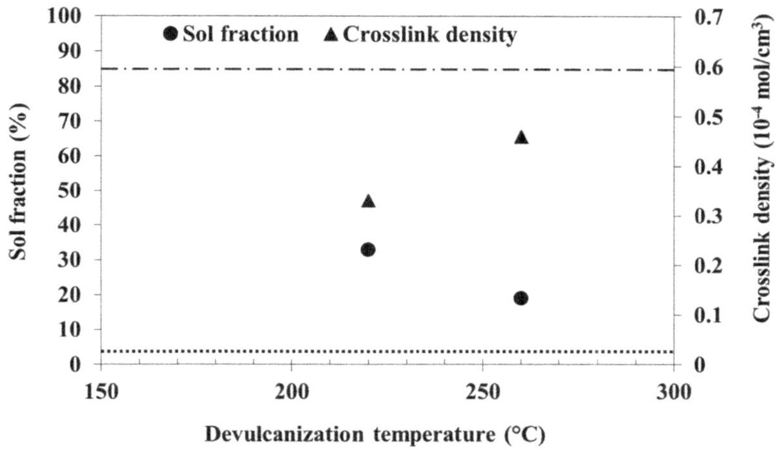

Figure 8.21 Sol fraction and crosslink density of BR de-vulcanizates *versus* de-vulcanization temperature; (.......) Sol fraction of untreated rubber vulcanizate; (–··–) Crosslink density of untreated rubber vulcanizate.

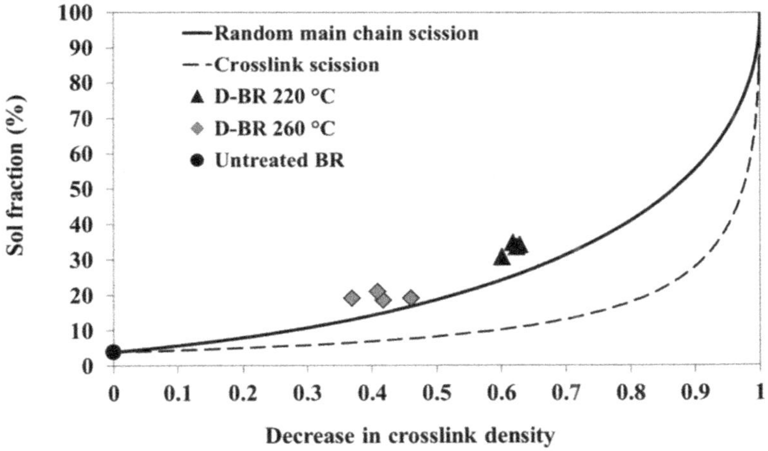

Figure 8.22 Relative decrease in crosslink density *versus* sol fraction of BR de-vulcanizates.

Chain scission occurs when breaking of the C–C bonds in the polymer chains is accompanied by hydrogen transfer, and results in chemically inactive polymer segments with a lower molecular weight.[57] The formation of active bond fragments presumably takes place when hydrogen transfer is not possible. Both hydrogen transfer as well as radical scavenging by the de-vulcanization aid of the active chain fragments are apparently more effective at a de-vulcanization temperature of 220 °C, leading to a higher sol fraction and a lower crosslink density compared to a temperature of 260 °C. For this polymer, the de-vulcanization temperature should be as low as possible.

8.5 De-vulcanization of NR

Thermochemical de-vulcanization of NR is commonly performed using DPDS as de-vulcanization aid. The sol fraction and crosslink density of NR de-vulcanizates as a function of the de-vulcanization temperature are shown in Figure 8.23. Almost 100% of the sol fraction of NR de-vulcanizates (D-NR) is observed at both de-vulcanization temperatures of 220 °C and 260 °C. Moreover, the crosslink densities of the de-vulcanizates treated at these temperatures are reduced to almost zero.[35,36] This indicates that breakdown of NR in this temperature range results in an almost complete destruction of the NR network. It has already been reported that the NR vulcanizate network can strongly be decreased in crosslink density at temperatures above 170 °C.[57] Moreover, DPDS was reported to be an effective de-vulcanization aid for NR. Temperature is the main governing factor and addition of DPDS can significantly enhance the effects.[57–59] The sol fraction of NR de-vulcanizates as a function of the relative decrease in crosslink density according to Horikx is shown in Figure 8.24.

As can be seen, the data points for NR de-vulcanizates reach the joint top point of both lines, main-chain and crosslink scission. This material is completely soluble, no network connections are remaining between the polymer molecules. Both NR de-vulcanizates had a very low viscosity and showed tacky behavior, another indication of a complete breakdown of the network. For this polymer, the de-vulcanization temperature is not critical within this

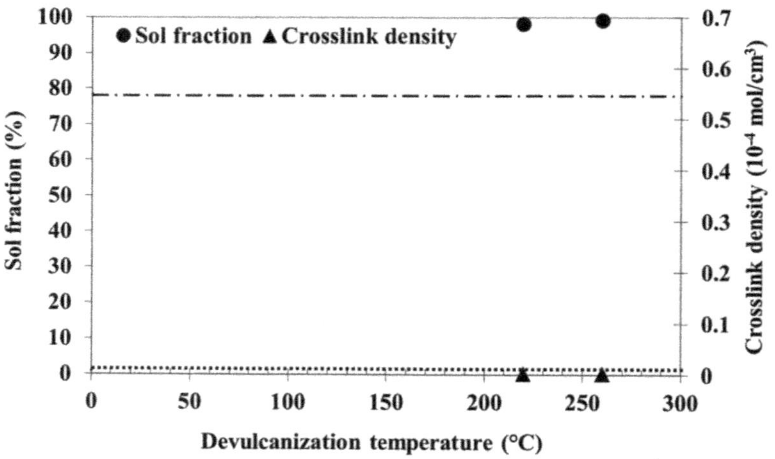

Figure 8.23 Sol fraction and crosslink density of NR de-vulcanizates *versus* de-vulcanization temperature; (.......) Sol fraction of untreated rubber vulcanizate; (– · –) Crosslink density of untreated rubber vulcanizate.

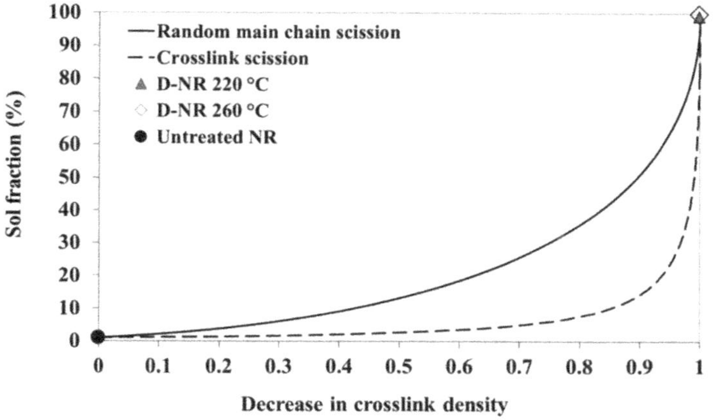

Figure 8.24 Relative decrease in crosslink density *versus* sol fraction of NR de-vulcanizates.

temperature range, as long as the polymer chains stay intact. The simplified reaction scheme proposed for breaking of a NR network by using DPDS is shown in Figure 8.5.

Aliphatic amines are also known to act as reclaiming agents for natural rubber. However, hexadecylamine is not working as a reclaiming agent for natural rubber based latex products. Instead of decreasing the viscosity and crosslink density of the rubber, it increases both properties relative to the thermal treatment, where no reclaiming agent is used. Figure 8.25 shows the increase in crosslink density with increasing hexadecylamine concentration for natural rubber reclamation.

Rajan *et al.* found that using hexadecylamine as a reclaiming agent resulted in the formation of additional crosslinks compared to the rubber reclamation without reclaiming agent.[14] This is explained by the presence of sulfur in the reclaimed rubber. During formation of crosslinks, polysulphide ions which were formed by cleavage of cyclic octasulfur are a crucial intermediate. Thus, these types of ions negatively influence the de-vulcanisation process when an amine is used as a reclaiming agent. Rubber reclaimed with hexadecylamine still has most of the polysulphidic crosslinks. This is explained by a complex formed by zinc-ions, which can be stabilized by amines as shown in Figure 8.26, thereby reducing the rate of crosslink scission.

8.6 De-vulcanization of CIIR

Thermochemical de-vulcanization of chlorinated butyl rubber (CIIR) was performed using DPDS as a de-vulcanization aid. Figure 8.27 shows the changes in sol fraction and crosslink density of CIIR de-vulcanizates as a function of de-vulcanization temperature. The relative decrease in crosslink density is shown in Figure 8.28. An increase in sol fraction with increasing de-vulcanization temperature was observed: the sol fractions were about

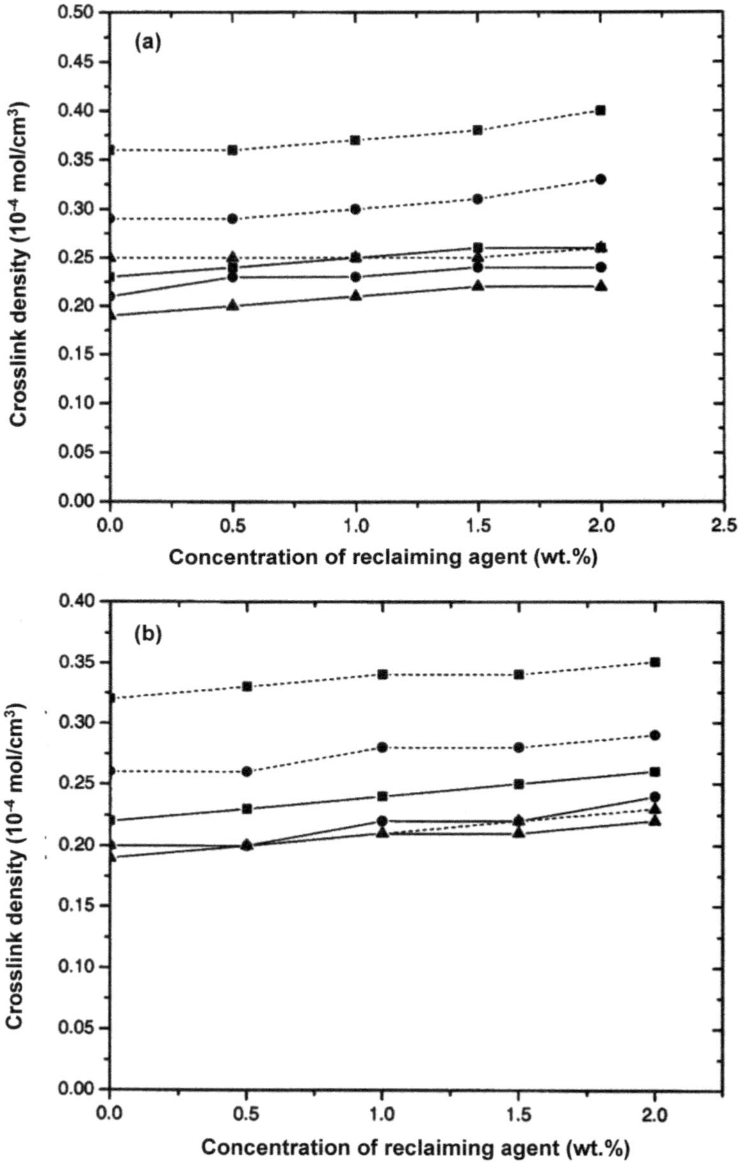

Figure 8.25 Crosslink density of natural rubber reclaim as a function of hexadecylamine concentration at reclaiming temperatures of (a) 170 °C and (b) 180 °C and at various reclaiming times; (■) 5 min; (▲) 7 min; (●) 10 min.

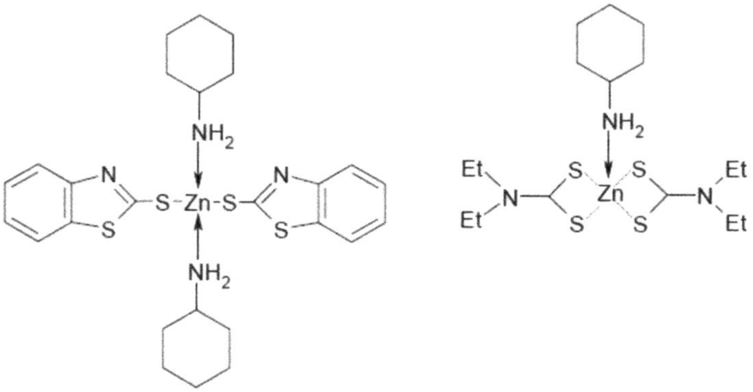

Figure 8.26 Hypothetical complex stabilization by amines.[60]

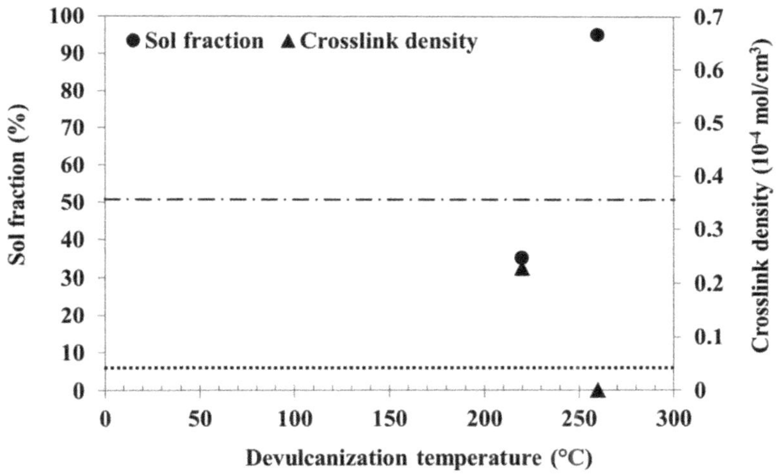

Figure 8.27 Sol fraction and crosslink density of CIIR de-vulcanizates *versus* de-vulcanization temperature; (.......): Sol fraction of untreated rubber vulcanizate; (– · –): Crosslink density of untreated rubber vulcanizate.

36% and more than 95% after de-vulcanization of CIIR at 220 °C and 260 °C, respectively. Consequently, the de-vulcanized CIIR at 260 °C had a very low viscosity and showed tacky behavior. However, in terms of crosslink density there was no significant decrease at a de-vulcanization temperature of 220 °C. It can clearly be noticed from Figure 8.28 that all experimental data are situated above the line of main-chain scission and at less than a 40% decrease in crosslink density compared to untreated CIIR. They are clustered at the left hand side which indicates that the crosslink density of the treated rubber is more or less the same as the crosslink density of the untreated

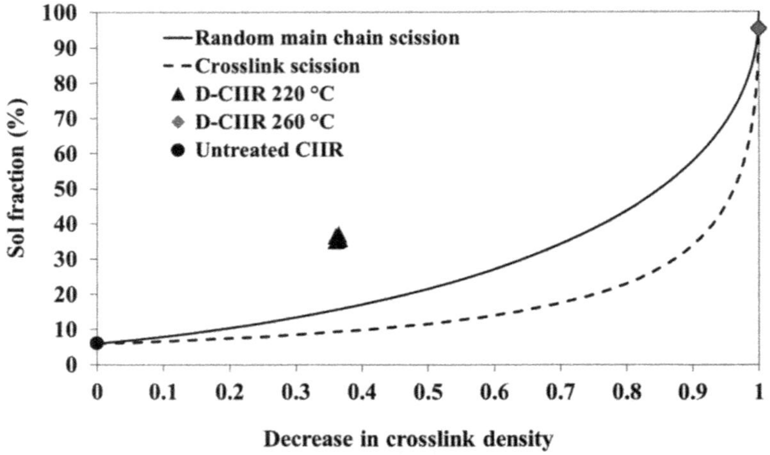

Figure 8.28 Relative decrease in crosslink density *versus* sol fraction of CIIR de-vulcanizates.

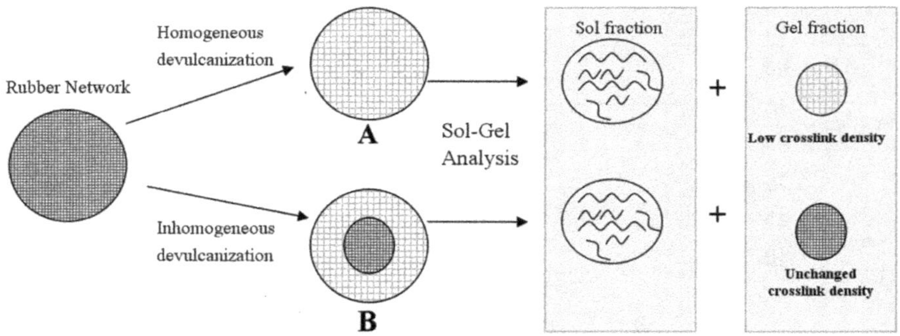

Figure 8.29 Schematic representation of two different de-vulcanization mechanisms, A: Homogeneous de-vulcanization and B: Inhomogeneous de-vulcanization.[36]

one. This may be attributed to a de-vulcanization mechanism which is not uniform and homogeneously distributed throughout the rubber particles, and for which the validity of the Horikx plot is limited. The two different de-vulcanization mechanisms, uniform and non-uniform, are schematically depicted in Figure 8.29.[38]

A limitation of the Horikx representation is that it assumes homogeneous breakdown of the vulcanized network throughout the particles (A). However, the results for de-vulcanized CIIR at 220 °C indicate a different mechanism, peeling off of the outer layers of the particles, while the inner cores of the particles stay more or less untreated at constant crosslink density (B). This inhomogeneity causes in actual practice a lower decrease in crosslink density at a particular sol fraction, than would have been obtained for homogeneous breakdown.

8.7 De-vulcanization of EPDM

For EPDM rubber, α-hydrogen-containing aliphatic amines, for example. hexadecylamine (HDA), are found to be very effective de-vulcanization aids.[7,13,63] The relative decrease in the EPDM crosslink density was reported to be dependent on the concentration of the amines, a higher concentration of amines leads to a stronger decrease in crosslink density. The use of HDA in a comparative study of two different vulcanization systems for carbon black filled EPDM-rubber was investigated, and HDA was found to be suitable as a de-vulcanization agent for both EPDM rubbers.[64] Dijkhuis *et al.* stated that de-vulcanized rubber from conventionally vulcanized EPDM, mainly polysulfidic in nature, shows a decrease in crosslink density with increasing HDA concentration and at low de-vulcanization temperatures.[11] After de-vulcanization at the lower limit of the experimental temperature window (*i.e.* 225 °C), the concentration of remaining di- and polysulfidic crosslinks is higher than the concentration of monosulfidic bonds, while at the upper temperature level (*i.e.* 275 °C), the concentration of monosulfidic bonds is highest, as shown in Figure 8.30. For efficiently vulcanized EPDM with primarily monosulfidic crosslinks, HDA again has a positive effect on the de-vulcanization efficiency at low temperatures of a maximum of 225 °C. At higher temperatures, the crosslink density increases with increasing concentrations of HDA. A comparative study between two different vulcanization systems was performed, and it was found that conventionally vulcanized EPDM de-vulcanizes to a larger extent by crosslink scission compared to the efficiently vulcanized material, which primarily shows main-chain scission. Both materials can be added to a virgin masterbatch in concentrations up to 50 wt% with a limited influence on the properties. This is a rather high concentration of recycled rubber compared to the maximum loading of 15 wt%, commonly known to be the practical limit.

The influences of three different types of de-vulcanization aids, HDA, DPDS and o'o'-dibenzamidodifenyldisulfide (BAFD) on the de-vulcanization efficiency of EPDM-rubber was investigated. Figure 8.31 shows the Horikx plots of de-vulcanized EPDM for all temperatures and de-vulcanization aids applied. As shown, addition of HDA results in de-vulcanized EPDM with the highest sol fractions and the largest relative decreases in crosslink density, as compared to DPDS and BAFD as de-vulcanization aids. DPDS and BAFD are not as effective as de-vulcanization aids and even increase the crosslink densities of the EPDM de-vulcanizates. The DPDS and BAFD require higher temperatures compared to HDA in order to generate plasticized and re-processable materials, and these are still difficult to reprocess due to their high Mooney viscosity. This appears to be due to a crosslinking effect of these two de-vulcanization aids. Addition of DPDS and BAFD increases the crosslink densities of the de-vulcanizates, while HDA lowers the concentrations of crosslinks. This is proven by the Horikx-plots which show for DPDS and BAFD a low relative decrease in crosslink density with hardly any generation of sol. Using these two as

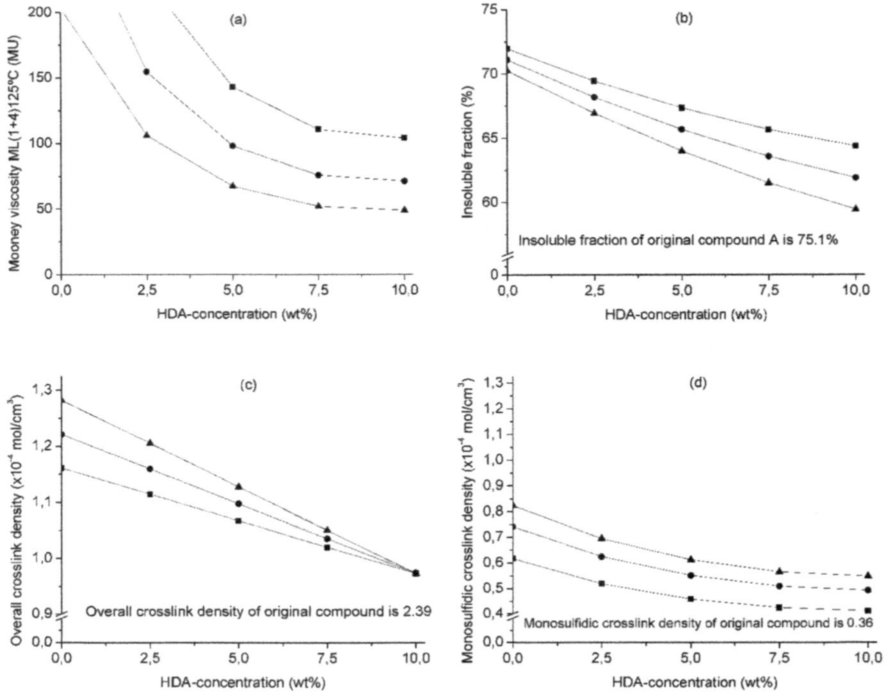

Figure 8.30 Influence of HDA-concentration on (a) Mooney viscosity; (b) insoluble fraction; (c) overall crosslink density and (d) monosulfidic crosslink density of de-vulcanized EPDM; (■) 225 °C; (●) 250 °C; (▲) 275 °C.[11]

de-vulcanization aids results in plasticized and re-processable de-vulcanizates only at the highest de-vulcanization temperatures in combination with high de-vulcanization aid concentrations. Even then, these de-vulcanizates have a high Mooney viscosity compared to ones obtained with HDA as a de-vulcanization aid.

DPDS cannot be used as a de-vulcanization aid for de-vulcanization of the mainly monosulfidic EPDM compound, because the obtained de-vulcanizates cannot be re-vulcanized again to result in good material properties. Addition of HDA- and BAFD-treated de-vulcanizates to virgin masterbatches results in both cases in re-vulcanizates with fully acceptable mechanical properties.

Due to the low amount of crosslinking sites available in EPDM (4 or 9 mol% at most), the most important factor in the recycling of EPDM-rubber is the balance between the breakdown of the vulcanized network and the ability to be revulcanized afterwards. When a de-vulcanization aid breaks all sulfur crosslinks, but does it in such a way that the crosslink sites are damaged, the recycled rubber cannot replace virgin rubber. This recyclate can then only be used in small amounts as a processing aid.

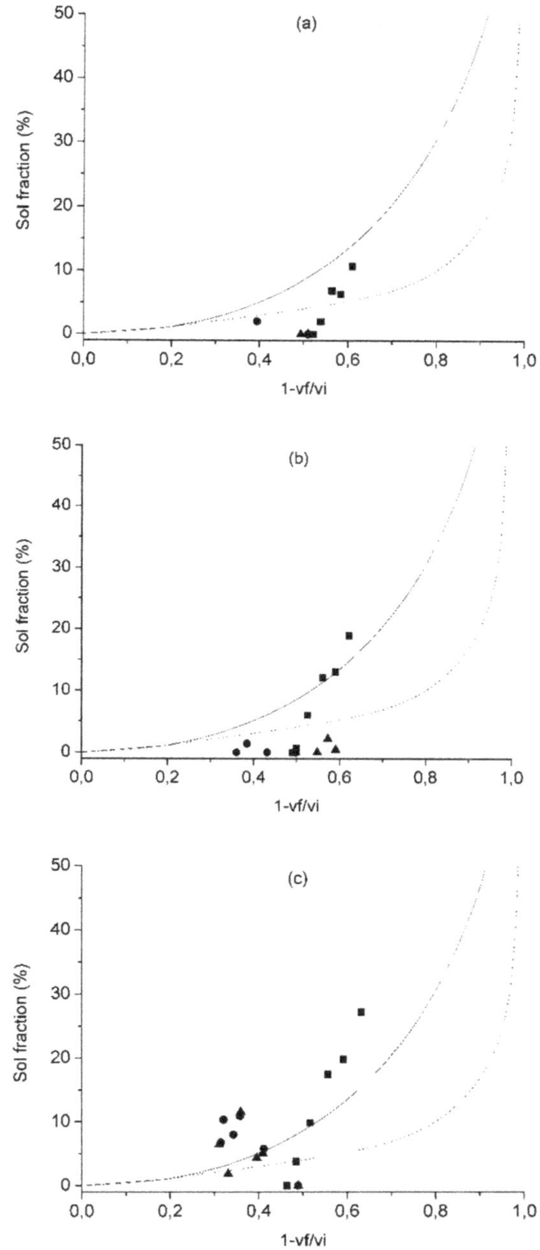

Figure 8.31 Fraction of sol generated during de-vulcanization *versus* the relative decrease in crosslink density. (—): only main-chain scission; (—): only crosslink scission; (a) 225 °C; (b) 250 °C and (c) 275 °C; (■) HDA; (●) DPDS; (▲) BAFD.[62]

8.7.1 Example of the Re-use of De-vulcanized Rubber: EPDM Roofing Foil

Addition of replasticized rubber into virgin rubber compounds is common practice. However, addition of de-vulcanized EPDM rubber to replace virgin EPDM is a relatively new concept. For sulfur-cured EPDM it was discovered previously that the use of an α-hydrogen containing amine as a de-vulcanization aid allows for the production of well plasticized de-vulcanizates.[13,61] They are characterized by a good re-vulcanization behavior, as the carbon–carbon unsaturation contained in the third monomer of EPDM is restored rather than de-naturated by the de-vulcanization process using these amines.

De-vulcanization of sulfur-cured EPDM building-profile scrap in a co-rotating twin-screw extruder with HDA as de-vulcanization aid results in a qualitatively excellent material.[60] It can be blended in practically all proportions with a virgin EPDM roof-sheeting compound, where the resulting re-vulcanizates all fulfill the most stringent industrial UEAtc (Union Européenne pour l'Agrément technique dans la construction) specifications.[63]

Increasing amounts of EPDM de-vulcanizate added to the virgin compound result in a decrease in vulcanization time and cure temperature for reversion-free vulcanization. The overall crosslink density shows a decrease with increasing de-vulcanizate content, while the ratio of mono- to di- and polysulfidic crosslinks increases (Figure 8.32). The insoluble fraction decreases with increasing de-vulcanizate content, indicating some loss of revulcanization ability of the de-vulcanizate relative to the virgin compound. Above a critical had concentration, the material starts vulcanizing at 125 °C readily, as next to being a de-vulcanization aid, HDA also has an activating effect on crosslinking. De-vulcanizate enhances the elasticity of uncured compounds, while it reduces the elasticity of the cured blends.

There are different specifications for EPDM roof-sheeting in use: EN 13956 (Norme Européenne), ASTM D 4637-04 and UEAtc, of which the UEAtc standard is the most stringent. For this reason, the revulcanizates were evaluated according to the UEAtc standard as detailed in Table 8.4.

The influence of de-vulcanizate content on the mechanical properties after re-vulcanization is depicted in Figure 8.33. Tensile strength and modulus at 300% strain decrease, while elongation at break and tear strength increase with higher de-vulcanizate concentrations. The decrease in tensile strength and increase in elongation at break with higher de-vulcanizate contents might be caused by a decrease in crosslink density, although different levels of filler reinforcement between the original building-profile compound and the roof-sheeting masterbatch may also play a role. In addition, an increasing ratio of mono- to di- and polysulfidic crosslinks with higher de-vulcanizate concentrations can also be held responsible for the reduction of tensile strength.[64] Another frequently reported observation is that de-vulcanizate particles act as fracture sites and thus lower the strength. Sample failure is initiated at the interfaces, simultaneously

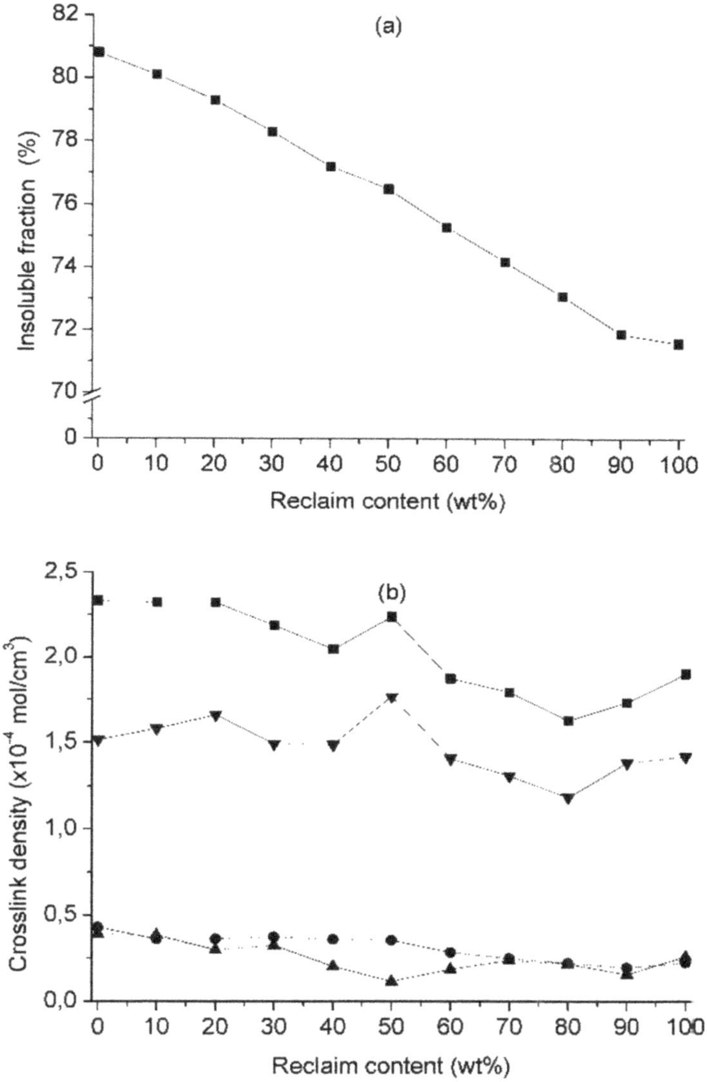

Figure 8.32 Influence of de-vulcanizate content on (a) insoluble fraction, and (b) crosslink densities of the revulcanizates; (■) overall; (●) monosulfidic; (▲) disulfidic and (▼) polysulfidic crosslink densities.[62]

at multiple points.[65] Still, another possible reason could be a plasticizing effect of the low molecular weight fragments contained in the de-vulcanizate. Under tear test conditions, the situation is different and propagation of cracks through the material is a critical step. The de-vulcanizate particles tend to stop or deflect the crack front, resulting in better tear strength values. All blends investigated in this study fulfill the UEAtc specifications,

Table 8.4 Specifications for EPDM roof-sheeting.

Property	UEAtc	EN 13956	ASTM D 4637-04
Tensile strength	≥ 8 MPa	≥ 6 MPa	≥ 9 MPa
Elongation at break	≥ 400%	≥ 300%	≥ 300%
Tear strength ISO 34 B with nick	≥ 20 kN m^{-1}	—	—

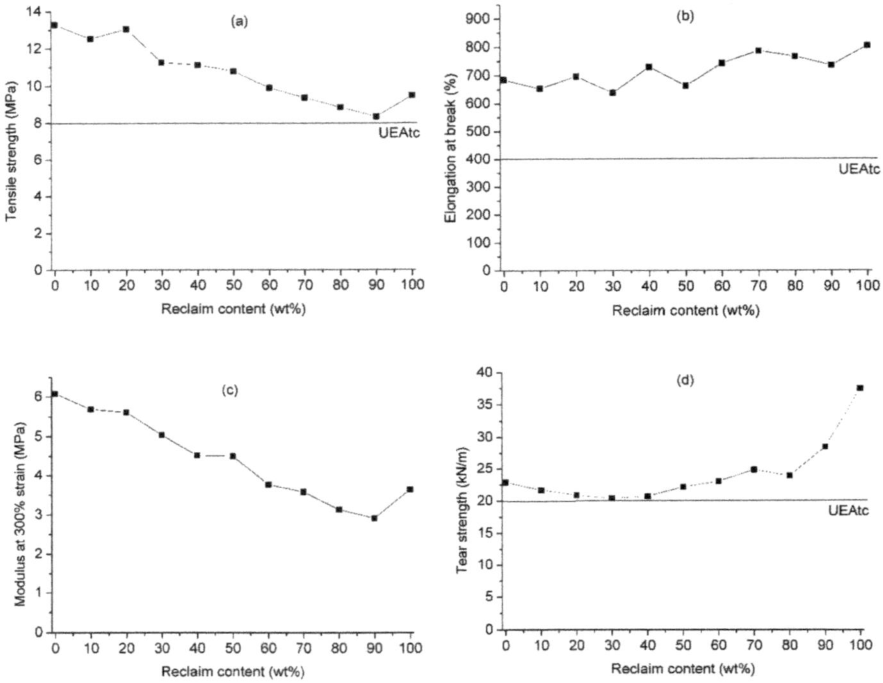

Figure 8.33 Influence of de-vulcanizate content on (a) tensile strength; (b) elongation at break; (c) modulus at 300% strain; (d) tear strength of the re-vulcanizates.[62]

with the proviso that the vulcanization in the present investigation was performed by compression molding for practical purposes, which is not a common procedure for EPDM roof-sheeting curing.

There is no general consensus about the temperature which should be taken to practically mimic the life-time of EPDM roof-sheeting, as the service temperatures spans from sub-zero temperatures during winter to temperatures as high as 80 °C in the summer. Life-time expectancy of compounds containing increasing amounts of the EPDM de-vulcanizates, to reach 250% elongation at break are shown in Table 8.5. Even at fulltime exposure of the roof-sheeting to a temperature of 80 °C, the material still shows acceptable life-times before reaching the threshold value of 250% elongation at break,

Table 8.5 Life-time expectancy of compounds containing increasing amounts of reclaim, to reach 250% elongation at break.

De-vulcanizate content (wt%)	Goodness of fit, R^2	Life-time expectancy in years at different temperatures					
		30 °C	40 °C	50 °C	60 °C	70 °C	80 °C
0	0.943	>100	53	17	6.1	2.3	0.9
10	0.930	>100	>100	39	13	4.4	1.6
20	0.927	>100	45	17	6.6	2.7	1.2
30	0.994	86.8	32	12	5.0	2.1	1.0
40	0.984	>100	79	26	9.1	3.4	1.3
50	0.933	64.9	3.1	1.7	1.0	0.6	0.4
60	0.984	>100	>100	>100	40	11	3.4
70	0.983	>100	96	30	10	3.7	1.4
80	0.994	>100	>100	91	26	7.7	2.5
90	0.996	>100	>100	39	13	4.5	1.7
100	← No failure →						

taking into account that such high temperatures only occurs on flat roofs during a limited period of time throughout the year. It has been stated though, that this aging method does not take into consideration the influence of exposure to all kinds of other weather conditions, such as (acid) rain, dirt, hail and excessive ozone attack. Only for the higher temperatures employed in this study, the threshold value of 250% in elongation is reached within the period of this study. If lower threshold values would have been chosen as practically realistic values, the times to reach a threshold value of for example 150% elongation at break would have required very long periods of aging, resulting in even significantly increased life-time expectancies of the compounds in roof-sheeting.

The life-time expectancy of virgin roof-sheeting compound with reclaim added is even extended relative to pure virgin compound. All blends perform well for very long periods of time before reaching the practical threshold value for elongation at break of 250%.

8.8 Concluding Remarks

The overview presented in this chapter has demonstrated that there are prospective ways to de-vulcanize various rubbers according to different de-vulcanization mechanisms. BR behaves more or less the same as SBR, in that they can be de-vulcanized but show chain recombination with increasing de-vulcanization temperatures. The de-vulcanization conditions optimized for SBR are also applicable for BR. For NR, the network is easily broken with a treatment temperature higher than 170 °C, but results in a combination of polymer and crosslink scission. NR turns out to be rather forgiving in terms of de-vulcanization conditions and can thus be treated in the same manner as SBR and BR. CIIR acts differently from the other types of rubber, network breakdown is enhanced by increasing the de-vulcanization temperature, it does not show recombination. However, at lower temperatures, the breakdown is not uniformly distributed throughout the particles; for a uniform breakdown higher temperatures are required. For EPDM-rubber, the balance between breakdown of the vulcanized network and the ability to be re-vulcanized afterwards is the main factor of consideration in order to be able to incorporate the recycled rubber into virgin rubber in appreciable amounts without losing too much of the properties.

The studies presented in this overview have demonstrated the perspectives of re-utilization of various de-vulcanized rubber types. Every type of rubber requires its own specific conditions to some extent, although sufficient commonalities were also found, allowing for de-vulcanization of blends of these rubbers in for example. sophisticated applications such as tires. However, for implementation in industrial practice, various additional factors will come into play to be considered for an economically feasible operation. Cleanliness and purity of the raw materials are important factors influencing final product properties. This also concerns regarding materials supply, which should be constant and of sufficient quantity. Extended studies are needed to scale the results up into an industrial operation and design continuous processes rather than batch processes. However, in the end, with all factors properly taken care of, it will result in applications in new products, which will then close the loop from waste rubber into valuable new applications.

References

1. J. A. Beckman, G. Crane, E. I. Kay and J. R. Laman, *Rubber Chem. Technol.*, 1974, **47**, 597.
2. M. Myhre and D. A. MacKillop, *Rubber Chem. Technol.*, 2002, **75**, 429.
3. M. Myhre, S. Saiwari, W. K. Dierkes and J. W. M. Noordermeer, *Rubber Chem. Technol.*, 2012, **85**, 408.
4. V. G. Gomes, *presented at the India Rubber Expo, IRE09*, January 2009, Kolkata, India.
5. H. Schnecko, *Kautsch. Gummi Kunstst.*, 1994, **47**, 885.
6. B. Adhikari, D. De and S. Maiti, *Prog. Polym. Sci.*, 2000, **25**, 909.
7. M. A. L. Verbruggen, PhD thesis, University of Twente, Enschede, The Netherlands, 2007.
8. H. Okamoto, *Int. Polym. Sci. Technol.*, 1980, **7**, T/59.
9. G. K. Jana and C. K. Das, *Polym. Plast. Technol. Eng.*, 2005, **44**, 1399.
10. G. K. Jana and C. K. Das, *Prog. Rubber Plast. Recycl. Technol.*, 2005, **21**(4), 319.
11. K. A. J. Dijkhuis, J. W. M. Noordermeer and W. K. Dierkes, *presented at a Meeting of the Rubber Division*, ACS, October 2008, Louisville, USA.
12. A. R. Jalilvand, I. Ghasemi, M. Karrabi and H. Azizi, *Prog. Rubber Plast. Recycl. Technol.*, 2008, **24**, 33.

13. M. A. L. Verbruggen, L. van der Does, J. W. M. Noordermeer, M. van Duin and H. J. Manuel, *Rubber Chem. Technol.*, 1999, **72**, 731.
14. V. V. Rajan, W. K. Dierkes, J. W. M. Noordermeer and R. Joseph, *Rubber Chem. Technol.*, 2005, **78**, 572.
15. A. Z. Kuzmiski, *Developments in Polymer Stabilization*, ed. G. Scott, Applied Science Publishers, London, 1981.
16. D. De, A. Das, D. De, B. Dey, S. C. Debnath and B. C. Roy, *Eur. Polym. J.*, 2006, **42**, 917.
17. D. De and D. De, *Mater. Sci. Appl.*, 2011, **2**, 486.
18. D. De, D. De and G. M. Singharoy, *Polym. Eng. Sci.*, 2007, **47**(7), 1091.
19. A. Charlesby, *Proc. R. Soc. London, Ser. A*, 1954, **222**, 60.
20. A. Charlesby, *J. Polym. Sci.*, 1953, **11**, 513.
21. M. M. Horikx, *J. Polym. Sci.*, 1956, **19**, 445.
22. M. Gordon, *Proc. R. Soc. London, Ser. A*, 1962, **268**, 240.
23. G. R. Dobson and M. Gordon, *J. Chem. Phys.*, 1964, **41**(8), 2389.
24. G. R. Dobson and M. Gordon, *J. Chem. Phys.*, 1965, **43**(2), 705.
25. P. J. Flory, *J. Am. Chem. Soc.*, 1941, **63**, 3096.
26. P. J. Flory and J. Rehner Jr., *J. Chem. Phys.*, 1943, **11**(11), 521.
27. V. V. Yashin and A. I. Isayev, *Rubber Chem. Technol.*, 2000, **73**, 325.
28. A. V. Tobolsky, *J. Appl. Phys.*, 1956, **27**, 673.
29. A. V. Tobolsky, D. J. Metz and R. B. Mesrobian, *J. Am. Chem. Soc.*, 1956, **72**, 1942.
30. K. Ono, *Angew. Makromol. Chem.*, 1997, **252**, 167.
31. F. Bjork, O. Dickman and B. Stenberg, *Rubber Chem. Technol.*, 1989, **62**, 387.
32. L. Mullins and A. G. Thomas, *J. Polym. Sci.*, 1960, **43**, 13.
33. L. Mullins and D. T. Turner, *J. Polym. Sci.*, 1960, **43**, 35.
34. R. N. Hader and D. S. le Beau, *Ind. Eng. Chem.*, 1951, **43**(2), 250.
35. S. S. Saiwari, PhD thesis, University of Twente, Enschede, The Netherlands, 2013.
36. S. Saiwari, W. K. Dierkes and J. W. M. Noordermeer, *Rubber Chem. Technol.*, 2014, **87**, 31.
37. S. Yamashita, *Int. Polym. Sci. Technol.*, 1981, **8**, T/77.
38. K. A. J. Dijkhuis, I. Babu, J. S. Lopullissa, J. W. M. Noordermeer and W. K. Dierkes, *Rubber Chem. Technol.*, 2008, **81**, 190.
39. S. L. Madorsky, "Thermal degradation of organic polymers", *Polymer Reviews*, ed. H. F. Mark and E. H. Immergut, Interscience Publishers, New York, U.S., 1964.
40. M. D. Sarkar, P. G. Mukunda, P. P. De and A. K. Bhowmick, *Rubber Chem. Technol.*, 1997, **70**, 855.
41. M. A. Golub, *Rubber Chem. Technol.*, 1978, **51**, 677.
42. M. A. Golub, *J. Polym. Sci., Polym. Lett. Ed.*, 1978, **16**, 253.
43. J. R. Shelton, *Rubber Chem. Technol.*, 1972, **45**, 359.
44. https://www.specialchem4adhesives.com (April 2013).
45. A. G. Ferradino, *Rubber Chem. Technol.*, 2003, **76**, 694.

46. N. M. Huntink, PhD thesis, University of Twente, Enschede, the Netherlands(2003).
47. H. Zweifel, "*Stabilization of Polymeric Materials*", Springer-Verlag, Berlin, Germany, 1998.
48. W. L. Hawkins, "*Polymer Degradation and Stabilization*", Springer-Verlag, Berlin, Germany, 1984.
49. H. H. G. Jellinek, "*Degradation and Stabilization of Polymers*", Elsevier, Amsterdam, the Netherlands, 1983.
50. W. Hofmann, "*Rubber Technology Handbook*", Hanser/Gardner Publications, Cincinnati, U.S., 1994.
51. G. Scott, "*Development in Polymer Stabilization-4*", Applied Science Publishers, London, U.K., 1981.
52. V. V. Rajan, PhD thesis, University of Twente, Enschede, The Netherlands(2005).
53. V. Yu Levin, S. H. Kim and A. I. Isayev, *Rubber Chem. Technol.*, 1997, **70**, 120.
54. J. Choi and A. I. Isayev, *Rubber Chem. Technol.*, 2011, **84**, 55.
55. A. I. Isayev, S. H. Kim and W. Feng, *Rubber Chem. Technol.*, 2005, **78**, 606.
56. J. S. Oh, A. I. Isayev and E. von Meerwall, *Rubber Chem. Technol.*, 2004, **77**, 745.
57. V. V. Rajan, W. K. Dierkes, J. W. M. Noordermeer and R. Joseph, *Rubber Chem. Technol.*, 2005, **78**, 855.
58. V. V. Rajan, W. K. Dierkes, R. Joseph and J. W. M. Noordermeer, *J. Appl. Polym. Sci.*, 2006, **102**, 4194.
59. A. R. Jalilvand, I. Ghasemi, M. Karrabi and H. Azizi, *Progr. Rubber Plast. Recycl. Technol.*, 2008, **24**(1), 33.
60. G. Heideman, PhD thesis, University of Twente, Enschede, The Netherlands(2004).
61. M. van Duin, J. W. M. Noordermeer, M. A. L. Verbruggen and L. van der Does, (to DSM) U.S. 6956065, 2005.
62. K. A. J. Dijkhuis, PhD thesis, University of Twente, Enschede, The Netherlands, 2008.
63. Union Européenne pour l'Agrément technique dans la construction, 1988.
64. A. I. Isayev, *Rubber Chem. Technol.*, 2003, **76**, 253.
65. T. D. Sreeja and S. K. N. Kutty, *Polym.-Plast. Technol. Eng.*, 2000, **39**(3), 501.

CHAPTER 9

Recycling of Latex Waste and Latex Products

A. R. AZURA* AND D. N. SYUHADA

School of Materials & Mineral Resources Engineering, Universiti Sains Malaysia, Engineering Campus, 14300, Nibong Tebal, Pulau Pinang, Malaysia
*E-mail: srazura@usm.my

9.1 Introduction

Latex is the starting material for a wide range of products. Latex concentrates are mixed with various compounding chemicals before being processed into products. Natural rubber (NR) latex is used extensively in the production of gloves, condoms, balloons, catheters, baby soothers, and dental dams.[1] The latex production method involves dipping formers for dipped products, foaming into foam or sponge, or extrusion into latex thread. Other than these products, latex can also be used in elastomeric thread, foam rubber, binder, adhesives, and many miscellaneous applications. Dipped products are made by dipping the formers into compound latex to enable deposition of a thin film of latex. In the glove industry, the common types of latex that being used include natural rubber (NR) latex, nitrile butadiene rubber (NBR) latex and chloroprene rubber (CR) latex. NR

Green Chemistry Series No. 59
Rubber Recycling: Challenges and Developments
Edited by Jin Kuk Kim, Prosenjit Saha, Sabu Thomas, Józef T. Haponiuk and M. K. Aswathi

Published by the Royal Society of Chemistry, www.rsc.org

latex is normally used to produce household gloves, while NBR and CR are used to produce medical or surgical gloves as the issue of natural rubber can caused allergies in consumers.

Latex foam is a suitable base material for cushioning in comfort products, such as mattresses, pillows, furniture, cushions, toppings over springs in automobile seats and rug padding. Initially the latex foam industry was developed by using NR latex. However, due to better improvement and price of synthetic latex, blends of styrene butadiene rubber (SBR) with NR latex foams are produced. Another alternative for NR latex is synthetic flexible polyurethane (PU) foam.[2] Foam products are usually made by incorporation of air into the compounded latex concentrate. The gelling agent is added to ensure the foamed latex is allowed to gel inside the mould.

While synthetic latex based on SBR or carboxylated styrene butadiene terpolymer (CSBR) has a major use in carpet backing and non woven fabric production. Lattices based on polychloroprene (CR) and NBR are used for specialist dipping product manufacture in which resistance to weather and oil is required. The competitive situation differs for polymers in latex form and products of dry rubber. Hence, excellent wet gel strength of NR latex makes it the preferred or only material for dipping and extrusion. However, the high mechanical stability of synthetic rubber lattices, especially SBR and CSBR, makes them more suitable for high speed carpet and fabric backing operations.[3]

9.2 Latex Waste

The high demand for latex goods and products give rise to latex waste accumulation worldwide. Latex waste can be divided into two categories, which are solid waste and liquid waste. Liquid latex waste includes expired raw latex and compounded waste latex. The compounded waste latex is defined as the liquid latex form that has been compounded according to a certain formulation, already used many times and also that obtained from the dipping line, especially after the dipping process has already stopped. It also includes the remaining used compounded latex in the production line which has exceeded the storage time, or due to improper care during the compounding and curing process. Hence, the resulting latex compound is not suitable to be used in the production of latex products as it would result in products with poor properties.[4] Expired raw latex is defined as unused raw latex which has already exceeded the storage time or does not fulfil the requirement to produce latex products in factories. Other than that, liquid latex waste is also obtained from latex paint industries.

Solid waste can be divided into two categories, which are rejected products and latex waste products from consumers. The rejected products come from the large number of defects encountered in articles produced by latex dipping, and are due to pinholes, craters and blisters, that are caused by air bubbles entrapped in the latex compound.[5] Due to the strict specifications for latex products, as much as 15% of the products can be rejected and these

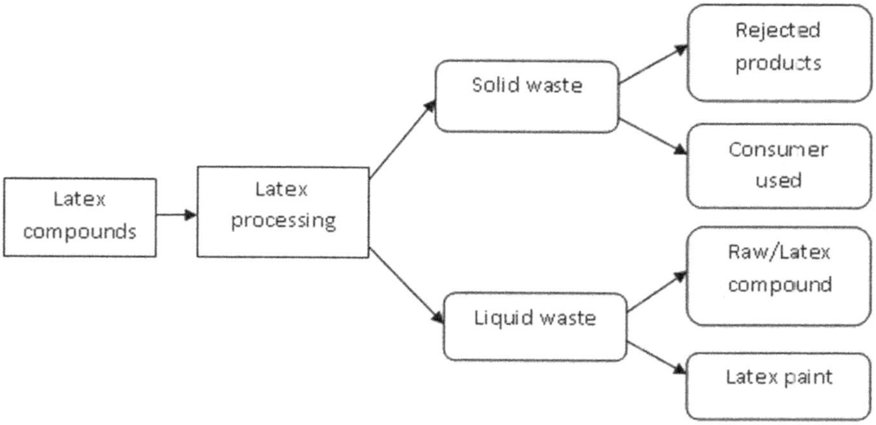

Figure 9.1 The categories of latex waste.

becomes rejected latex products from industry.[6] The expanding market for manufactured latex products such as gloves and prophylactics[7] therefore increases the solid latex waste. In these industries, a high percentage of latex waste products is due to the requirements to ensure high quality products are produced, and due to the unstable nature of latex. The used latex products from consumers, in particular gloves from household utilities, are also defined as scrap latex products.[8] Figure 9.1 elaborates on the categories of latex waste.

According to ISO 14 000, hazardous waste is a material that is corrosive, poisonous or harmful in any other way. It may exist in either solid, gas, liquid or slurry form. The disposal of untreated hazardous waste in landfill sites is normally prohibited. The waste needs to be treated before it proceeds through the disposal process.[9] Moreover, the authorities also have a restriction that liquid latex waste cannot be drained off to the environment without purification. Latex waste needs to be coagulated with acid and neutralized before it is drained off. The coagulants are needed to wash out the acid and to properly dry the products before disposal as part of the requirements from the authorities. Due to these reasons, production factories face a major challenge to manage the economic and environmental issues for latex waste treatment.

9.3 Recycling of Liquid Latex Waste

9.3.1 Laminated Mould Cleaning

Mould cleaning is the process of the removal of built-up deposited or residue material on the mould surface. It is an important process in product manufacturing industries to obtain a good quality of finished product. The cleaning frequency depends on the application of the components being manufactured, the types of compound being processed and the temperature of vulcanization.[10] A lot of cleaning techniques are used, such as manual scouring with steel wool and abrasives, wire brushing, grit blasting, vapor blasting and chemical cleaning in salt, acid or alkali baths.[11] The first two techniques are suitable to be used for simple moulds and the rest can be applied to any size or design of the mould.

Another technique for mould cleaning is through the use of a uncured rubber compound containing amine generating chemicals, which are available to strip the compound and release agent residues during curing of the rubber compound. The mould is either totally filled by the cleaning rubber compound or in the case of larger moulds such as tire moulds, with a veneer over a conventional rubber stock. The dirt is then removed after the rubber cleaning compound has cured, which becomes bonded into the rubber surface during the vulcanization cycle. By using these cleaning rubbers, the downtime for the mould is minimal as the mould does not have to be removed from the machine.[12] However, the use of uncured rubber compound has a limitation which depends on the operator's skill to obtain a homogenous distribution of compounding ingredients and cleaning agents. The usage of a high amount of cleaning agent has also become one of the limitations of the uncured rubber compound which is not economic.

Recently, a new substitution for the mould cleaning process has been found, which is utilization of a laminated NR latex film.[13] NR latex films were prepared as a base material for the mould cleaning compound. The NR latex films were laminated to increase the thickness as a thin film. If the NR latex film is used without lamination, it will cause a problem during the mould cleaning process as the excess of the film may remain on the mould, which creates an additional cleaning process. In order to solve the thin NR latex films and utilize the waste NR latex, the lamination of NR film and NR waste film was carried out.

NR latex compound with mould cleaning agent and NR latex waste were cast onto a glass plate. The NR latex sheet obtained from casting was then laminated using latex waste film as the core layer and compressed using compression moulding to obtain the latex composite. A fresh layer of NR latex film with cleaning agent monoethanolamine (MEA) is used as the outer layer of the laminated latex film while the waste natural rubber (WNR) latex is used for the core layer, as shown in Figure 9.2, for the arrangement of the laminated film before compression moulding. The usage of cleaning agent is only limited to the outer layer of the thin NR latex films, which reduced the

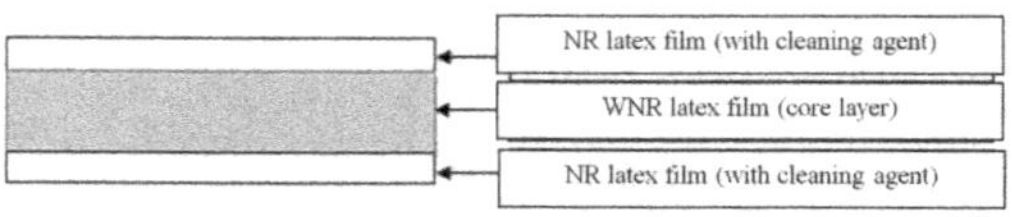

Figure 9.2 The arrangement of laminated NR latex films.

amount of cleaning agent usage compared to the uncured dry rubber compound. It is a green and environmental friendly mould cleaning compound because the NR latex waste from industry can be used as a core layer and the usage of the mould cleaning agent is minimized as it is only used in the NR latex film (outer layer). The effects of different cleaning agents and different mould temperatures on the tensile properties of the NR latex film were investigated in order to obtain good cleaning effectiveness of the mould cleaning compounds.

Two types of cleaning agents were used, MEA and diethylene glycol. The effect of the cleaning agent on tensile strength and tear strength of the NR latex film is shown in Figure 9.3.[13] From the results, the MEA cleaning agent gives a better tensile strength, which can be related to a compatible cleaning agent to be used in NR latex. The tear strength of the NR latex film also gives a similar trend to tensile strength, where a MEA cleaning agent provides a high tear strength compared to a control sample and diethylene glycol (DEG) cleaning agent.

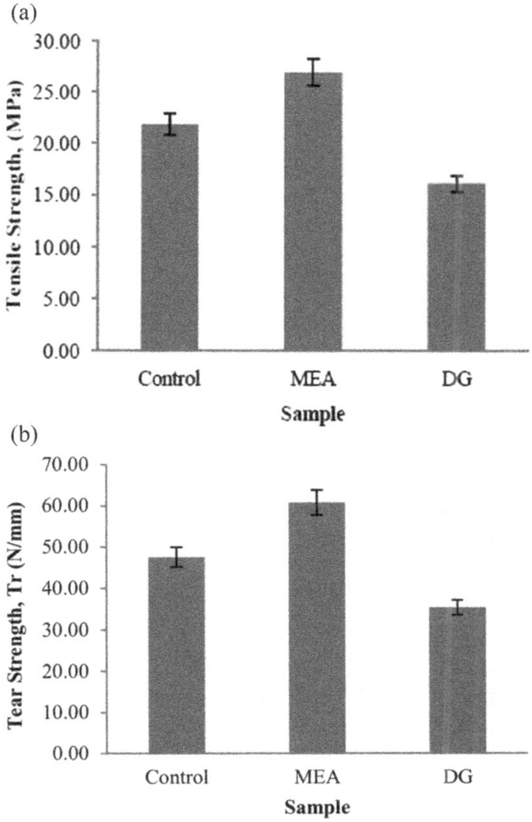

Figure 9.3 (a) Tensile strength and (b) tear strength for different cleaning agents in NR latex film.[13]

Table 9.1 Effect of moulding temperature on tensile strength of the laminated film with different types of core layer.[13]

	Tensile strength, MPa	
Temperature	With filler	Without filler
120 °C	9.0	8.4
140 °C	9.4	10.2
160 °C	10.4	12.2

In order to understand the effect of the moulding temperature on the properties of laminated films, different moulding temperatures at similar moulding times are carried out. Table 9.1 presents the relationship between the moulding temperature and tensile strength of the laminated cleaning compound with different types of core layer, one without filler and the other one with filler.[13] From the results it can be seen that the tensile strength of the laminated compounds increased as the moulding temperature increased. Although the increase in temperature caused degradation to the rubber sheet, which decreased the tensile strength, but as the films were produced as lamination films, stress transfer occurs which contributes to an increase in the tensile strength. The moulding process at 160 °C gives better strength to laminate films which have a similar temperature to that normally run during the mould cleaning process.

The cleaning effectiveness test was performed for laminated NR latex and laminated NBR latex films. This test does not have any specific guidelines due to the variation of dirt for all cleaning parts as we used a different existing available mould in the laboratory. The test is carried out to check the effectiveness of the laminated films as a mould cleaning compound. The mould cleaning process is carried out at the same time as the compressin moulding process. The effectiveness of the mould cleaning was assessed after the moulding process was finished by visual inspection of the dirt on the laminated films. A fresh layer of the NR latex films with the cleaning agent MEA is used as the outer layer of the laminated latex film, while WNR latex is used for the core layer. The arrangement of the laminated latex film is shown in Figure 9.4.

After compression moulding, the features of the laminated latex film can be observed as shown in Figure 9.5. The flow ability of the fresh NR latex is higher than WNR latex. Hence, when subjected to compression moulding, the NR latex film is squeezed out and flows to occupy the mould space, while the WNR latex film did not flow as well as the fresh NR latex (outer layer) due to the waste latex already being cured and the cross linking chain restricts the waste NR film from being squeezed out to occupy the mould space. The better surface compatibility between fresh NR latex film and WNR latex film is due to the similar rubber used, but also depends on the temperature of the moulding. The fresh NR latex film

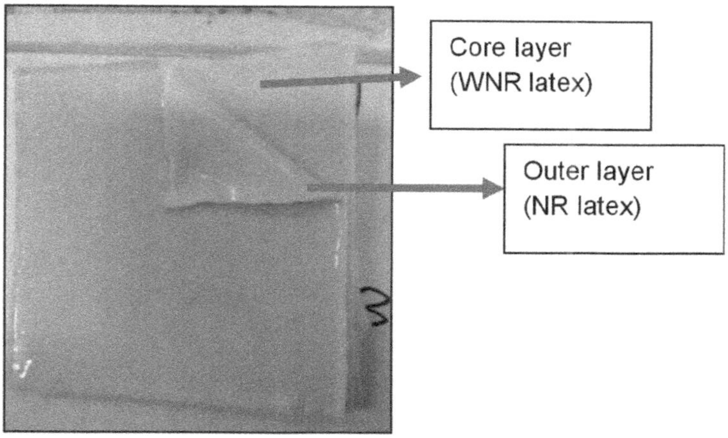

Figure 9.4 NR latex laminated film before compression moulding with WNR latex film as the core layer.

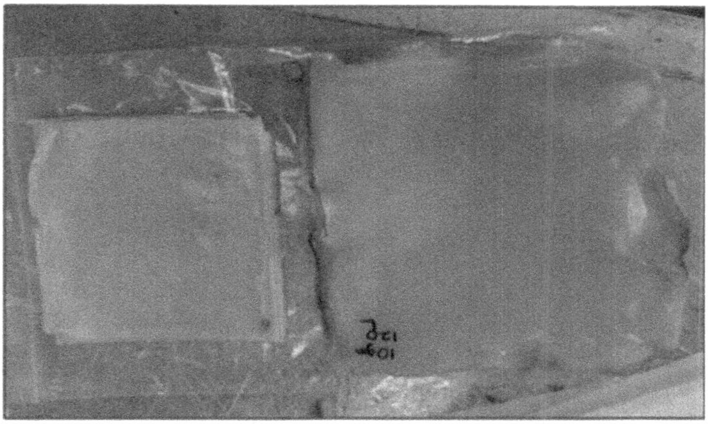

Figure 9.5 NR Latex laminated film before and after compression moulding with a NR latex sheet as the core layer.

can be used in WNR latex film to form a laminated latex film. The surface compatibility is determined from the tensile test. A good surface compatibility is observed from good tensile strength without delamination occuring in the films.

Figures 9.6–9.8 show the effectiveness of mould cleaning of the NR latex laminated film. L0 in Figure 9.6 is a laminated latex film with an outer layer of latex sheet without cleaning agent (MEA). From observation, the mould cleaning ability can be said to be zero when compared to the laminated film with cleaning agent. This shows that the cleaning agent plays an important role

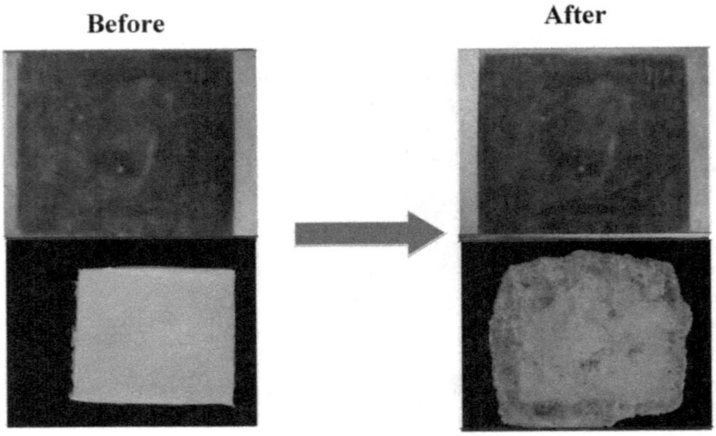

Figure 9.6 Mould and L0 NR latex films (without cleaning agent) before and after compression moulding.

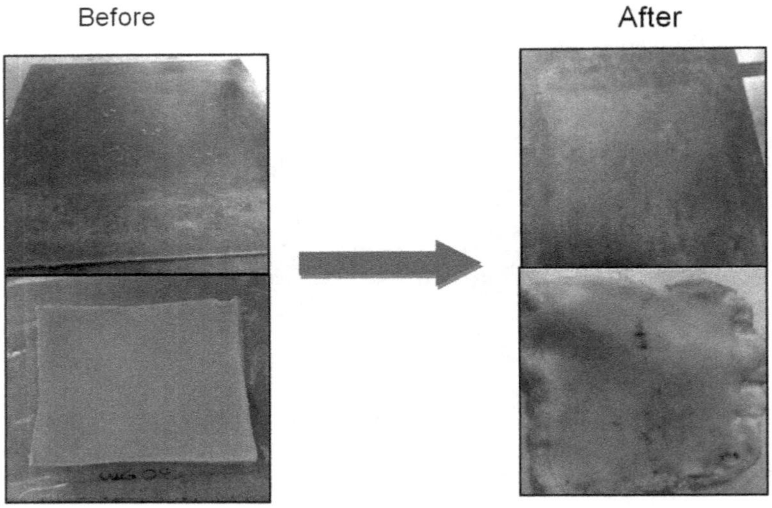

Figure 9.7 Mould and LN1 before and after compression moulding.

in the mould cleaning process. From Figures 9.7 and 9.8 it can be observed that the effectiveness of the mould cleaning ability is reduced from LN1 and LN5. This is influenced by the decrement of the thickness of the fresh NR latex film (outer layer) which contains the cleaning agent. LN1 consists of only fresh NR latex, which contributes to the good flow ability of the films to clean the mould and the content of the mould cleaning agent is also high in this sample because there is no waste latex used in this sample. Sample LN5 consists of a small amount of fresh NR latex used as an outer layer and high waste NR latex as the core layer. This caused the flow ability of the sample to

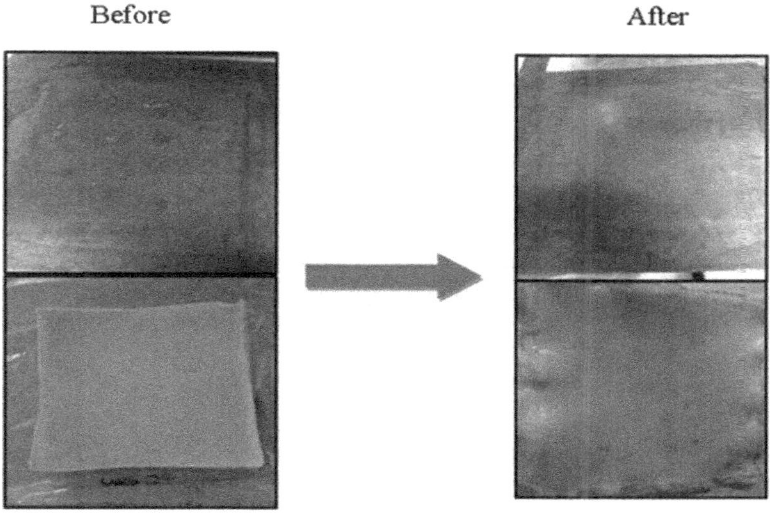

Figure 9.8 Mould and LN5 before and after compression moulding.

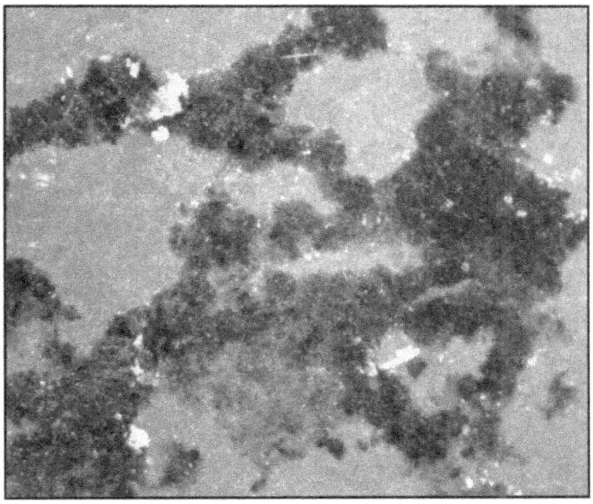

Figure 9.9 40x magnified surface of NR latex laminated film after mould cleaning with an optical microscope.

not be as good as the LN1 sample and the content of the cleaning agent in the sample is also low, thus the effectiveness of mould cleaning is reduced from LN1 compared to LN5.[14]

An optical microscope was used to magnify the surface of the latex laminated films after the mould cleaning process. The dirt on the surface of the latex laminated films is shown in Figure 9.9 and can be found in all samples with a different distribution of the area of the dirt area. The effectiveness of the mould cleaning is slightly influenced by the thickness of the laminated film, but is actually independent of the thickness of the latex layer, as long as the layer is added with the cleaning agent.[15] Therefore, it can be concluded that the thickness of the fresh latex films is not directly affected by the effectiveness of the mould cleaning ability. The flow and distribution of the cleaning agent in the fresh latex is the main reason for the effectiveness of the mould cleaning process, which is indirectly influenced by the thickness of the fresh latex layer. The usage of fresh latex is important because it can flow easier than waste latex due to the ability to cover the surface of the laminated films and because of greater surface contact with the mould. Hence, more of the mould can be cleaned by the laminated latex films.

9.3.2 Outdoor Cleaning

The main focus of this project was to prepare a NR latex cleaning compound for outdoor application using waste NR latex. The utilization of waste NR latex is expected to resolve the disposal problem of liquid waste NR latex. The outdoor cleaning compound is produced using waste NR latex as a cleaning compound which indirectly converts the liquid waste into solid waste after the application. The solid waste is easier to treat and dispose of compared to disposal of liquid latex waste.

For outdoor cleaning applications, three different types of latex are used: (1) high ammonia (HA) latex compounds/fresh compound; (2) subnormal latex/pre-vulcanized latex which has already expired; and (3)waste latex compound supplied from a latex factory. Table 9.2 shows the tensile strength of the three different cleaning latex compounds with the addition of different loading of MEA and DEG cleaning agents. Generally, increased addition of MEA cleaning agent loading caused a reduction in the tensile strength and increments of elongation at break of the latex compound. This is due to a function of MEA cleaning agents, which also function as diluting agents. Hence, the excess additions of MEA cause destabilising of the latex compound which will affect the product formation. The addition of MEA loading increases the stability of latex and if

Table 9.2 Tensile strength of latex cleaning compounds with the addition of MEA and DEG.

	Tensile strength, MPa						
	Control sample	MEA Loading, phr			DEG loading, phr		
		1	3	5	1	3	5
HA latex	24.9	26.2	27.2	20.6	22.6	20.4	19.7
Latex waste	19.0	17.0	18.6	18.2	15	13.2	13.0
Sub normal latex	17.6	14.0	13.7	12.4	15.2	15.7	14.4

Table 9.3 Tear strength of latex cleaning compound with the addition of MEA and DEG.

	Tear strength, N mm^{-1}						
	Control sample	MEA loading, phr			DEG loading, phr		
		1	3	5	1	3	5
HA latex	48.8	45.3	49.1	40.8	46.9	41.7	49.5
Latex waste	65.0	53.0	53.4	47.2	63.4	55.7	51.4
Subnormal latex	23.6	21.6	19.0	18.6	18.6	18.1	17.5

the latex can be shaped into a product, it will result in a softer product with lower strength.

For DEG cleaning agent, the addition of DEG showed a decrease in the tensile strength of all the NR latex cleaning compounds. Higher loading of DEG, resulted in a lower tensile strength of the NR latex film. As observed in Table 9.2, NR latex compounds without DEG have the highest tensile strength compared to DEG added NR latex compounds. DEG can be used as antifreeze and de-icing fluid, similar to MEA, excess addition of DEG will cause the problem of destabilising the latex and will weaken the bonding between the cleaning agent and latex particles, thus resulting in a film with low tensile strength.[13]

Table 9.3 shows the tear strength of the NR latex compounding. Overall, the tear strength of NR latex cleaning compounds with MEA and DEG have a similar trend. The addition of MEA and DEG do not appear to make much difference to the tear strength of NR latex cleaning compounds. Nevertheless, it can be noted that as the loading of the cleaning agent increased, the tear strength of the NR latex cleaning compound decreased. This is because excess cleaning agent in the NR latex compound will work as a plasticizer in latex and weaken the bonding between the cleaning agent and rubber particles,[16] thus resulting in low tear strength.

From Table 9.3, it was observed that waste NR latex cleaning compound either with added MEA or DEG has a higher tear strength compared to the control compound. The waste NR latex compound was received as a readily compounded NR latex from the factory with unknown compounding ingredients. It is believed that the waste NR latex compound has had a filler added as one of the compounding ingredients. The presence of filler particles helps to deviate the crack propagation and therefore, the filled film exhibited knotty tearing behaviour.[17] The unfilled films are easy to tear as there was no filler to divert the crack, thus the crack propagated straight across the tear specimen.

Table 9.4 shows the cleaning effectiveness of the NR latex cleaning compound for HA, NR waste latex and subnormal latex. The result shows that all the NR latex cleaning compounds were able to attract dirt from the dirty surfaces. Waste NR latex cleaning compound showed the best cleaning efficiency

Table 9.4 Cleaning effectiveness of NR latex cleaning compounds on outdoor surfaces.

	BEFORE	AFTER	
		Surface	Film
HA latex compound/3 layers			
Waste latex compound/3 layers			

Subnormal latex compound/3 layers			

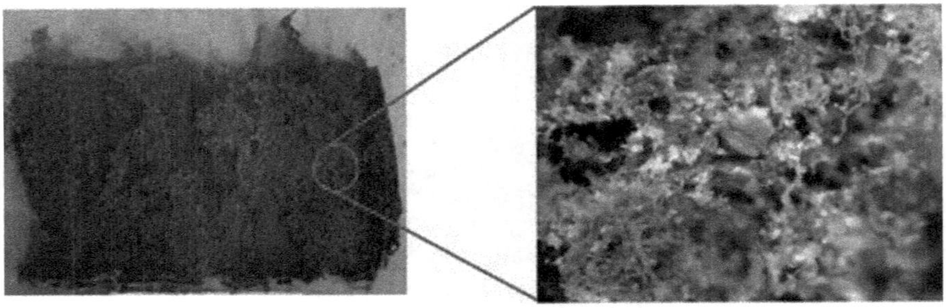

Figure 9.10 Dirt on the surface of waste NR latex cleaning compound film under 40x magnification.

compared to HA and the subnormal NR latex cleaning compound. This is due to the presence of filler in the waste latex, which in the latex industries the filler normally used is calcium carbonate. Calcium carbonate at certain loadings has a tendency to lower the surface tension between the adhesive and the substrate leading to an increase in the wettability of the adhesive on the substrate.[18] Good wettability enhances mechanical interlocking and anchorage of the adhesive in pores and irregularities in the adherent. Thus, the presence of calcium carbonate in waste latex increases the adhesion of waste latex cleaning compound on the dirt and hence contributes to the excellent cleaning efficiency of the waste NR latex cleaning compound. Figure 9.10 shows the ability of NR latex compounds to remove the dirt from outdoor dirty surfaces.

Other than that, the NR latex cleaning compounds were observed to clean the dirty surface better with a higher loading of MEA and DEG. An i ncrease in the layers of the NR latex cleaning compound applied onto the dirty surfaces was also found to improve the cleaning of the surfaces, as more NR latex cleaning compound was used to cover the dirty surfaces. The waste NR latex cleaning compound with added MEA or DEG performed the best in cleaning of the dirty surfaces. This is probably due to the presence of filler in the waste NR latex cleaning compound which increased its ability to attract more dirt. Hence, it can be concluded that the waste latex compound from the latex industry has the potential to be reused as a latex cleaning compound.

9.3.3 Former Cleaning

Proper cleaning of formers is an important factor in the successful production of products through the latex dipping process. During the dipping process, it is essential that the former surface should be uniformly wetted by any coacervant solution which is used, as well as by the latex, the latter requirement is especially important in single dipping processes.[5] The presence of grease or impurities on the former surface must be avoided because it can lead to thinning or the complete absence of deposit on the products. It is also important that particles of dust and dirt be excluded from the former, because this will promote the formation of pinholes and local stress concentration and hence result in rejected products. In general, the thinner the produced products, the greater the importance of former cleanliness during the glove production process.

This project aims to re-use the latex waste compound from the industry with the addition of cleaning agent as a NR latex cleaning compound. The NR latex cleaning compound is intended to be used for an industrial cleaning purpose, such as cleaning the formers used in the production of latex dipped articles. It works as an alternative to replacing the former cleaning process in which acid and alkaline solutions are usually used in industry. This common practice of cleaning and processing in the latex industries can cause corrosion/porosity to occur and result in the former becoming worn out because of the acid used. Corrosion/porosity can cause microcracks, flaws, and voids inside metal and porcelain which can result in stress concentration. In addition, corrosion/porosity of the former the the latex glove industries can mean the former needs to be changed every 6 month, which most of the latex glove industries try to avoid.[19]

The formers may be dusted to prevent adhesion to itself and other formers after removing products from the former. First of all, the weight of dust is measured on the former surface before and after different repeated cycles in the dipping process. Table 9.5 shows the weight of dust before and after the dipping cycle. From the results, it can be seen that the weight of dust increases for every dipping cycle. The weight of dust on the former affected the surface/porosity of the former when immersed in the latex tank and the NR latex cannot fully coat the surface of the former. The formation of dust in the long run can reduce the properties of the final product, which causes defects like pinholes and uneven thickness of the final product.

After the weight of each former used in the dipping process is measured, the former is subjected to a cleaning process prior to continuing dipping cycles. For this purpose, the former was subjected to different methods of former cleaning processes which included the acid and alkali method, HA latex/fresh latex compound, industry latex waste and control latex waste with the addition of a cleaning agent. Table 9.6 shows the percentage dust weight

Table 9.5 The weight of dust before and after different cycles in the dipping process.

	Before dipping process (g)			After dipping process (g)		
Methods Cycle	10	20	30	10	20	30
Acid + alkali	0.03	0.01	0.06	0.26	0.19	0.12
HA + MEA	0.09	0.03	0.02	0.25	0.18	0.13
Industry latex waste + MEA	0.07	0.07	0.04	0.26	0.16	0.09
Control latex waste	0.02	0.03	0.05	0.28	0.19	0.12

Table 9.6 Percentage of dust weight loss of different cleaning methods.

	Percentage dust weight loss/Dipping cycles		
Types of samples	10	20	30
Acid + alkali	93%	86%	56%
HA NR latex + MEA	100%	100%	77%
Industry latex waste + MEA	96%	100%	89%
Control latex waste	92%	82%	67%

Figure 9.11 Microstructure of dust stucks on the surface of the latex film cleaning compound (control latex waste without MEA) after repeated dipping cycles.

loss after the former was subjected to different cleaning methods. From the results, the most effective method to clean the former without damaging/corroding the former is by using the NR latex cleaning compound which the sequences follow by HA latex/fresh latex, industrial latex waste and control waste latex. The percentage of weight loss is better compared to the acid–alkali method which indicated its suitability to be used as a substitute to the conventional process using acid and alkali. It was found that the addition of MEA increases the effectiveness of the cleaning method. Hence, latex waste compound can be re-used in former cleaning, which is cost effective and does not damage the former.

The deposition of dust on the latex film surface of the NR latex cleaning compound based on fresh HA latex, industry latex waste and control latex waste was investigated based on polarized optical microscope (POM) analysis. The results obtained showed the ability of the NR latex as a cleaning compound to remove the deposit from the former surfaces. Figure 9.11 illustrates the microstructure of dust sticking to the surface latex film cleaning compound made from control latex waste after repeated cycles of the dipping process. Actually, the dust on the surface of the latex cleaning compound films is calcium carbonate which had been used to aid the stripping of NR latex films from the former. The particle size of calcium carbonate is <30 μm. Figures 9.12 and 9.13 show that the particle size of calcium carbonate is less than 30 μm.

Figure 9.12 Microstructure of dust stuck on the surface of the latex film cleaning compound (industry latex waste with MEA) after several dipping cycles.

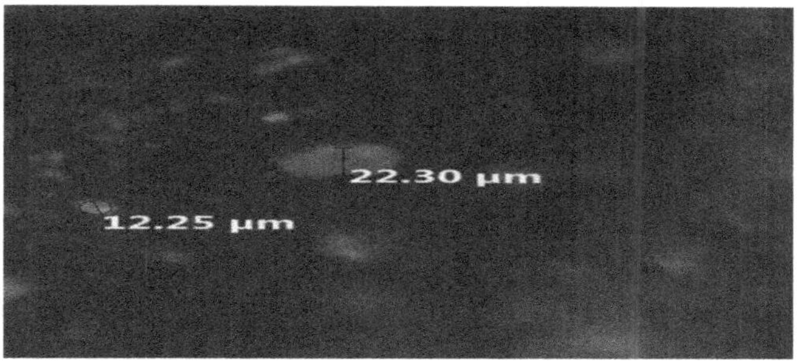

Figure 9.13 Calcium carbonate stuck on the surface of the NR latex film cleaning compound after several dipping cycles using fresh HA latex with a cleaning agent (MEA).

9.3.4 Blending of Waste NR Latex

The increasing consumption of NR latex caused the waste rubber formed in latex-based industries to reach 15% of consumption. Hence, an attempt was made to blend waste natural rubber latex (WNRL) with polystyrene (PS) in order to utilize and add value to the waste. The blends were prepared by using a solution and melt mixing technique. According to the work by Boondamneon,[20] solution mixing for WNRL/PS does not create a continuing film unless WNRL is treated with NR latex. The blending is done by using 5% (w/v) of chloroform solution.

From the results, the 70/30 TWL/PS (treated latex waste/PS) blend gave the highest mechanical properties compared to other composition and NRL/PS blends. The introduction of 15 phr of styrene graft natural rubber (SNR) as a compatibilizer proved the best mechanical and morphological properties, as shown in Figure 9.14. However, there is a high probability of phase separation in solution blending, hence melt-mixing was carried out. Based on the study using melt mixing with a Haake internal mixer, WL/PS showed the highest tensile strength and thermal stability, comparable to the treated WL/PS. When comparing the WNRL/PS blend and NR/PS blend, the results show

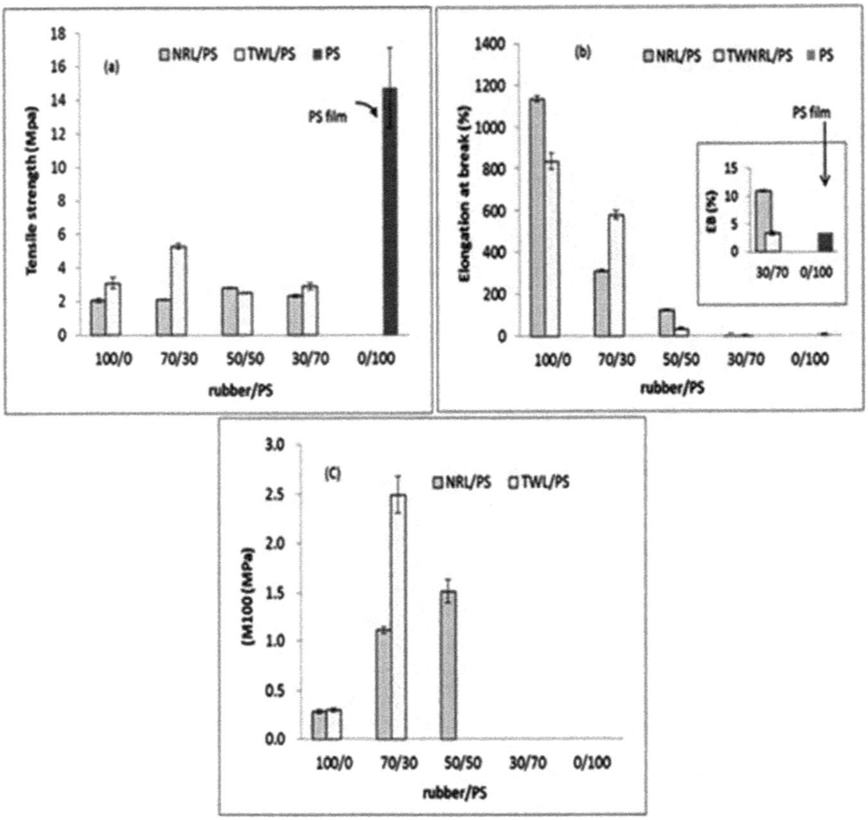

Figure 9.14 Mechanical properties of NR LATEX/PS TWL/PS blends: (a) tensile strength (b) elongation at break, and (c) M100.[20]

that the WRNL/PS blend gave higher tensile strength and thermal stability compared to the NR/PS blend at a similar composition. The introduction of a styrene-butadiene-styrene copolymer (SBS) gave the best overall properties compared to SNR.[21] The blends based on WRNL and PS could be applied to shoe applications such as the sole and so forth.

9.3.5 Recycling of Latex Paint

Latex waste paint (WLP) alone constitutes 12% of the total hazardous waste and has been accumulating since the inception of its collection policy. It is projected that the collected WLP will grow substantially for two reasons: (a) current collection programs result in the capture of only 10% to 30% of the total hazardous waste; public education programs can raise this level significantly, and (b) increasing of environmental regulations on volatile organic compounds (VOCs) which increase the production and use of latex paints relative to other solvent-based paints, consequently increasing the generated WLP.[22] Industrial paint sludge, water treatment sediment, leftover latex paint, and other chemically similar materials are typically considered wastes, however, such wastes can be used as substitute raw materials in the manufacturing of paint or other products. For example, about 50% of leftover latex paint can be remanufactured and used in recycled content paint products. The remaining 50% of leftover latex paints, plus industrial paint sludges and water treatment sediments, are routinely disposed of in incinerators at a high cost and untold future liability.

The study conducted by Nehdi and Sumner[23] aimed to investigate the benefits of recycling latex waste paint in concrete with a special focus on concrete sidewalks. The latex waste paints are used as a partial replacement for virgin latex and for mixing water. Based on their study, the incorporation of latex waste paint and concrete mixture was proven to improve workability, high flexural strength, lower chloride penetrability, better resistance to deicing salt surface scaling and is more economic. This is because they required less water-reducing and air-entraining admixtures. Another study on latex waste paint in concrete conducted by Mohammed *et al.*[24] investigated a variable percentage of latex waste paint that was replaced in mixtures and compared it to a mixture containing no latex and a mixture containing 15% latex. The study found that the latex waste paint mixture demonstrated improved properties over normal concrete at similar properties as the SB mixtures. The study focused on concrete for sidewalks and non-structural elements.

Reused paint has the ability to be high quality and comparable to new paint. Nevertheless, this method requires filtering with the addition of additives. Segala[22] studied a technique to recycle paint known as processed latex pigment. Processed latex pigment is collected from automobile manufacturing plants, entrainment production, maintenance companies, and professional painters. It is a patented method for recycling industrial paint sludge, water-treatment sediment, and leftover latex paint. Processed latex paint is shipped to Portland cement plant to use as an additive for special cement or as a raw material for the kiln feedstock. Processed latex paint can also be used to replace silica, aluminium, iron and calcium carbonate.

9.4 Recycling of Latex Products

9.4.1 Reclaiming of Latex Waste Products

In the process of recycling rubber products, the first step is a preliminary treatment, such as cutting, milling and so on. Sometimes there are impurities present, which need to be removed, such as steel cord. Then the waste rubber is cured by mechanical methods, chemical methods, freezing and crushing. Eventually, the waste rubber is made into rubber chips or powdered rubber. Manufacturing powdered rubber is the main recycling route for waste rubber because it provides economic and social benefits, which are: (a) it will conserve consumption of facilities and energy; (b) conservation of the mollifier and activating agent used in desulfurizing; (c) a decrease in the cost of the product by blending fine powdered rubber with raw rubber; (d) production of fine powdered rubber does not cause environmental pollution; and (e) the content of fine powdered rubber in products is more than that of powdered rubber; moreover, it has better mechanical behaviour. Thus, the application prospects for powdered rubber are quite wide.[25]

The recycling of scrap and used rubber has been practiced by the rubber industry for many years, almost entirely by conversion to reclaimed rubber (RR).[26] Even though the incentives are to reduce compound cost and to conserve raw materials and energy, the use of reclaim constitutes only a very small percentage of raw rubber consumption. This is mainly due to the deterioration in technical properties.[27] In contrast to reclaimed rubber, scrap latex products contain rubber hydrocarbon of a very high quality which is only lightly cross-linked. Due to the unstable nature of the latex compound and the strictest specifications in the quality of latex products such as condoms and examination gloves, the rejection in the latex industry comes to as high as 10 to 15% of the rubber consumed. The rubber Research Institute of India has developed a process for reclaiming latex products, as has the rubber Research Institute of Malaysia.[28]

Nowadays, researchers pay more attention to scrap latex rejects compared to reclaimed rubber as the use of reclaimed rubber was limited owing to three main reasons:

1. Easy reclamation was not possible in the case of waste tyres due to the presence of steel belts or plies.[29]
2. The properties of reclaimed rubber were inferior due to its degradation during reclamation.[30]
3. The various components of a tyre or blend may not respond in the same way to reclaim.[26]

Scrap latex rejects or latex product waste represents a source of high quality rubber hydrocarbon, making it a suitable candidate for generating reclaimed rubber of superior quality.[31] The Rubber Research Institute of India reported a method for reclaiming latex products. Latex product such as a powdered waste condom was reclaimed as filler in raw rubber. In this process, waste condoms were powdered initially by passing them through a hot two roll mill at 80–90 °C to a size of about 40 mesh. The powdered material was mixed with 10 phr of naphthenic oil and 1 phr of pentachlorothiophenol (PCTP) on a cold mill. The resulting compound was heated in an air oven at 140 °C for 30 min. The reclaimed rubber obtained by this process was found to form a smooth band on the mill and contained about 82% of rubber hydrocarbon. However, it was found that only a small amount of this reclaim could be added to raw rubber without adversely affecting the mechanical properties. This is mainly due to the degradation of the rubber hydrocarbon during the reclaiming process.

Kumar *et al.*[32] revealed a possible technique to examine both recycle rubber-virgin rubber properties, either increasing the effective surface area by reducing the granulate size or by surface modification of granulates in order to improve the adhesion between the matrix and granulates. Dierkes *et al.*[33] studied the influence of different amounts of reclaim on the properties of a virgin compound in terms of mechanical and dynamic viscoelastic properties, crosslink density and sulphur distribution between the matrix and the reclaim. They summarized that the mechanical and viscoelastic properties are affected by the following factors: (a) presence of gel in reclaim; (b) adhesion of the reclaim to the matrix; (c) particle size of the reclaim; (d) sulphur distribution between the matrix and reclaim; and (e) crosslink density and distribution. The increase in concentrations of reclaim will reduce the molecular weight of the sol fraction in the reclaim and the presence of the crosslinked gel, thus reducing the tensile strength. This is due to the reclaim particles acting as an obstacle to stress transmission within a continuous matrix, resulting in an initiation of failure expressed by lower tensile strength. The stress accumulates on the interface between the reclaimed particle and the matrix, and the fracture starts from this point. Increasing concentration of the reclaim leads to the increasing of weak spots, thus reducing the tensile strength of the system.

9.4.2 Latex Waste Products as Filler

In rubber industries the waste rubber was recycled by remoulding, as a filler for virgin material, blending with other virgin materials, reclaiming to reduce the crosslinked molecules and used as a new material again. Recycled gloves were also used as potential filler in natural rubber compounding.[34] It has been observed that as the filler contents increase, the curing characteristics, such as the scorch time and cure time decrease for powder gloves but not for chlorinated gloves. In addition, for compounds with calcium carbonate as the base filler, the physical properties, tensile strength, 300% modulus, elongation at break, tear strength, % of crack growth and abrasion index change insignificantly with the glove rejects loading. Thus, 40 phr of glove rejects loading is recommended for obtaining good properties and extrusion quality.

Ismail *et al.*[35] investigated the comparison properties of recycle rubber powder (RRP), carbon black and calcium carbonate filled natural rubber compounds. They found that tensile modulus and hardness increases with increased carbon black loading, whereas elongation at break, resilience and swelling properties showed an opposite trend. According to their finding, tensile strength of the RRP filled natural rubber compound gradually increases up to 10 phr and starts to deteriorate at higher filler loading. The increase of tensile strength at below 10 phr is due to the smaller particle size and uniform distribution of RRP in NR compounds. At higher filler loading, the weak interaction and bonding between the rubber powder particles and rubber matrix is responsible for the deterioration in tensile strength. The other properties such as tensile modulus, hardness, elongation at break, resilience, and swelling percentage showed small changes (increase or decrease) with an increase in RRP and calcium carbonate loading in NR compounds. The results of this study discuss the effect of RRP loading on the tensile modulus of NR compounds. It can be observed that M100 and M300 increase with the increasing of RRP loading. This is due to the RRP behaving like a rigid particulate filler as it has a higher modulus then the NR matrix.

There are wide varieties of successful applications such as in cement/concrete where the addition of the recycled rubber reduces vibration transmission/cracking and improves fracture resistance. Other than that, crumb rubber is effective at absorbing heavy metals and organic solvents such as benzene, toluene and other organic solvents, which are used in sludge treatment plants. Asphalt containing crumb rubber is more durable, has a better resistance to thermal and reflective cracking, reduces the level of noise generated by vehicles travelling on it, and provides a more comfortable ride. Flooring, walkway tiles and sport surfaces such as running or jogging tracks constitute a growing market for recycled rubber.[36]

Mathew *et al.*[30] developed a cost-effective technique for the reuse of latex waste rubber in epoxidised natural rubber (ENR). The powdered rubber is prepared in a fast rotating toothed-wheel mill. The mixing of ENR and filler were carried out using a two-roll mixing mill. From the results, the cure characteristics such as the optimum cure time, rheometric scorch time and induction are found to decrease with increasing of the concentration of latex waste filler. Due to the presence of the unreacted accelerator zinc diethyldithiocarbamate (ZDEC) in the latex waste, the tensile strength of the mixing shows better properties at finer particles of latex waste.

Another report discussed the chemical modification of latex thread waste and glove waste into the preparation of thermoplastic elastomers by blending the waste with high density polyethylene (HDPE). The mixing of glove waste was performed on a Brabender plasticorder at 150 °C and 50 rpm. The mix was then sheeted on a mixing mill. The tensile strength of the blend reduced with the increase of rubber due to the drop in crystallinity. The tensile strength of the waste blend shows higher properties compared to NR/HDPE, which may occur due to the partially crosslinked nature of the waste.[37]

Noriman *et al.*[38] conducted research into the characteristics and properties of SBR/recycled acrylonitrile-butadiene rubber (SBR/NBR_R). From their research, the curing characteristics such as the scorch time and cure time decreased with increasing of the NBR_r blends. This is due to the presence of cross-linked precursors and unreacted curative in the recycle rubber.[39]

Other than that, they found that the processability for the blends becomes more difficult with increasing NBR_r due to the presence of cross-linked NBR_r and other additives. The tensile strength results of the blend indicated that the increasing of the recycled rubber has reduced the tensile strength of the blends. This is due to the incompatibility of the blends. In order to overcome this problem, the blends are incorporated with compatibilizer to improve the properties. The compatibilizers that have been used are *trans*-polyoctylene rubber (TOR)[40] and epoxidised natural rubber (ENR-50).[41] Both of the compatibilizers have almost similar tensile strength. The incorporation of the compatibilizer with the blends improved the tensile strength compared with the blend without compatibilizer.

9.5 Conclusions

Latex is the starting material for a wide range of products which are used extensively for the manufacture of dipping products, particularly in glove manufacturing. The increasing demand of latex products year by year increases the production of latex waste. Latex waste can be divided into two categories, solid waste and liquid waste. Solid latex waste is normally subjected to incineration, or disposed in landfill, which creates an ecological problem. Thus, researchers have provided another alternative by recycling of the solid waste by reclaiming the solid waste, blending and also using it as a filler in the rubber industry. Liquid latex waste can be divided into two types, latex paint and latex waste compound. Disposing of liquid waste is more difficult since it has to be converted into a solid state before it can be disposed, which requires a high cost to the manufacturer. However, latex waste paint can be recycled through addition in concrete, which provides better workability to the cement industry. It has also been used as processed latex paint where it can replace other fillers such as silica and calcium carbonate in the cement industry. The latex compound waste can be utilized as cleaning compounds, such as laminated mould cleaning, outdoor cleaning and former cleaning compounds. This gives a new added value to the latex waste which converts the liquid based materials into useful products in solid waste which are easier to dispose.

References

1. E. Yip and P. Cacioli, The manufacture of gloves from natural rubber latex, *J. Allergy Clin. Immunol.*, 2002, **110**(2), S3–S14.
2. H. M. Lim and H. Amir, Polyurethane and natural rubber latex blend foam rubber, *J. Rubber Resour.*, 2011, **14**, 41–50.
3. A. K. Bhowmick, *Rubber Products Manufacturing Technology*, CRC Press, New York, 1994.
4. H. Hamidah, A comparative study on the effect of compounded latex and diffusion of curatives on tensile properties of natural latex films, *Jurnal Teknologi Proses*, 2006, 27–30.

5. D. C. Blackley, *High Polymer Latices: Science and Technology*, Chapman & Hall, London, UK, 1997.
6. E. Abraham, B. M. Cherian, P. A. Elbi, L. A. Pothern and S. Thomas, *Recent Developments in Polymer Recycling. Kerala, India*, Transworld Research Network, 2011, pp. 47–49.
7. N. D. Moung, Natural rubber latex: then and now, in *Latex & Synthetic Polymer Dispersions 13-14 March 2012*, Smithers, KL Concorde Hotel, Kuala Lumpur, 2012, pp. 1–14.
8. S. G. Reena and J. Rani, The Utilization of Waste Latex Products in Styrene-Butadiene Rubber, *Iran. Polym. J.*, 1996, **96**, 1026–1265.
9. A. S. Morris, *ISO 14000 Environmental Management Standards; Engineering and Financial Aspect*, Johd Wiley & Sons, England, 2004.
10. J. A. Lindsay, *Rubber Injection Molding; A Practical Guide*, Rapra Technology Ltd., United Kingdom, 1999.
11. Rapra Technology Limited, *Rubber Machinery Sourcebook*, iSmithers Rapra Publishing, Shawbury, 2000, p. 20.
12. R. B. Simpson, *Rubber Basics*, Rapra Technology Limited, Shawbury, 2002.
13. M. S. Mufidah, A. R. Azura and A. Baharin, Effect of Cleaning Agent on Tensile and Swelling Properties of Natural Rubber Latex Film, *Adv. Mater. Res.*, 2014, **858**, 46–51.
14. M. M. Sidek, A. A. Rashid and B. Azahari, Effect of Different thickness of core layer on tensile properties of laminated Natural rubber latex, *Adv. Mater. Res.*, 2014, **1024**, 259–262.
15. A. A. Rashid, M. M. Sidek and T. S. Chean, Effect of lamination layers on tensile properties and mould cleaning efficiency of laminated NBR latex composites, *Adv. Mater. Res.*, 2014, **1024**, 189–192.
16. W. George, *Handbook of Plasticizers*, ChemTec Publishing, Toronto, 2004.
17. R. P. Brown, *Physical Testing of Rubber*, Chapman & Hall, London, 1996.
18. B. T. Poh, P. G. Lee and S. C. Chuah, Adhesion Property of Epoxidized Natural Rubber (ENR)-based Adhesive Containing Calcium Carbonate, *eXPRESS Polym. Lett.*, 2008, **2**(6), 398–403.
19. M. Herrmann, R. Rottenegger, J. Tinschert and R. Marx, The effect of corrosive environment on the porcelain-to-metal bond–a fracture mechanics investigation, *Dent. Mater.*, 1992, **8**(1), 2–6.
20. O. Boondamneon, A. R. Azura, M. Ohshima, S. Chuayjuljit and A. Ariffin, Recycling waste natural rubber latex by blending with polystyrene- characterization of mechanical properties, *Int. J. Mod. Phys.: Conf. Ser.*, 2013, **6**, 391–396.
21. O. Boondamneon, A. R. Azura, M. Ohshima, S. Chuayjuljit and A. Ariffin, Effect of latex treatment on properties of WNRL/PS blends, *Adv. Mater. Res.*, 2012, **488–489**, 478–482.
22. L. M. Segala, Recycling of nonhazardous industrial paint sludge, nonreusable leftover latex paint, and similar materials, *Met. Finish.*, 2003, **101**(3), 38–40.

23. M. Nehdi and J. Sumner, Recycling latex waste paint in concrete, *Cem. Concr. Res.*, 2003, **33**, 857–863.
24. A. Mohammed, M. Nehdi and A. Adawi, Recycling latex waste paint in concrete with added value, *ACI Mater. J.*, 2008, **105**(4), 367–374.
25. Y. Fang, Z. Maosheng and Y. Wang, The status of recycling of waste rubber, *Mater. Des.*, 2001, **22**, 123–127.
26. A. Acetta and J. M. Vergnaud, Upgrading of Scrap Rubber Powder by Vulcanization without New Rubber, *Rubber Chem. Technol.*, 1981, **54**, 302.
27. R. P. Burford and M. Pittnlu, Characterization and performance of powdered rubber, *Rubber Chem. Technol.*, 1982, **55**, 1233.
28. R. S. George and R. Joseph, The utilization of latex waste products in styrene-butadience rubber, *Iran. Polym. J.*, 1996, **5**, 199–205.
29. P. Rajalingam, J. Sharpe and W. E. Baker, *Rubber Chem. Technol.*, 1993, **66**, 664.
30. G. Mathew, R. P. Singh, N. R. Nair and S. Thomas, Recycling of natural rubber latex waste and its interaction in epoxidised natural rubbers interaction in epoxidised natural rubber, *Polymer*, 2001, 2137–2165.
31. V. V. Rajan, W. K. Dierkes, R. Joseph and J. W. M. Noordermeer, Science and technology of rubber reclamation with special attention to NR-based latex waste products, *Prog. Polym. Sci.*, 2006, **31**(9), 811–834.
32. P. Kumar, Y. Fukahori, A. G. Thomas and J. J. C. Busfield, Recycled rubber: the rubber granulate-virgin rubber interface, *Rubber Chem. Technol.*, 2007, **80**, 24–39.
33. W. K. Dierkes, V. V. Rajan and J. W. M. Noordermeer, Application of NR-based latex reclaim: the link between structure and properties, *Rubber Chem. Technol.*, 2007, **80**, 40–60.
34. O. Patarapaiboolchai, T. Tulayapitak and I. Sriwang, Recycled gloves as a filler, *J. Sci. Technol.*, 2002, 701–706.
35. H. Ismail, R. Nordin and A. Noor, Cure characteristics. Tensile properties and swelling behavior of recycled rubber poder-filler natural rubber compounds, *Polym. Test.*, 2002, **21**, 565–569.
36. W. Klingensmith and B. Rodgers, Natural rubber ad recycled materials, in *Rubber Compounding: Chemistry and Applications*, ed. B. Rodgers, Marcel Dekkar, Inc, New York, 2004, ch. 1.
37. S. Rajalekshmi and R. Joseph, Studied on the rheology and the mechanicalpropertiesofthermoplasticelastomerfromlatexproductwasteandhigh density polyethylene, *J. Elastomers Plast.*, 2002, **314**, 313–322.
38. N. Z. Noriman, H. Ismail and A. Rashid, Curing characteristics, mechanical and morphological properties of styrene butadiene rubber/virgin acrylonitrile-butadiene rubber (SBR/NBRv) and styrene butadiene rubber/recycled acrylonitrile-butadiene rubber (SBR/NBRr) blends, *Polym.-Plast. Technol. Eng.*, 2008, **47**, 1016–1023.
39. T. D. Sreeja and S. K. N. Kutty, Cure characteristics and mechanical properties of natural rubber/reclaimed rubber blends, *Polym.-Plast. Technol. Eng.*, 2000, **39**(3), 501–512.

40. N. Z. Noriman, H. Ismail and A. Rashid, Characteristic of styrene butadiene rubber/recycled acrylonitrile-butadiene rubber (sbr/nbrr) blends: the effect of epoxidized natural rubber (ENR-50) as a compatibilizer. material characterization, *Polym. Test.*, 2010, **29**, 200–208.
41. N. Z. Noriman, H. Ismail and A. Rashid, Properties of styrene butadiene rubber/recycled acrylonitrile-butadiene rubber (SBR/NBRr) blends: effect of addition trans-polyoctylene (TOR), *J. Appl. Polym. Sci.*, 2012, **126**, E56–E63.

CHAPTER 10

Recycling of Rubber Blends for Durable Construction

SATHISH KUMAR PALANIAPPAN[a], RAJASEKAR RATHANASAMY*[b], SAMIR KUMAR PAL[a] AND GANESH CHANDRA NAYAK[c]

[a]Department of Mining Engineering, Indian Institute of Technology Kharagpur, West Bengal, 721 302, India; [b]Department of Mechanical Engineering, Kongu Engineering College-Erode, Tamil Nadu, 638 052, India; [c]Department of Applied Chemistry, Indian Institute of Technology (Indian School of Mines) Dhanbad, Jharkhand, 826 004, India
*E-mail: rajasekar.cr@gmail.com

10.1 Introduction

Generally, the blending of two or more types of polymers is a useful technique for the preparation and development of materials with properties superior to those of the individual constituents.[1,2] Mostly, polymer blends are incompatible and this plays an important role in the modern polymer industry not only for the development of new materials, but also for practical recycling purposes. In general, the final performance of a polymer blend is determined by the polymer compatibility, dispersion, stiffness and the phase morphology of each component.[3–10] The accumulation and disposal of polymer wastes such as the products of plastics, rubbers and synthetic rubbers has become a global issue. Polymer waste with a crosslink structure is difficult to recycle as they are neither fusible nor soluble. A critical degradation

Green Chemistry Series No. 59
Rubber Recycling: Challenges and Developments
Edited by Jin Kuk Kim, Prosenjit Saha, Sabu Thomas, Józef T. Haponiuk and M. K. Aswathi

Published by the Royal Society of Chemistry, www.rsc.org

method is needed to reshape the polymer waste with this crosslink structure under extreme temperature and pressure.[11,12] This has contributed to the large accumulation of polymer waste, especially waste from tires. Considering the social reasons, economic and environmental advantages, polymer recycling is now a subject of considerable importance.[13–16] Three main approaches that have been used to extract value from waste materials are as follows: In the first approach, polymers can either be recycled chemically through degradation or depolymerization chemistry. Secondly, they can be recycled physically through thermal reshaping into objects of lesser value than the original, but still useful in selected applications. Finally, waste polymers can simply be burned to extract their energetic value.[13]

Elastomers (rubbers) form a unique class of the materials which possess high elastic deformation and are able to resume their original shape after deformation. Usually elastomers have inadequate mechanical and thermal properties.[17] The production of rubber materials rises every year. A number of possible applications of various forms of waste rubber, either on its own or combined with other materials, in broad disciplines have been studied and reported.[3,18] The purpose of blending the rubber is to improve the physical and mechanical properties as well as to modify the processing characteristics and to reduce the cost of the final product. However, most rubber based polymer blends are immiscible and usually exhibit phase separated morphology and poor interfacial adhesion between the phases.[1,19] The interfacial adhesion can be improved by introducing a third component into an immiscible binary system that will either interact chemically with both the phases or will have a specific interaction with one phase and a physical interaction with the other.[1,20] This phenomenon can be justified as an improvement in the compatibilization of the blend. Commonly, compatibilization can lead to a finer phase structure and enhanced interfacial adhesion.[1,21] Waste rubber can be generated during the manufacturing process of the product and also by disposal of post-consumer products.[22,23] Recycled rubber can be generalized as any rubber waste that has been converted into an economically useful form, such as reclaimed rubber, ground rubber or reprocessed synthetic rubber.[13,24] The manufacturing of powdered rubber has obvious economic and social benefits, as it decreases the cost of the product by blending fine powdered rubber with raw rubber.[3,13,25] Blending recycled rubber with other materials has also been an attractive alternative in waste rubber recycling.[26]

Due to the large volumes of scraps associated with rubber production, the old practice of land filling has been unable to keep up with production because of rising land costs, environmental concerns and fire risks occurring at these sites.[15,27,28] Considering that vulcanized rubber is not easily biodegradable, recycling methods must be proposed in order to solve this environmental problem and recover a material with high commercial value.[15,29] However, recycled latex has become a focus of attention compared to reclaimed rubber due to the lightly cross-linked and high quality nature of the rubber hydrocarbon.[13,30] In Malaysia, a study on waste tires shows that 8.2 million or approximately 57 391 tons of waste tires were generated annually.[11,31] A lot of things can benefit from the use of waste rubber products. Therefore, many researchers have done studies on forming waste tire rubber into new products. Among these was the research on a new compound using recycled rubber by Sekhar Research Innovations (SRI).[32] The compound, called SRI compound was formulated from recycled tires and waste rubber from making new tires, rethreaded tires or manufacturing automotive parts.[11] Recycled rubber based blends find application in playground surfaces, recycled rubber flooring, sporting mats,[33] automotive, aerospace and consumer products and other engineering products. Environmental legislation will help to promote recycling, but long-term technological developments are needed.[34]

10.2 Recycling of Rubber Based Blends for Durable Construction

Amir *et al.* described the thermal and rheological behavior in blends of recycled rubber (RR) and natural rubber (NR).[11] Tire thread was synthesized by optimizing the content of RR and standard malaysian rubber (SMR). Recycled rubber from lorry tires and two natural rubbers, such as SMR-L and SMR-10 were used. The samples were subjected to differential scanning calorimetry (DSC), X-ray diffraction (XRD), capillary rheometery and Mooney viscometer analysis. The blends of SMR-10/RR exhibited a higher thermal stability and Mooney viscosity compared to SMR-L/RR. However, both the blends showed similar shear viscosities and shear rates. Ahmad *et al.* studied the compatibility of sulfur cured NR/recycled acrylonitrile butadiene rubber (NBRr) blends with 10 phr of epoxidized natural rubber (ENR-50) as a compatibilizer using cure characteristics, mechanical properties and scanning electron microscopy (SEM) analysis. The compatibilized blends showed a lower scorch time, cure time, higher torque, enhanced mechanical properties, increased hardness, better adhesion and difficulty in processing compared to uncompatibilized samples.[35] Da Silva, Araujo, and De Melo analyzed the rheological properties and FTIR spectroscopy of compatibilized polyamide 6 (PA-6)/recycled styrene butadiene rubber (SBRr) blends using melt mixed Haake curves. Maleic anhydride grafted polyethylene samples exhibit the best results compared to graphitized polypropylene blends.[36] The prepared samples were subjected to mechanical tests (tensile and impact), dynamic mechanical thermal analysis (DMTA), DSC and SEM analysis. Reduced impact strength and elastic modulus of the compatibilized blends were observed compared to pure PA-6. Ciro *et al.* analyzed the rheometric, thermogravimetric (TGA), DSC, SEM, density, melt flow index and tensile behavior on the effect of particle size (850 μm) of the RR in recycled polypropylene/polyethylene (RPP)/RR blends. An increase in RPP content enhanced the maximum torque. The presence of fillers in both the polymers increased the blend densities, whereas addition of RR affected the crystallinity of RPP. The DSC results displayed two melting temperatures for RPP (LDPE and PP).[37]

Azmi *et al.* adopted a two roll mill method to prepare blends of SBR/virgin chloroprene rubber (CRv) and SBR/recycled CR (CRr) with a CRr size range from 0.3–0.7 mm.[38] The authors analyzed the effect of the tensile and morphological properties of both the blends. The results depicted that all the blend ratio of SBR/CRr showed a lower cure time, higher scorch time, minimum torque, tensile strength, tensile modulus and elongation at break compared to blends of SBR/CRv. SEM analysis of the 50SBR/50CRr blend showed improved adhesion and better dispersion compared to 50SBR/50CRv. Sonnier *et al.* used γ irradiation on recycled high density polyethylene (rHDPE) and ground tire rubber (GTR) powder forming free radicals, leading to chain scission within rubber particles, crosslinking of matrix and co-crosslinking at the interface.[39] The irradiated blends exhibit improved mechanical performances, better adhesion and strong elongation of GTR particles upon elongation compared to uncompatibilized rHDPE/GTR blends. Naskar *et al.* used rGRT as a replacement for ethylene propylene diene monomer terpolymer (EPDM) in a dynamically vulcanized EPDM/acrylic-modified HDPE blend. The 60EPDM/40m-HDPE blend behaved like a thermoplastic elastomer and finally, it was observed that 50 wt% of EPDM can be replaced by rGRT without deterioration of the technical properties. TEM studies displayed m-HDPE as a continuous phase and EPDM as a co-continuous phase.[40] Hrdlicka, Kuta and Hajek prepared blends using an internal mixer which behaved like thermoplastic elastomers, containing low density polyethylene (LDPE) and recycled waste tire rubber.[41] The very good recycling ability and higher mechanical performances for the best composition blends were observed on both DCP and sulfur cured systems. Malarvizhi, Senthil and Kamaraj analyzed the feasibility and technical performances of stone matrix asphalt (SMA) with recycled crumb rubber (CR) and LDPE as additives.[42] The increase in the addition of rubber content indicated improved engineering properties and long term performance of the SMA mixtures.

Al-Malki *et al.* developed thermoplastic elastomers based on different weight ratios of HDPE and GTR. The prepared blends were subjected to mechanical, rheological and morphological analysis to understand the compatibility between the blends.[43] Boondamnoen *et al.* blended a 50 : 50 ratio of waste natural rubber latex (WNRL) as a dispersed phase with polystyrene (PS) as a matrix for recycling. The prepared blends displayed higher tensile strength, Young's modulus, crosslinking density and lower elongation at break compared to dynamically vulcanized natural rubber (NRv)/PS and NR/PS blends. The WNRL/PS blend required higher energy to break compared to (NRv)/PS and the NR/PS blends were analyzed using SEM and TEM blend morphology.[44] Kim, Lee and Balasubramanian developed ultrasonically treated and untreated thermoplastic elastomers based on waste tire rubber as a dispersed phase and two polyolefins (MA-*g*-PP and LDPE) as the continuous phase.[26] However, the untreated samples gave good mechanical properties compared to treated samples. Similarly, addition of MA-*g*-SEBS as a compatibilizer showed the best results for both untreated and treated samples. Yahya *et al.* investigated the effect of particle sizes of rNBRg on the curing characteristics, tensile properties and physical properties of NR/SBR/

rNBRg blends.[45] The samples were prepared using a two roll mill at room temperature with different particle sizes. A reduction in the particle size of rNBRg decreased the scorch time, cure time, tensile strength, blend resilience and minimum torque, whereas rigidity, hardness, maximum torque and crosslink density increased. Ismail and Suryadiansyah described that different weight ratio of RR were prepared through a sanding process and used to develop PP/RR blends. Similarly, a series of blends using NR were also prepared. The distribution of recycled rubber particle size is displayed in Figure 10.1. The prepared samples depicted a higher tensile strength and Young's modulus, and a lower elongation at break and stabilization torque. SEM analysis confirmed that higher energy may be required to cause failure on PP/RR blends compared to PP/NR samples.[46]

Navarro *et al.* reported the thermal and rheological properties of modified bitumen blends on the effect of GTR and RPE.[47] The addition of neat or both GTR and RPE into the blends enhanced the thermo-rheological behavior, mechanical performances and in-service temperatures. The author developed these compounds to resolve the waste disposal problem and to improve the quality of blends for roofing and waterproofing applications. Boondamnoen *et al.* treated NRL with waste NRL to improve the desired properties and blended with PS using a Haake internal mixer. Measuring the torque while mixing, the storage modulus and tan delta from rheometer were recorded for both the untreated and treated blend samples. The prepared samples were subjected to TGA, DSC, degradation studies and morphological analysis. Treated WNRL displayed a lower mixing torque, good processability, lower storage modulus and cross-linked molecules. The presence of two Tg confirmed through DSC explained that both the polymers are immiscible. The lower energy required to break the molecules of the TWNL/PS blend showed less degradation temperature.[48]

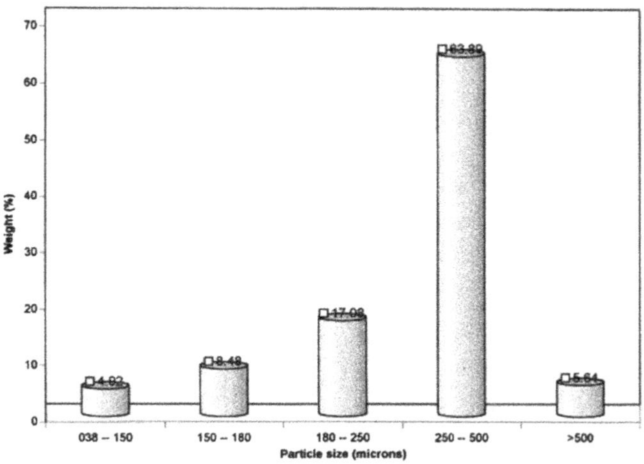

Figure 10.1 Distribution of recycle rubber particle size. Reproduced from ref. 46 with permission from Elsevier, Copyright 2002.

Sangwichien, Sumanatrakool and Patarapaiboolchai used carbon black and reclaimed rubber to analyze the curing characteristics and mechanical properties of NR/PP blends in-order to improve the tensile behavior. Increasing amounts of both the fillers enhanced the tensile strength, scorch time and decreasing the cure time of the NR/PP blends. As a result, reclaimed rubber render improved elasticity and tensile mechanical performances.[49] Noriman *et al.* investigated the weathering test, tensile strength, elongation at break, tensile modulus, SEM and retentions of SBR/NBRr blends with and without *trans*-polyoctylene rubber (TOR) as a compatibilizer. Better dispersion of NBRr particles in the SBR matrix of the tensile fractured surface were observed through SEM images, minimizing the degradation process.[50] Recently, the authors Noriman, Ismail and Rashid[51,52] reported the comparison properties of virgin and NBRr of SBR/NBRr blends. Hayeemasae and Ismail created a value added NR material with R-EPDM to prepare NR/R-EPDM blends. ZnO nanoparticles as a curing activator were added onto it and the thermo-mechanical performances of the prepared blends were investigated. The synthesized ZnO nanoparticles with a dimension ranging from 10–40 mm have no impurities to improve their thermal stability and dynamic mechanical properties of the blends.[53]

Phinyocheep, Axtell and Maiseaumsook investigated the mechanical properties of PP filled recycled rubber dusts. The size reduction process in rubber recycling was avoided due to these rubber dusts. PP wax and ethylene vinyl acetate wax were used to develop extruded and injection molded compounds. The amount of rubber dust addition influenced the tensile properties and impact strength was higher for injection molded samples compared to extruded blends.[54] Ismail and Suryadiansyah developed thermoplastic elastomers through melt mixing by blending PP, NR and RRP. The effect on partial replacement of rubber on mechanical properties, swelling behavior, torque development and morphological properties of the prepared blends were investigated in detail.[55] An increase in the RPP content enhanced the tensile strength, Young's modulus and swelling resistance and displayed a decreasing trend for torque and elongation at break. Grigoryeva *et al.* investigated TGA, DSC, DMA, wide angle X-ray scattering (WAXS), mechanical testing and rheological measurements for developed recycled LDPE and butadiene rubber based thermoplastic elastomers. The addition of reactive polymers does not prevent the crystallization and this was confirmed through WAXS studies. A decreased melting point and improved compatibility depict enhanced mechanical properties. However, dual phase continuity morphology was observed for all the prepared samples.[56]

Supri analyzed the effect of fly ash loading on morphological properties of recycled poly(vinyl chloride)/acrylonitrile butadiene rubber blends.[57] SEM analyses of the tensile fractured surfaces showed weak filler matrix interactions. An increase in the loading of fly ash content deteriorated the blends mechanical properties. This may be due to the presence of more impurities in fly ash making them chemically inactive. Mangaraj followed a cost effective and efficient method for recycling of rubber waste by blending ground rubber with thermoplastic and thermoset polymers. The author effectively adopted reactive and non-reactive compatibilization in the blends to develop co-continuous type morphology for improved mechanical properties.[58] Noriman, Ismail and Rashid studied TGA, DSC and fatigue life of SBR/NBRr blends with and without TOR. Addition of TOR onto the blends showed a rougher surface with many tear lines, good interaction between the matrices, altering of the calorimetric profile, increased thermal stability and fatigue life.[59] Yahya *et al.* investigated tensile properties and morphology of different ratios of NR/SBR/rNBRg blends. The author used discarded waste nitrile gloves in this study to solve some environmental issues. The increase in addition of rNBRg depicts decreased tensile properties, resilience, crosslink density and increased hardness for the blends. The blends with 50/30/20 ratio showed optimum mechanical properties. However, morphological observations indicate poor dispersion of rNBRg particle onto the blends. The random dispersion of rNBRg at 40 phr showed a significant drop in the hardness value.[60]

Zulkepli, Ismail and Rashid analyzed the effect of different particle sizes of NBRr on the curing characteristics, and mechanical and morphological properties of SBR/NBRr blends.[61] The addition of NBRr decreased the scorch time and cure time of the blends. The presence of cross-linking precursors and unreacted curative decreased the particle sizes of NBRr. This shows better mechanical properties, high hardness, cross-linking density, decreased resilience, increased contact surface area and efficient interfacial interaction. Dekic *et al.* substituted recycled rubber powder with particle sizes ranging from 0.1–1.0 mm onto the NR/SBR compounds to analyze the wear behavior, hardness, tensile properties and morphological characteristics. Addition of RPP shows reduced tear strength for all the blend ratios.[62] Ramarad *et al.* reviewed the incorporation of waste tire rubber onto the polymer matrices. The composition of waste tire manufacturing is tabulated in Table 10.1.

The researchers also surveyed the methods of waste tire downsizing and the importance of WTR characterization and modifications. Figures 10.2 and 10.3 depict flowcharts for waste tire utilization and representation of devulcanization or reclaiming. The different methods used for devulcanization or reclaiming are presented in Table 10.2.

Table 10.1 Composition of materials used in tire manufacturing. Reproduced from ref. 63 with permission from Elsevier, Copyright 2015.

Type of tire	Car/passenger	Truck
Rubber/elastomer	41–48	41–45
Carbon black	22–28	20–28
Metal/steel	13–16	20–27
Textile	4–6	0–10
Additives	10–12	7–10

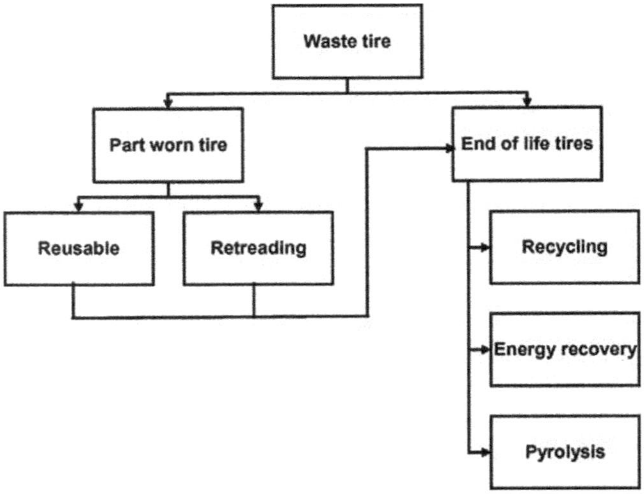

Figure 10.2 Flow of waste tire utilization. Reproduced from ref. 63 with permission from Elsevier, Copyright 2015.

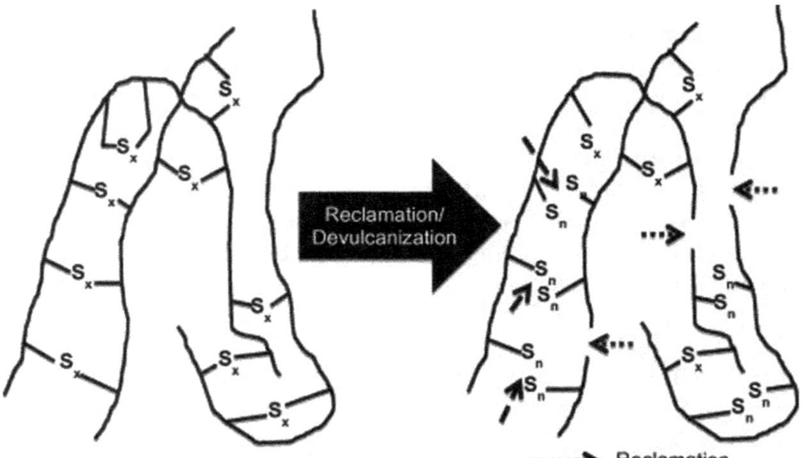

Figure 10.3 Schematic representation of devulcanization and reclaiming. Reproduced from ref. 63 with permission from Elsevier, Copyright 2015.

The effects of WTR size, rubber content (Figure 10.4), loading, modification, compatibilization and crosslinking on the mechanical, thermal and rheological properties of neat rubber and blends were reported. The difference in the microstructure of polymer blends with GTR and RTR is displayed in Figure 10.5. The study concentrated on the dynamic mechanical properties, thermal, aging and swelling properties of WTR based polymeric blends.[63]

Table 10.2 Tire rubber reclaiming devulcanization processes. Reproduced from ref. 63 with permission from Elsevier, Copyright 2015.

Physical	Chemical	Microbial
Mechanical	Radical scavengers	Aerobic reactions
Thermo-mechanical	Nucleophilic additives	Anaerobic reactions
Microwave	Catalyst systems	Bacterial
Ultrasonic	Chemical probes	Fungi

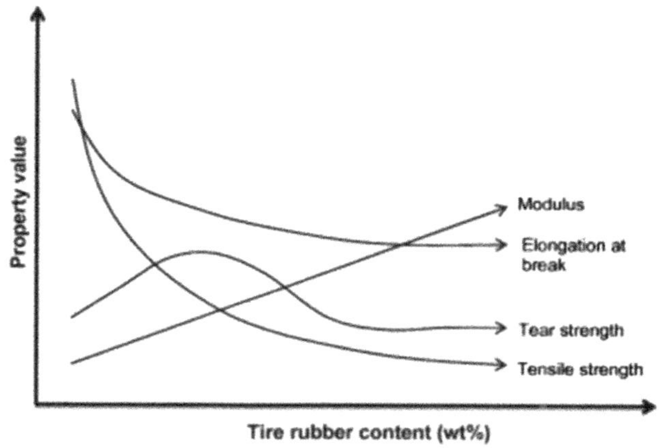

Figure 10.4 Effect of tire rubber content on mechanical properties of rubber blend. Reproduced from ref. 63 with permission from Elsevier, Copyright 2015.

Noriman, Ismail and Rashid investigated resilience (Figure 10.6), cure characteristics, mechanical properties, FTIR (Figure 10.7), DSC and SEM to determine the compatibility of SBR/NBRr blends with 10 phr of ENR-50 as a compatibilizer.[1] The addition of ENR-50 lowers the scorch time, cure time, minimum torque, elongation at break and rebound resilience of the blends compared to uncompatibilized samples.

It also depicts the enhanced tensile strength, tensile modulus, crosslink density and hardness at all blend ratios. SEM and DSC results confirmed better adhesion and improved compatibility of compatibilized SBR/NBRr blends. Noriman and Ismail developed SBR/NBRr blends reinforced with CB/Sil hybrid filler treated with and without silane coupling agent (Si69). Improved tensile properties, cure characteristics and good dispersion of filler onto the prepared blends were observed. The addition of filler content shows increased technical properties.[64]

Husseinsyah, Yeng and Ken analyzed the tensile properties and morphology of CR incorporated rPP/CR blends. Enhanced tensile strength, modulus of elasticity and decreased elongation at break for the prepared blends were

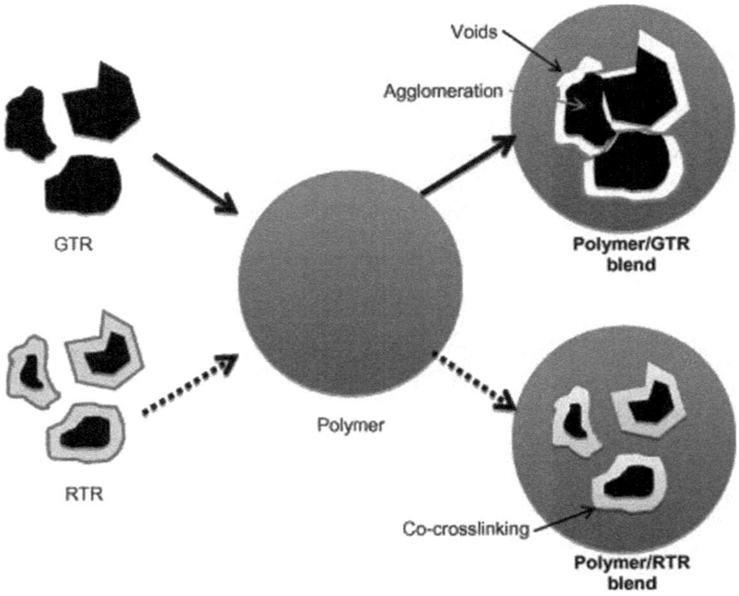

Figure 10.5 Schematic diagram showing the difference in microstructure of polymer blends with GTR and RTR. Reproduced from ref. 63 with permission from Elsevier, Copyright 2015.

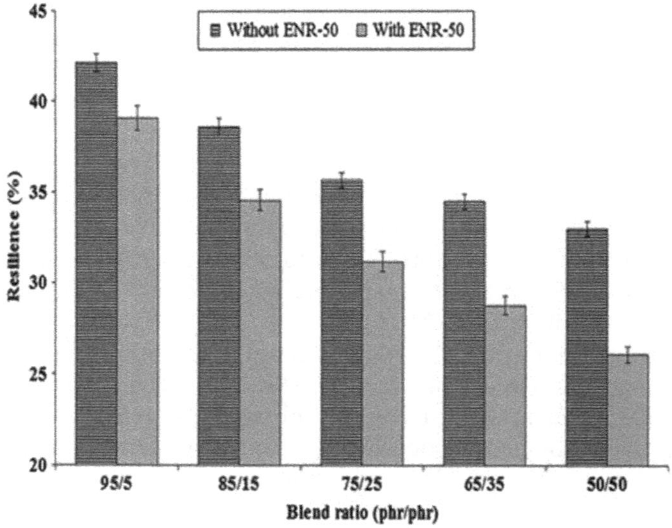

Figure 10.6 The resilience of SBR/NBRr blends with and without ENR-50. Reproduced from ref. 1 with permission from Elsevier, Copyright 2010.

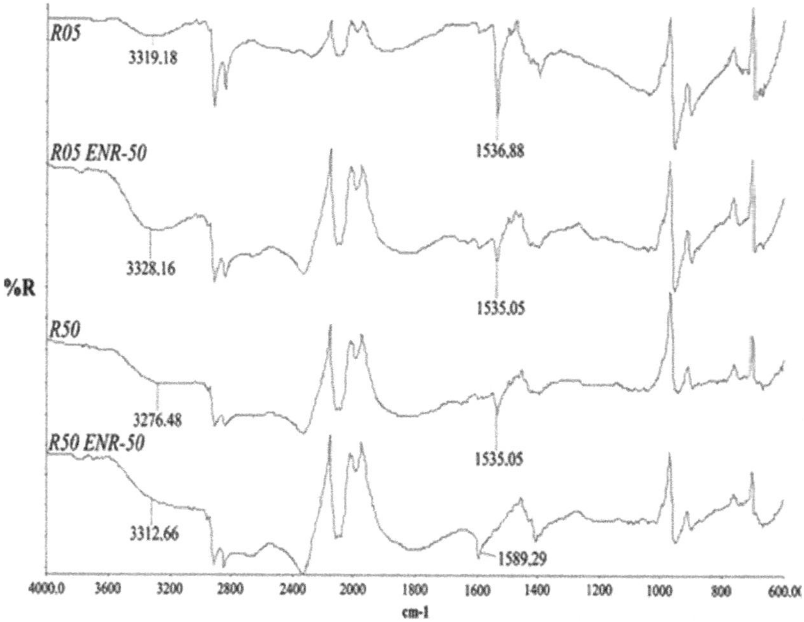

Figure 10.7 FTIR analysis of SBR/NBRr blend with and without ENR-50. Reproduced from ref. 1 with permission from Elsevier, Copyright 2010.

observed. Morphological images shows incompatibility between the polymers whereas CR particles pulled out from the rPP surfaces.[65] Luo and Isayev mainly addressed and analyzed the mechanical properties, rheology, morphology, thermal stability and melting behavior of GRT/polyolefin blends.[66] Various cure systems and compatibilizers were used to make the polymers more compatible. Finally, a phenolic resin cure system and maleic anhydride grafted PP compatibilizer results improved the Young's modulus and tensile strength of the GRT/PP blend system. Roche N reported the dynamic mechanical properties of neat EVA alone, GTR/EVA blends and GTR/EVA porous blends. The Williams Landel Ferry theory of expanding the dynamic properties over a wide range of frequencies was adopted to calculate the Young's modulus and loss factor for the blends. As a result, addition of GTR shows improved chain mobility of the polymer, loss factor and impact strength.[67] Colom *et al.* analyzed the mechanical properties, SEM analysis and particle size of reused tire powder reinforced recycled HDPE based composites. The effect of chemical acid pretreatment for example, H_2SO_4, HNO_3 or a mixture of both, and the particle size of the powder on the rubber were reported in detail.[68] The developed surfaces adhered to better mechanical interactions which improved the mechanical behavior of the blends.

Hassan *et al.* reported improvement in tensile strength, thermal stability, melting temperature and electrical resistance, but, decreased elongation at break, elastic modulus and micro-structural changes of waste polyethylene (WPE)/recycled waste rubber powder (RWRP) blend composites.[69] SEM micrographs depict the presence of some holes and voids around the talc particles in the polymer matrix. Grigoryeva *et al.* explained the functionalization of LDPE or HDPE with MAH and GTR with MAH or acrylamide *via* chemical or irradiation techniques.[70] The interfacial interaction of the prepared blends, grafting degree and molecular mass distribution was analyzed through FTIR and size exclusion chromatography (SEC). In addition, the authors extended their research by developing a dynamically vulcanized thermoplastic elastomer based on PE/GTR and EPDM rubber. Their phase structural changes and improved technical properties were reported through DSC, DMA and mechanical testing results. Galpaya focused on developing a dynamically vulcanized TPE based on PP and vNBR or rNBR to overcome the existing problem of disposing nitrile rubber waste materials.[71–73] Blend morphology, FTIR, tensile properties, swelling behavior and natural weathering resistance of the prepared blends were investigated in detail.[72] In addition, a polymer degradation study was performed using electron beam irradiation. The effect of physical properties and morphology of the diglycidyl ether of bisphenol-A epoxy resin (DGEBA) or MA-*g*-PP, epoxy resin (EP) or NR latex modified rNBR ($rNBR_{LM}$) as a compatibilizer on the prepared TPE were determined.[74,75] As a result, improved oil resistance was observed for DGEBA and MA-*g*-PP blends compared to $rNBR_{LM}$. The obtained results were also compared with the electron beam irradiated samples.[76]

Noriman and Ismail analyzed the effect of ENR-50 on the thermal properties, weathering resistance and fatigue life of SBR/NBRr blends.[77] TG thermograms of compatibilized blends show a lowered thermal stability, fatigue life and reduced char residue whereas improved interfacial adhesion, tensile properties and weathering resistance were observed compared to blends without ENR-50. Similarly, Noriman and Ismail described the effect of electron beam irradiation of the same blends with and without the polyfunctional monomer, trimethylopropane triacrylate (TMPTA).[33] Improved fatigue life and decreased tensile properties were observed due to the effect of polymer degradation. Salleh, Ismail and Ahmad investigated the minimum torque, maximum torque, scorch time and cure time due to the addition of vCR or rCR and CB onto the NR/CR blends. The results of rCR shows a pronounced effect compared to vCR loading. Elongation at break and swelling percentage of NR/CR blends decreased for both vCR and rCR addition, whereas tensile strength and fatigue life increased for 25 phr of vCR loading and decreased further.[78]

Li reported the microstructural changes and physical properties of blends containing recycled EPDM rubber, EMA, LDPE and thermoplastics.[79] The effect of a reactive and non-reactive compatibilizer on the prepared blends was extensively reported. Scaffaro *et al.* conducted research on the prepared blend based on recycled polyethylene and GTR. The results showed that the compound exhibited fairly good properties in terms of mechanical and rheological analysis with a relatively low concentration of GTR and proper thermal conditions.[80]

10.3 Conclusion

Recycling is one of the most important tools to reduce environmental pollution. Recycling, in a way, reduces the financial burden of a country by preserving valuable resources. Recycling of rubber based blends will soon be an accelerated field which will remove environmental and social impacts while contributing towards sustainable manufacturing. Therefore, recycling of waste rubber from used tires is becoming crucial and the speed at which the necessary technologies are developing may not be adequate. Recycling of rubber is the best process of disposing unwanted rubber, and is practiced in most developing countries. Reclaimed rubber offers technical properties equivalent to virgin rubbers. Rubber blends recycling helps to produce many useful products that can be used in asphalt, pavements, concrete and so forth.

References

1. N. Z. Noriman, H. Ismail and A. A. Rashid, *Polym. Test.*, 2010, **29**, 200.
2. F. Findik, R. Yilmaz and T. Koksal, *Mater. Des.*, 2004, **25**, 269.
3. G. Markovic, O. Veljkovic, M. M. Cincovic, V. Jonanovic, S. S. Jovanovic and J. B. Simendic, *Composites, Part B*, 2013, **45**, 178.
4. T. Laosee, P. Phinyocheep and F. H. Axtell, *J. Sci. Soc. Thailand*, 1998, **24**, 251.
5. M. Gordana, O. Veljkovic, S. S. Jovanovic, M. M. Cincovic and J. B. Simendic, *Chem. Ind. Chem. Eng. Q.*, 2011, **17**(3), 315.
6. M. Forrest, *Recycling and Re-use of Waste Rubber*, Smithers Rapra, 2014, ISBN: 978-184735-684-0.
7. J. Jose, Recycling of polymer blends, in: *Recent Developments in Polymer Recycling*, Transworld Research Network, 2011, p. 187, ISBN: 978-81-7895-524-7.
8. S. K. De, A. I. Isayev and K. Khait, *Rubber Recycling*, Taylor and Francis Group, 2005, ISBN: 0-8493-1527-1.
9. J. E. Mark, B. Erman and C. Michael Ronald, *The Science and Technology of Rubber*, Elsevier, 2013, ISBN: 978-0-12-394584-6.
10. V. L. Shulman, *Tyre recycling*, Rapra Rev. Rep., 2004, ISBN: 1-85957-489-0.
11. A. Amir, *J. Technol.*, 2015, **76**(5), 89.
12. C. Fengzhen and Q. Jialin, *Waste Manage.*, 2003, **23**, 463.
13. F. Sadaka, I. Campistron, A. Laguerre and J. F. Pilard, *Polym. Degrad. Stab.*, 2013, **98**, 736.
14. M. Awang, H. Ismail and M. A. Hazizan, *Polym. Test.*, 2008, **27**, 93.
15. B. Adhikari, D. De and S. Maiti, *Prog. Polym. Sci.*, 2000, **25**, 909.
16. F. A. Aisien, F. K. Hymore and R. O. Ebewele, *Environ. Monit. Assess.*, 2003, **85**, 175.
17. A. Shakun, J. Vuorinen, M. Hoikkanen, M. Poikelispaa and A. Das, *Composites, Part A*, 2014, **64**, 49.
18. D. A. Baeta, J. A. Zattera, M. G. Oliveira and P. J. Oliveira, *Braz. J. Chem. Eng.*, 2009, **26**(1), 23.

19. B. G. Soares, M. S. M. Almeida and P. I. C. Guimaraes, *Eur. Polym. J.*, 2004, **40**, 2185.
20. A. Tedesco, R. V. Barbosa, S. M. B. Nachtigall and R. S. Mauler, *Polym. Test.*, 2002, **11**, 21.
21. Z. Horak, V. Fort, D. Hlavata, F. Lednicky and F. Vecerka, *Polymer*, 1996, **37**(1), 65.
22. H. S. Ismail and A. M. M. Yusof, *Polym. Test.*, 2004, **23**(6), 675.
23. M. C. Bignozzi and F. Sandrolini, *Cem. Concr. Res.*, 2006, **36**(4), 735.
24. R. Schaefer and R. A. Isringhaus, Reclaimed rubber, in *Rubber Technology*, ed. M. Morton, Van Nostrand Reinhold, New York, 3rd edn, 1987, p. 505.
25. D. Sunday, S. Ogunniyi and M. Mureyani, *Iran. Polym. J.*, 2001, **10**(3), 143.
26. J. K. Kim, S. H. Lee and M. Balasubramanian, *Polymer*, 2006, **16**(4), 263.
27. A. Zanchet, L. N. Carli, M. Giovanela, R. N. Brandalise and J. S. Crespo, *Mater. Des.*, 2012, **39**, 437.
28. A. I. Isayev and B. Sujan, *J. Elastomers Plast.*, 2006, **38**, 291.
29. Y. Fang, M. Zhan and Y. Wang, *Mater. Des.*, 2001, **22**, 123.
30. R. S. George and J. Rani, *Iran. Polym. J.*, 1996, **5**, 199.
31. T. S. Kumar, *Degree of Doctor of Philosophy Thesis, Waste Tyre Management in Malaysia*, Putra University, Malaysia, 2006.
32. Y. Geoffrey, *Turn Old Tyres into New Ones*, 2011, available: http://www.thestar.com.my/story.aspx/?file=%2f2011%2f2%2f1%2flifefocus%2f7767404&sec=lifefocus.
33. N. Z. Noriman and H. Ismail, *Mater. Des.*, 2011, **32**, 3336.
34. Y. Yang, R. Boom, B. Irion, D. J. V. Heerdenm Pieter Kuiper and H. D. Wit, *Chem. Eng. Process.*, 2012, **51**, 53.
35. H. S. Ahmad, H. Ismail and A. A. Rashid, *Sains Malays.*, 2015, **44**(6), 835.
36. D. F. D. Silva, E. M. Araujo and J. A. D. Melo, *J. Aerosp. Technol. Manage.*, 2013, **5**(2), 231.
37. E. Ciro, J. Parra, M. Zapata and E. A. Murillo, *Ing. Cienc.*, 2015, **11**(22), 173.
38. A. A. Azmi, N. N. Zulkepli, M. Razif Nordin, M. M. A. Abdullah, I. Sandu, M. A. A. Mohd Salleh and I. Hanafi, *Mater. Plast.*, 2013, **50**(3), 175.
39. R. Sonnier, E. Leroy, L. Clerc, A. Bergeret and J. M. Lopez Cuesta, *Polym. Degrad. Stab.*, 2006, **91**, 2375.
40. A. K. Naskar, A. K. Bhowmick and D. S. De, *Polym. Eng. Sci.*, 2001, **41**(6), 1087.
41. Z. Hrdlicka, A. Kuta and J. Hajek, *Polymer*, 2010, **55**(11–12), 832.
42. G. Malarvizhi, N. Senthil and C. Kamaraj, *Int. J. Sci. Res. Publ.*, 2012, **2**(10), 1.
43. Z. T. Al-Malki, E. A. Al-Nasir, M. N. Khalaf and R. K. Zidan, *Open J. Polym. Chem.*, 2013, **3**, 99.
44. O. Boondamnoen, M. Ohshima, A. R. Azura, S. Chuayjuljit and A. Ariffin, *Int. J. Mod. Phys.: Conf. Ser.*, 2012, **6**, 391.
45. N. Z. N. Yahya, N. N. Zulkepli, H. Ismail, M. A. A. Mohd Salleh and R. Santiagoo, *Key Eng. Mater.*, 2016, **673**, 151.
46. H. Ismail and Suryadiansyah, *Polym. Test.*, 2002, **21**, 389.

47. F. J. Navarro, P. Partal, F. J. Martinez Boza and C. Gallegos, *Polym. Test.*, 2010, **29**, 588.
48. O. Boondamnoen, A. A. Rashid, M. Ohshima, S. Chuayjuljit and A. Ariffin, *ASEAN Eng. J. Part B*, 2013, **2**(2), 53.
49. C. Sangwichien, P. Sumanatrakool and O. Patarapaiboolchai, *Chiang Mai J. Sci.*, 2008, **35**(1), 141.
50. N. Z. Noriman, H. Ismail, I. Lokman Hakim, A. A. Azrem and S. T. Sam, *Aust. J. Basic Appl. Sci.*, 2013, **7**(5), 308.
51. N. Z. Noriman, H. Ismail and A. A. Rashid, *Polym.-Plast. Technol. Eng.*, 2008, **47**, 1016.
52. N. Z. Noriman, H. Ismail and A. A. Rashid, *Iran. Polym. J.*, 2009, **18**(2), 139.
53. N. Hayeemasae and H. Ismail, *Int. J. Polym. Anal. Charact.*, 2015, **20**, 514.
54. P. Phinyocheep, F. H. Axtell and T. Maiseaumsook, *Elastomer*, 1999, **34**(3), 212.
55. H. Ismail and Suryadiansyah, *Polym.-Plast. Technol. Eng.*, 2002, **41**(5), 833.
56. O. Grigoryeva, A. Fainleib, O. Starostenko, A. Tolstov and W. Brostow, *Polym. Int.*, 2004, **53**, 1693.
57. H. Supri, *J. Technol. Process.*, 2006, **5**(1), 58.
58. D. Mangaraj, *Rubber Chem. Technol.*, 2005, **78**(3), 536.
59. N. Z. Noriman, H. Ismail and A. A. Rashid, *Adv. Polym. Technol.*, 2012, **31**(2), 100.
60. N. Z. N. Yahya, N. N. Zulkepli, H. Ismail, M. A. A. Mohd Salleh and R. Santiagoo, *Adv. Environ. Biol.*, 2013, **7**(12), 3731.
61. N. N. Zulkepli, H. Ismail and A. Rashid, *Iran. Polym. J.*, 2009, **2**(18), 148.
62. P. S. Dekic, D. I. Temeljkovski, B. Rancic and S. Nusev, *J. Sci. Ind. Res.*, 2012, **71**, 295.
63. S. Ramarad, M. Khalid, C. T. Ratnam, A. Luqman Chuah and W. Rashmi, *Prog. Mater. Sci.*, 2015, **72**, 100.
64. N. Z. Noriman and H. Ismail, *J. Appl. Polym. Sci.*, 2012, **124**, 19.
65. S. Husseinsyah, C. M. Yeng and P. W. Ken, *Appl. Mech. Mater.*, 2015, **754–755**, 192.
66. T. Luo and A. I. Isayev, *J. Elastomers Plast.*, 1998, **30**, 133.
67. N. Roche, M. N. Ichchou, M. Salvia and A. Chettah, *J. Elastomers Plast.*, 2011, **43**, 317.
68. X. Colom, J. Canavate, F. Carrillo and J. J. Sunol, *J. Appl. Polym. Sci.*, 2009, **112**, 1882.
69. M. M. Hassan, R. O. Aly, J. A. Hasanen and E. S. F. El Sayed, *J. Appl. Polym. Sci.*, 2010, **117**, 2428.
70. O. Grigoryeva, A. Fainleib, J. Grenet and J. M. Saiter, *Rubber Chem. Technol.*, 2008, **81**(5), 737.
71. G. Dona and D. Galpaya, Processing characteristics, physical properties and morphological study of polypropylene (PP)/recycled acrylonitrile butadiene rubber (rNBR) blends, PhD thesis, Universiti Sains Malaysia, August 2009.

72. H. Ismail, D. Galpaya and Z. Ahmad, *Polym.-Plast. Technol. Eng.*, 2009, **49**(1), 110.
73. H. Ismail, D. Galpaya and Z. Ahmad, *Polym.-Plast. Technol. Eng.*, 2009, **48**(4), 440.
74. H. Ismail, D. Galpaya and Z. Ahmad, *Polym.-Plast. Technol. Eng.*, 2010, **49**(11), 1150.
75. H. Ismail, D. Galpaya and Z. Ahmad, *Polym. Test.*, 2009, **28**(4), 363.
76. H. Ismail, D. Galpaya and Z. Ahmad, *J. Vinyl Addit. Technol.*, 2010, **16**(2), 141.
77. N. Z. Noriman and H. Ismail, *J. Appl. Polym. Sci.*, 2012, **123**, 779.
78. S. Z. Salleh, H. Ismail and Z. Ahmad, *J. Polym. Eng.*, 2013, **33**(9), 803.
79. J. Li, Blends of EPDM rubber/thermoplastics, PhD thesis, Lulea University of Technology, 2008.
80. R. Scaffaro, N. Tzankova Dintcheve, M. A. Nocilla and F. P. La Mantia, *Polym. Degrad. Stab.*, 2005, **90**(2), 281.

CHAPTER 11

Recycling of Rubber Composites and Nanocomposites

RAMAKRISHNAN SHANMUGAM[a], SATHISH KUMAR PALANIAPPAN[b], RAJASEKAR RATHANASAMY*[a], KRISHNAMURTHY KASILINGAM[a] AND GANESH CHANDRA NAYAK[c]

[a]Department of Mechanical Engineering, Kongu Engineering College-Erode, Tamil Nadu, 638 060, India; [b]Department of Mining Engineering, Indian Institute of Technology Kharagpur, West Bengal, 721 302, India; [c]Department of Applied Chemistry, Indian Institute of Technology (Indian School of Mines) Dhanbad, Jharkhand, 826 004, India
*E-mail: rajasekar.cr@gmail.com

11.1 Introduction

Recycling is the process of changing waste materials into new products with the aid of appropriate technology. Recycling plays a predominant role in sustainable manufacturing. The chain of activities used for managing wastes is as follows: reduce, reclaim and recycle. Recycling is one of the most important tools to reduce environmental pollution. Glass, metal, plastic, textiles, paper, some electronics and biodegradable product waste, such as food, garden waste, are the most common recyclable materials at the current time.

Green Chemistry Series No. 59
Rubber Recycling: Challenges and Developments
Edited by Jin Kuk Kim, Prosenjit Saha, Sabu Thomas, Józef T. Haponiuk and M. K. Aswathi

Published by the Royal Society of Chemistry, www.rsc.org

A huge amount of energy has been utilized to make products from raw materials. Recycling requires much less energy and therefore helps to preserve natural resources, reduce air pollution (from incineration) and water pollution (from land filling) by minimizing the need for "conventional" waste disposal. Landfill sites occupy a significant area, as well as releasing harmful chemicals and greenhouse gases from refuse. The process of making products from raw materials is cheaper compared to recycled products. Recycling, in a way, reduces the financial burden of a country by preserving valuable resources.

Polymers are used broadly in several fields due to their excellent properties such as good mechanical performance, low density, low cost and comfort of processing.[1–3] Elastomers are polymers which are characterized by long-range elasticity and become useful when cross-linked to the preferred extent. According to a survey in 2006, the consumption of rubber worldwide is about 21.5 metric tons. Among the total consumption, natural rubber (NR) accounts for 43.1% and the synthetic rubber (SR) accounts for the remaining (56.9%).[4]

In recent years, researchers have focused on new alternatives for conventional polymers and as a result, polymer nanocomposites (materials with nanosized reinforcement) have been developed. The incorporation of nanofillers in the polymer matrix increases the physical properties due to a high surface to volume ratio. As pure rubber lacks an adequate modulus and strength, an additional reinforcing phase is essential for the practical use of rubber materials. Rubber is normally reinforced with fibers, carbon black (CB), and silicates. The concept of a nanofiller reinforced elastomer was demonstrated by the incorporation of nanoparticles in a rubbery polymer matrix such as clay into natural rubber.[5,6] The major application of rubber is in automobile industries (tires and tubes). They are also used in hoses, gaskets, oil seals, and so forth.

The nanocomposites can be classified into three types based on the number of dimensions of the dispersed particles in the nanometer range, as follows:

(i) Iso-dimensional nanofillers.
(ii) Elongated structure: two dimensions in the nano scale.
(iii) Nanocomposites in sheet form in which only one dimension is in nanometer scale.

Every year an enormous amount of waste polymers are produced from various sources such as households, industry, agriculture and so on. The effective treatment of waste polymers is done by the recycling process, as it overcomes the disadvantages of conventional processes (combustion or burying underground) such as formation of fumes, dust, toxic gases in the air and polluting underground water. Reclamation and recycling of plastics/rubber nanocomposites can be done through reuse, mechanical recycling, and chemical recycling.[7] Primary, mechanical, chemical or feedstock recycling and energy recovery are some other techniques to recycle waste polymers.[8,9]

Recycling of rubber is the best process of disposing of unwanted rubber, practiced in most developing countries. Reclaimed rubber offers technical properties equivalent to virgin rubbers. Rubber recycling helps to produce many useful products that can be used in asphalt, pavements and concrete.

11.2 Various Nanofillers

Complete dispersion of conventional fillers (CB and mineral fillers) is inhibited by their large structure. Therefore, strong shear fields and filler modification are required to ensure fine dispersion of conventional fillers in elastomers. Conventional fillers in the range of 20–40% should be added to obtain specific properties. The same properties can be obtained by adding a very low quantity of nanofiller (3–5 wt%). Nanoscience or nanotechnology is rightly considered as a revolutionary science comprising many disciplines. The term nano denotes the range of 1–100 nm. Nanofillers are particles added to composite materials to lower the consumption of more expensive binder materials to achieve superior properties of the composite material. The most commonly used nanofillers are carbon black, fly ash, carbon nanotube, nanosilica, nanoclay, layered double hydroxide (LDH), fullerenes, graphene and so on. Carbon black is a material produced by the incomplete combustion of heavy petroleum, and mainly used as the reinforcing filler in tires and other rubber products. It is also used as a color pigment for paints, inks and plastics. Fly ash, also known as flue-ash, is produced during coal combustion in power plants and is placed in landfill. A small amount is usually recycled and it is used in cement, asphalt concrete and metal matrix composites. Nano silica has the advantages of small particle size and narrow particle distribution. It has a large number of hydroxyl groups and shows high reflectivity to long wave, visible light and ultraviolet rays. Used widely in the field of adhesives, batteries, cosmetics, glass, plaster, paints, steel and chemical fibers, nanoclays are layered mineral silicates and are organized into various types such as bentonite, hectorite, halloysite, kaolinite and montmorillonite, depending on their morphology and chemical composition. Layered double hydroxides are layered crystalline materials which consist of positively charged layers and weakly bound charge-balancing anions located in the interlayer region. LDH possess flexibility in organic modifiers and modification processes. Carbon nanotubes (CNTs) are allotropes of carbon with a cylindrical nanostructure. They show excellent strength and electrical properties and are effective conductors of heat. There are two main types of nanotubes, namely single-walled nanotubes (SWNTs) and multi-walled nanotubes (MWNTs). A fullerene is a carbon molecule with at least 60 atoms of carbon in the shape of a hollow sphere, ellipsoid, tube, and many other forms. Fullerenes are identical in structure to graphite with pentagonal or sometimes heptagonal rings. Fullerene are widely used in microelectromechanical systems. Graphene is a single, tightly packed layer of carbon atoms that are bonded together in a hexagonal honeycomb lattice. Graphene has many extraordinary properties such as being the thinnest, lightest, and strongest material, and a good conductor of heat with unique levels of light absorption.

11.3 Recycling of Rubber Nanocomposites

Isayev discussed the harmful effects of waste rubber on cross-linked elastomers and the probable ways of recycling them.[10] The current hurdles in recycling rubber/waste tires outweigh the potential of using waste rubber as raw material. Various methods for recycling waste rubber and used tires such as grinding, reclaiming and devulcanization and the possibility of using recycled rubber as a raw material have been detailed previously. It has been indicated that retreading is an economical process that conserves energy. The recycling of rubber vulcanizates can be achieved by reclaiming technology, wherein the scrap tire rubber is made into a recyclable form with desirable properties.[10]

Isayev reported the current trends in recycling, reclaiming and devulcanization of natural and synthetic isoprene rubbers.[11] The devulcanization process in particular is a challenging process as it involves breaking down of sulfur induced 3D cross-links. Initially, waste rubber was incinerated to reclaim metal and energy. Due to environmental constraints, the chemical processing gained attention in spite of its inherent limitations. Many mechanical, thermo-mechanical and cryo-mechanical methods were proposed to fragment or shred rubber but devulcanization was still a big challenge. For the recycling of rubber, physical or mechano-chemical methods may be employed. In the physical process, the rubber is ground into small pieces with little effect on the chemical bonding. In the latter process, the thermoset elastomer is converted into a reprocessable and revulcanizable compound, similar to virgin rubber. The rubber reclaimed from mechanical grinding can be used in road filling applications. The micronized rubber was found to be compatible with virgin rubber. Compared to chemical, biotechnological and microwave methods, it was found that the ultrasonic method is the most appropriate for the effective recycling of NR.[11] The above mentioned study mainly relates to the processes concerned with the recycling of tires. The process includes a series of steps designed to reduce a rubber tire into a pulverized material. This process involves first shredding the tires or recyclable material into chips. The chips are then cooled to cryogenic temperatures at which said pieces become brittle. The brittle pieces are then pulverized in the second step. The pulverized material is then screened, removing large chunks of rubber from the mixture. Finally, the remaining fiber and magnetic material is separated out of the pulverized material, through the use of a magnetic separator and a vibrational (fiber) separator.[12]

Hamad *et al.* reviewed the current trends in recycling polymeric waste *viz.* traditional polymers and blends. The effects of recycling on the mechanical and physical properties of the recycled materials has been explored.

Mechanical recycling was found to be a better method to reclaim rubber compared to complicated chemical treatment. Incorporation of a smaller quantity of virgin polymers in the presence of a compatibilizer improved the properties of waste rubber considerably. The blending of PLA and PET waste was enhanced by addition of ABS. It was concluded that mechanical recycling, particularly blending, resulted in better properties of the composites.[13,14]

Abraham *et al.* reported the current advances in the recycling of rubber waste with a particular emphasis on the waste latex reclamation process.[15] Due to the exponential growth of the rubber industry, recycling of rubber waste has become a challenge in the environmental, social and economic domains. Waste latex rubber (WLR) constitutes around 10–15% of the total rubber used. This large quantity of WLR is the result of the rigid quality norms and the unstable nature of the latex compound. It has been reported that reclaiming of waste rubber could be a better solution to the disposal problem. Reclaiming is the process of breaking down the rubber vulcanizates with the aid of mechanical and chemical techniques. Normally 5–10% of reclaimed rubber can be added to raw rubber without altering the properties. In certain applications the quantity of reclaimed rubber may be up to 40%. However, the proportion of reclaimed rubber is restricted to 2–5% in radial tires. The reclamation process involves breaking of cross links in the vulcanized rubber. The mechanical reclaiming process is carried out in an open two-roll mixing mill at elevated temperatures. Extreme molecular weight collapse is reported at high temperatures (>200 °C). Addition of reclaimed rubber increased the cure time, modulus, abrasion loss, and hardness, with a decrease in tensile strength, elongation at break, tear strength and resilience. An alternative process using heat and chemicals has also been reported. A combinational process that involves breaking of carbon-to-carbon bonds and sulfur cross links is also employed for the reclamation of rubber. The scrap rubber can be ground in the cryo-mechanical process. It has been concluded that use of reclaimed rubber reduced the cost of the finished product with regard to energy conservation.[15,16]

Fang, Zhan and Wang emphasized the significance of recycling rubber and discussed the various methods and techniques for recycling rubber. In particular, powder rubber (PR) was discussed in detail. The importance of rubber recycling has been stressed from the ecological, economic and social perspectives. The recycling of rubber involves preliminary treatment and curing, followed by crushing. The waste rubber is then made into rubber powder. The different varieties of PR are listed as sundry PR and tire PR, rough PR, tiny PR, fine PR, ultra fine PR, and so forth. To enhance the adhesion strength of PR, it is subjected to activation and modification. The activation is usually carried out by making the particles highly polar and adding cross linked polymers. In spite of its size limitation, PR finds applications in a variety of applications such as a modifying agent for thermoplastics, elastomers to improve impact strength, and so on.[17,18]

Zhang *et al.* discussed the different technologies used for waste rubber recycling, particularly waste ground rubber tire (WGRT). As WGRT is a thermoset material, recycling is not a viable option. However, powder utilization has been proposed as a suitable technique for using WGRT effectively. When WGRT is incorporated in thermoplastics, thermoplastic elastomers (TPEs) are obtained with better properties. Waste polypropylene (WPP) is usually chosen along with WGRT to prepare TPEs. A very weak bond was found between WPP and WGRT. Either vulcanization or partial vulcanization may improve the molecular entanglement between the two. Furthermore, incorporation of a compatibilizer and bitumen was found to increase the adhesion between WPP and WGRT.[19,20]

Yang *et al.* reported on the various aspects pertaining to recycling of composites. The advent of development has expanded the usage of polymers in very large quantities; however, recycling of composites is not widely carried out owing to the heterogeneous nature of the composites. The different methods/techniques suitable for recycling are mechanical recycling, thermal recycling and chemical recycling. Recycling can be carried out in two forms: either material recycling or product recycling as re-use. However, due to limited demand and high processing costs, very few technologies exist for reclaiming the polymers. It is thus concluded that greater attention should be given to the recycling of composites.[21,22]

Reijnders discussed the recycling of elastomeric nanocomposites in detail with a special emphasis on particulate releases. As recycling is crucial for the environmental performance of the products, it is of great importance. The environmental performance can be measured based on the number of times a product can be recycled with minimal inputs. It was found that nanoparticles improve biodegradability. The major types of recycling methods used in the case of rubber tires are retreading, recycling and combustion. The rubber powder thus obtained can be used in a wide variety of applications. Rubber granulate can be used along with virgin rubber providing the modulus of the granulate is similar to that of the virgin. Thermal recycling is carried out in kilns where energy is partially recovered. Retreading is also preferred due to the high price of new tires and mineral oils. From the resource cascade point of view, retreading comes first, followed by usage of rubber powder/granulate and lastly, incineration. Pyrolysis of tires may also be carried out to obtain carbon black, oil and methane.[23]

Adhikari, De and Maiti summarized the various methods/techniques for reclamation and recycling of rubber. Reclaiming was proposed as one of the most viable alternative for effectively using waste rubber. The reclaiming process falls into two broad categories namely, the physical reclaiming process and chemical reclaiming process. Physical reclaiming was found to be better than the chemical reclaiming process. In the physical reclaiming process, the 3D cross-links in rubber are broken down with the help of external energy. The rubber chain is thus transformed into small fragments which can be easily blended with virgin rubber. The external energy is supplied by a two-roll mixing mill and milling. In the thermo-mechanical reclaiming process, the vulcanizate is suspended in solvent and ground to fine powder. The ground rubber is used in the civil engineering fields, tires, hoses, belts and mechanical goods, wire and cable. The advantages of reclaimed rubber are easy breakdown and mixing time, low power consumption, and better curing and aging characteristics.[24]

Raghavan highlighted the strategies used to recycle/reuse tires in an effective way. It has been pointed out that the simplest, yet effective, strategy is to produce tires with high resilience and durability. Despite the fact that tires have a longer life expectancy of about 80 000 miles, the handling of wastes as a solid form presents a huge challenge for the industry and society at large. The following methods were proposed to reuse/recycle/reclaim rubber in an energy-efficient, cost-effective and environmental friendly way. In retreading, a new tire the tread is placed over the worn out tire. In splitting, the bead wire is removed and the tread part is also cut from the tire. In rubber reclaiming, the covalent carbon–sulfur bond is broken, that is, devulcanized. Steam devulcanization is a common method along with microbial devulcanization. In pyrolysis, the chemical bonds of rubber are completely ruptured with the aid of heat. Scrap or crumb rubber can also be used in Portland cement/concrete. Rubberized asphalt and rubberized concrete are also gaining attention, owing to their superior performance and desirable properties. It was concluded that the strength of rubberized concrete can be improved significantly when granular rubber is replaced by fibrous rubber.[25]

11.4 Reclamation of Rubber Composites/Waste Tires

Marković *et al.* prepared polybutadiene rubber (BR) with different quantities of chlorosulphonated polyethylene rubber (CSM) and a fixed quantity (50 phr) of waste rubber powder (WRP). The samples were prepared to laboratory scale in a two-roll mill and the physical–mechanical properties were also evaluated. A slight increase in torque was reported upon incorporation of CSM, along with WRP. The decrease in cure rate index (CRI) was attributed to the difference in viscosity between BR and CSM. The highest value of tensile strength was obtained at 50 phr loading of CSM. Upon further loading, a decrease in the tensile strength value was observed. The rubber exposed to gamma irradiation exhibited better crosslinking. Therefore, incorporation of WRP in BR enhanced the properties to a significant level.[26]

Vladuta *et al.* investigated the effects of HDPE, LDPE, and TiO_2 nanoparticles on the compatibility of PET-rubber nanocomposites. The compounds were prepared from waste rubber and PET by the compression molding process. The thermophysical properties of various recycled materials are listed in Table 11.1. It was found that mixture processing of rubber faced serious hindrance due to the thermodynamic immiscibility of polymers. Additives, if added in small amounts, improved the mechanical strength of the nanocomposites significantly and reduced the interfacial tension between the

Table 11.1 The thermophysical parameters for the recycled material. Reproduced from ref. 27 with permission from Elsevier, Copyright 2015.

Recycled material	T_m [°C]	ΔH_m [J g^{-1}]
PET	248.83	40.64
Rubber	22.93	0.90
LPDE	121.81	0.32
HDPE	161.81	0.62

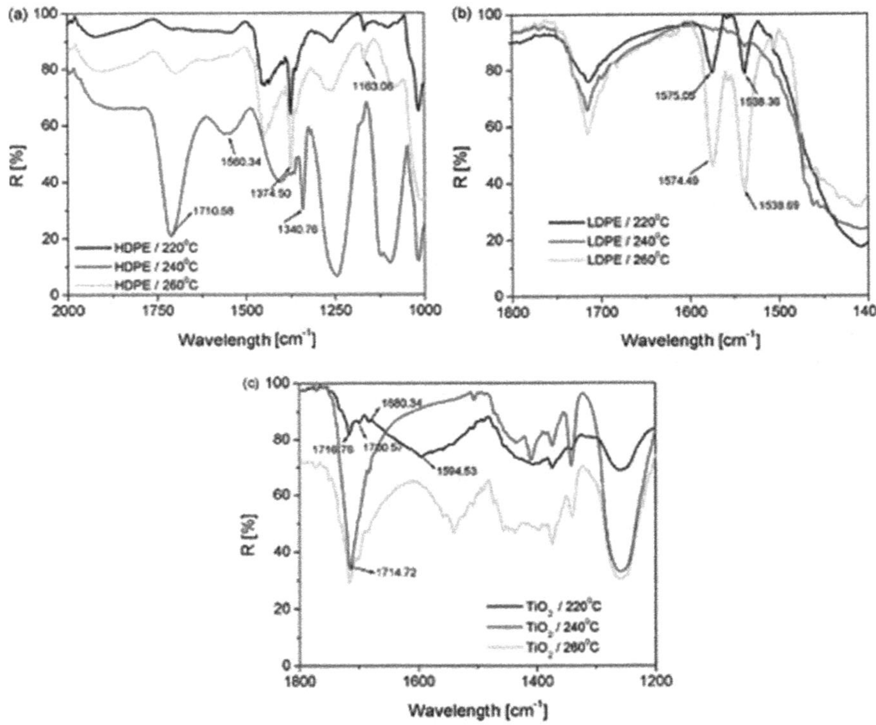

Figure 11.1 FTIR spectra for the composites with: (a) HDPE; (b) LDPE; and (c) TiO_2. Reproduced from ref. 27 with permission from Elsevier, Copyright 2015.

components. The biggest interfacial change was observed when HDPE was incorporated in the composites (Figure 11.1), whereas a larger degree of porosity was obtained upon adding LDPE. From Figure 11.2, it is evident that the composites were smooth and uniform. The addition of TiO_2 nanoparticles reduced the degree of crystallinity and simultaneously augmented the surface roughness.[27]

Noriman prepared styrene butadiene rubber (SBR)/NBRr blends using the two-roll mill method with five different compositions. The effects of ENR-50 as a compatibilizer on the physical and mechanical properties of the SBR/NBRr

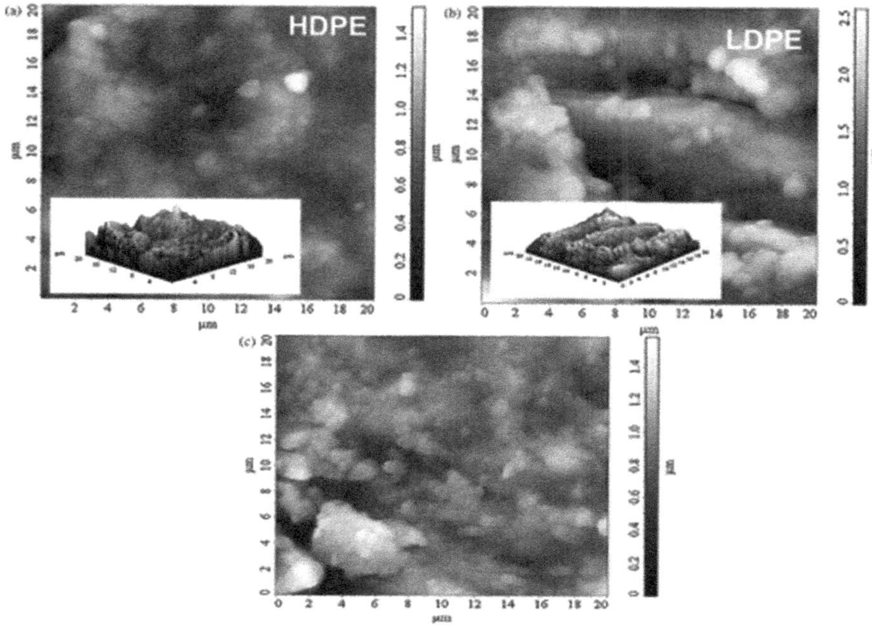

Figure 11.2 AFM micrograph for the samples with: (a) HDPE; (b) LDPE; and (c) TiO_2. Reproduced from ref. 27 with permission from Elsevier, Copyright 2015.

Table 11.2 Characteristics of materials. Reproduced from ref. 28 with permission from Elsevier, Copyright 2015.

Materials	Description	Source
Styrene butadiene rubber (SBR)	1502	Bayer (M) Ltd
Recycled acrylonitrile-butadiene rubber (NBRr)	Size: S1-117–334 μm	Juara One Resources Sdn, Bhd, Bukit Mertajam, Penang, Malaysia
Epoxidized natural rubber (ENR-50)	Grade EPOXYPRENE 50 specific gravity; 1.03	Malaysian Rubber Board
Carbon black	N330	Malayan Carbon (M) Ltd
N-cyclohexyl-2-benzothiazyl sulfenamin de (CBS), zinc oxide, stearic acid, sulfur and processing oil		Anchor Chemical Co (M) Ltd

blends was studied. The characteristics of the materials used for the study are shown in Table 11.2. To determine the compatibility of prepared blends cure characteristics and mechanical properties, Fourier-transform infrared spectroscopy (FTIR), differential scanning calorimetry (DSC) and scanning electron microscopy (SEM) were performed. The cure characteristics resulted in lower scorch time and cure time compared with pure blends without

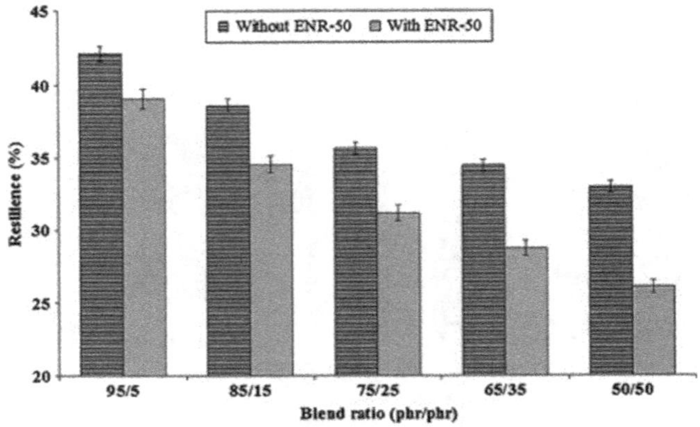

Figure 11.3 The resilience of SBR/NBRr blends with and without ENR-50. Reproduced from ref. 28 with permission from Elsevier, Copyright 2015.

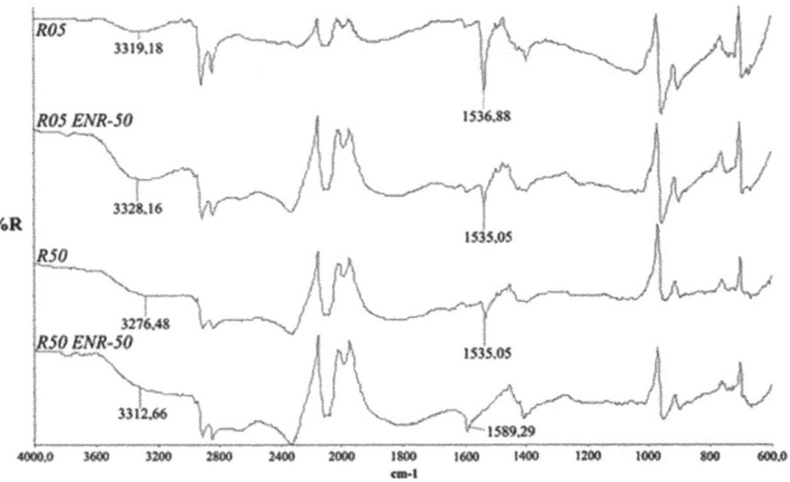

Figure 11.4 FTIR analysis of SBR/NBRr blend with and without ENR-50. Reproduced from ref. 28 with permission from Elsevier, Copyright 2015.

compatibilizer. The lower minimum torque and higher maximum torque were obtained for better processability of blends with the addition of a compatibilizer. An increase in the cross link density tends to give an enhancement in hardness for compatibilized samples. The DSC and SEM results displayed improvement in the compatibility of the prepared blends, better dispersion and adhesion. FTIR revealed the oxiraneandisoprene group of the NBRr/SBR blends and showed better compatibility with ENR-50 (Figure 11.4). An increase in the tensile strength, Young's modulus, and a decrease in the elongation at break for the compatibilizer SBR/NBRr were obtained. The rebound resilience property was found to decrease with an increase in NBRr content (Figure 11.3). The crack resistance also exhibited considerable

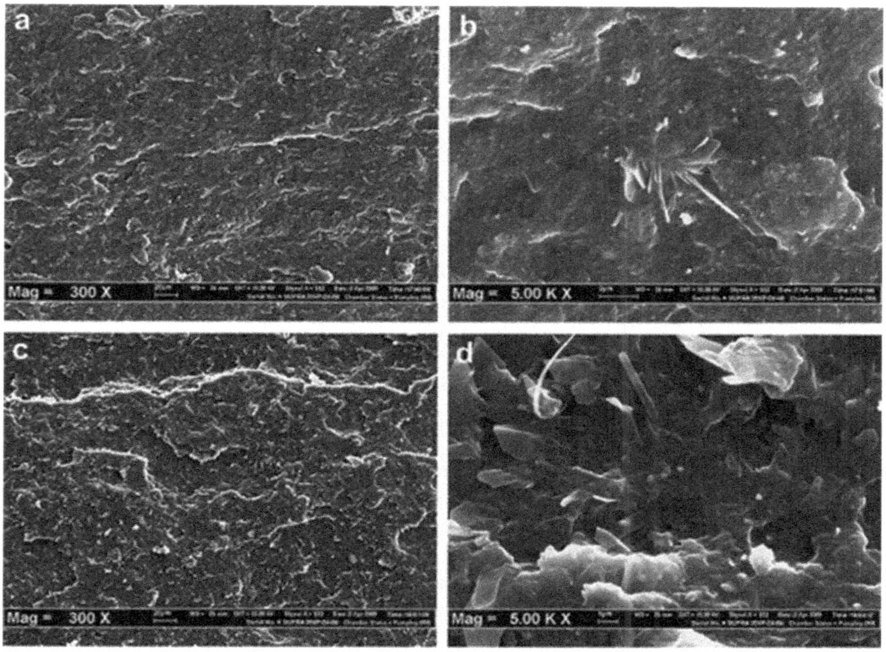

Figure 11.5 SEM micrographs showing tensile fracture surface of SBR/NBRr blends at 50/50 blend ratio (a) and (c) with ENR-50; (b) and (d) without ENR-50 at 300X magnifications and 5.00X magnification. Reproduced from ref. 28 with permission from Elsevier, Copyright 2015.

Table 11.3 Composition of the tested materials. Reproduced from ref. 29 with permission from Elsevier, Copyright 2015.

	LDPE (wt%)	GTR (wt%)	EVA (wt%)
PE	100	0	0
EVA	0	0	100
PEGTR30	70	30	0
PEGTR30EVA20	50	30	20

improvement due to better dispersion of the NBRr in the SBR matrix (Figure 11.5).[28]

Meszaros *et al.* explained that there are still some problems in the recycling of polymers, especially in recycling of cross-linked rubber materials. The composition of the tested materials is listed in Table 11.3. Ground tire rubber (GTR) is applied as a recycled material to overcome the lack of compatibility between the cross-linked rubber and thermoplastic matrix using the electron beam (EB) irradiation technique. GTR was incorporated into PE and PE/EVA matrices. The results show a beneficial effect on the polymer-GTR interfacial connection. EB irradiation shows an enhancement in the tensile strength, elongation at break and a positive effect on the impact

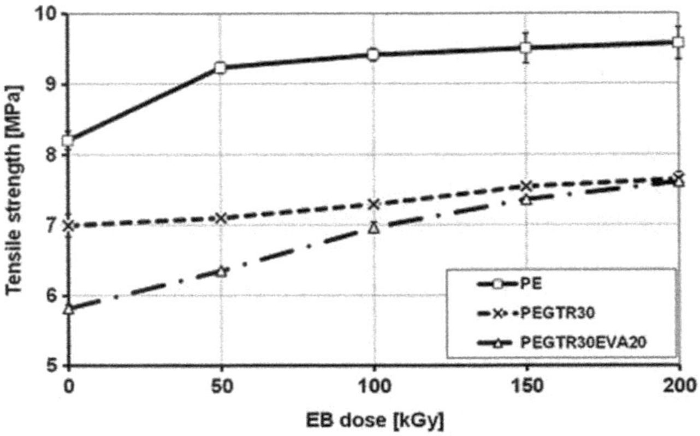

Figure 11.6 Tensile strength of the GTR containing thermoplastics as a function of the EB dose. Reproduced from ref. 29 with permission from Elsevier, Copyright 2015.

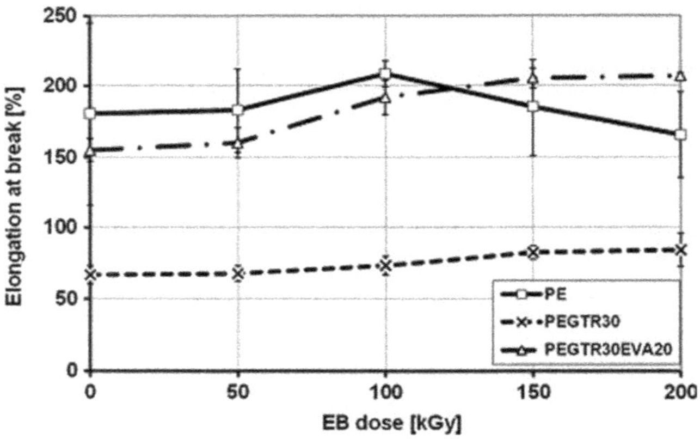

Figure 11.7 Elongation at break of the GTR containing thermoplastics as a function of the EB dose. Reproduced from ref. 29 with permission from Elsevier, Copyright 2015.

strength and dumping properties (Figures 11.6–11.9). The same trend was reported for perforation energy and Shore D hardness of the samples (Figures 11.10 and 11.11). Dynamic mechanical analysis (DMA) confirmed the EVA and EB compatibilization effect on the GTR and polymer matrix.[29]

Mohaved *et al.* reported that various automotive parts are manufactured using asethylene-propylene–dieneterpolymer (EPDM) rubber. Recycling of EPDM is a major problem. To overcome these drawbacks, devulcanization of waste EPDM automotive parts was done with heating and high pressure steam. Devulcanizing agents such as mercapto-benzothiazole sulfide (MBTS) and tetramethyl-thiuramdisulfide (TMTD), aromatic and aliphatic oils were

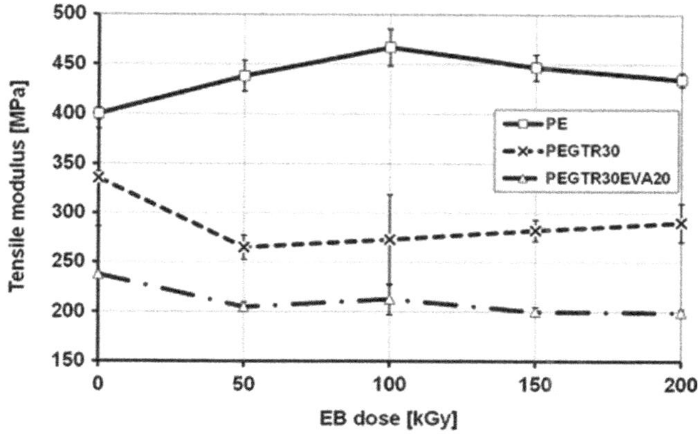

Figure 11.8 Tensile modulus of the GTR containing thermoplastics as a function of the EB dose. Reproduced from ref. 29 with permission from Elsevier, Copyright 2015.

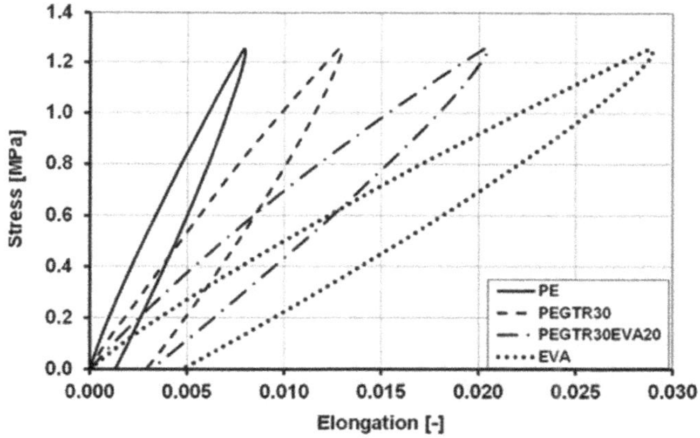

Figure 11.9 Hysteresis cycles of the recycled rubber containing thermoplastics after 150 kGy EB treatment. Reproduced from ref. 29 with permission from Elsevier, Copyright 2015.

used. The portion of EPDM rubber in automotive parts was replaced and revulcanized using a SEM vulcanization system. Replacement shows no effect on scorch, cure times, crosslink density, rate of cure, viscosity, hardness, compression set and modulus. The curing characteristics of the rubber compounds are shown in Table 11.4. Thus, reclaimed rubber used in a low percentage shows an adverse effect. The SEM micrograph shows the extent of filler dispersion in the rubber matrix (Figure 11.12). The results conclude that a new effective recycling route for waste EPDM powder could be in automotive strips.[30]

Aly investigated the mechanical and thermal properties of irradiated waste polyethylene (WPE)/Acrylonitrile–butadiene rubber (NBR) blends

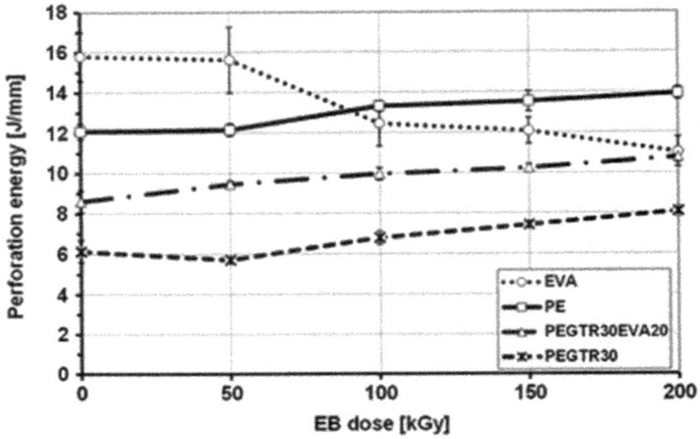

Figure 11.10 Fracture (perforation) energy of the recycled rubber containing thermoplastics as a function of the EB dose. Reproduced from ref. 29 with permission from Elsevier, Copyright 2015.

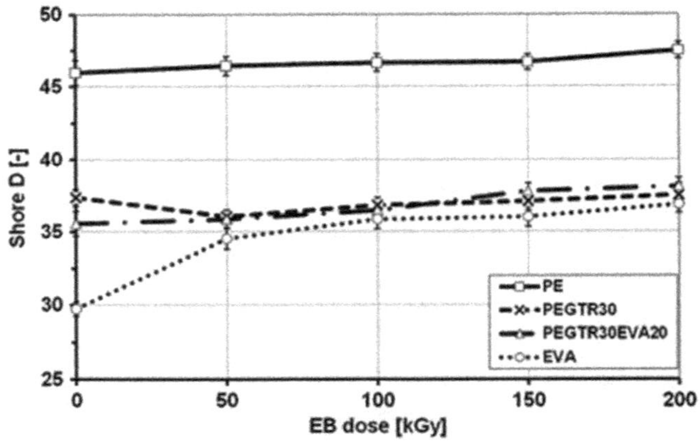

Figure 11.11 Hardness (Shore D) of the recycled rubber containing thermoplastics as a function of the EB dose. Reproduced from ref. 29 with permission from Elsevier, Copyright 2015.

by varying the content of NBR. The blends were prepared by the compression molding technique and were then subjected to gamma irradiation. The mechanical properties of irradiated blends were compared to that of unirradiated blends. A significant increase in the swelling percentage was reported at 50% loading of NBR. Upon irradiation, the swelling percentage decreased due to crosslinking. In the case of tensile strength, the irradiated blends exhibited higher values compared to unirradiated ones. The elastic modulus of the WPE/NBR blends decreased with the addition of NBR due to agglomeration. However, an increase in elongation was found to be a result of the increased NBR content. It was concluded that

Table 11.4 Curing properties of the rubber compounds. Reproduced from ref. 30 with permission from Elsevier, Copyright 2015.

Compound no.	Scorch time (t_{s1}) (m:s)	Cure time (t_{90}) (m:s)	Δtorque (dN m)	CRI (S^{-1})	Mooney viscosity (MU)
Reference	0:25	2:06	41.0	0.99	57.1
B-3-1	0:23	1:53	40.0	1.11	56.6
B-3-2	0:19	1:10	24.6	1.96	54.2
B-3	0:18	0:36	13.11	5.56	32.2
B-4-1	0:23	1:56	40.2	1.08	57.1
B-4-2	0:20	1:12	23.0	1.92	49.5
B-4	0:17	0:36	15.2	5.26	40.3
B-5-1	0:22	1:56	42.2	1.06	58.8
B-5-2	0:17	1:14	27.4	1.75	58.6
B-5	0:15	0:36	16.8	4.76	35.2
B-6-1	0:26	1:56	40.7	1.11	59.7
B-6-2	0:21	1:10	22.3	2.04	49.9
B-6	0:20	0:40	16.1	5.00	37.5

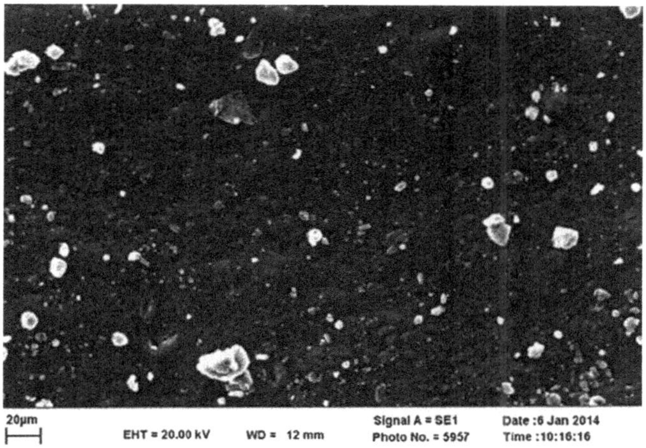

Figure 11.12 SEM micrograph showing the interior of the reference compound (RC) after freeze-fracture. Reproduced from ref. 30 with permission from Elsevier, Copyright 2015.

irradiation improved the tensile strength and elongation at break, but had no effect on elastic modulus. A remarkable enhancement in thermal stability also resulted from higher NBR content.[31,32]

Han *et al.* prepared biodegradable composites with poly(L-lactic acid) and recycled carbon fiber (PLLA/RCF). The compounds were prepared using the melt extrusion process. The effect of the silane coupling agent on the interfacial adhesion between RCF and PLLA was investigated. Significant improvements were reported in the crystallization behavior and mechanical properties of the compounds. The surface treatment of RCF ensured homogeneous distribution of RCF in the PLLA matrix. Addition of just 3 wt% of RCF improved the tensile and flexural properties of the composites considerably. This trend continued with further addition of RCF into the PLLA matrix. However, improvement in Izod impact strength was obtained only at 11 wt% of RCF and a negative trend was observed with further addition of RCF. Nucleation capacity in PLLA was reported after incorporation of RCF. Furthermore, remarkable improvements were observed in the heat resistance, processability and storage moduli of the PLLA/RCF composites.[33,34]

Bendjaouahdou and Bensaad studied the effects of organoclay on the properties of TPEs-polypropylene and natural rubber (PP/NR) blends. The samples were prepared using the melt extrusion technique. The mechanical properties were investigated along with the morphological analysis. Reduction of tensile strength and shore A hardness were observed in PP with addition of 10 wt% of NR. However, elongation at break, thereby ductility, increased with the addition of NR. This trend was attributed to the soft nature of NR in the rigid PP phase. Maximum tensile strength, hardness, and toughness were obtained at 3 wt% organoclay loading of PP/NR composites. The increase in the properties of the blend resulted from the intercalation of the PP/NR chains between the organoclay platelets. The interfacial adhesion between PP and NR was improved with the addition of organoclay. The incorporation of organoclay in the blend reduced the average rubber inter-particle distance. However, at further loading of organoclay (>3 wt%), the rubber domain size increased and weak interfacial adhesion resulted from droplet like dispersion of the filler. Better properties were observed at 3 wt% loading of filler in TPE.[35]

Shakun *et al.* explored the possibility of using nanodiamonds (ND) as fillers in soft rubbers. The ND powders are normally obtained by detonation synthesis, laser ablation and irradiation processes and do forth. The surface atoms of ND were found to contain a large number of highly-polar chemically reactive groups. Currently, NDs are obtained from the decomposition of explosives devoid of oxygen. The nanostructured explosives are found to influence the primary particle and aggregate size of NDs. Sometimes, the detonation yielded diamond blends (DB) consisting of ND and graphite structures. When added to elastomers, DB did not improve the properties significantly and acted like grease in the solid state. The incorporation of DB decreased the tensile strength, but improved the tearing and abrasion resistance. The addition of DB in isoprene rubber increased the fatigue resistance and strength. However, abrasion resistance remained unaltered. When DB was incorporated in SBR, better cohesive strength was obtained. The tear and abrasion resistance properties of NBR improved significantly when DB was added in small quantities (0.14–1.67 wt%). Direct incorporation of ND in silicone rubber led to agglomeration and certain special dispersion routes were explored. The NDs were also employed to strengthen recycled rubber. The incorporation of DB/NDs in TPE improved their properties considerably.[36]

Maragiannis *et al.* investigated the reasons behind the breaking of blades in tire strip cutting machines. Tire recycling involves three processes: strip cutting, making the strips into small pieces, making the pieces into powder. The cutting blade is normally made of AISI D2 steel. It was found that the blade failed after 3 months, even though the effective life of blade was nearly one year. The crack initiation was noted at the blade base *viz.* front connection hole and then propagated to the backside. As a result, the blade failed due to a brittle fracture. It was found that heat treatment did not induce cracking in the tool. As D2 steel contains carbides and tempered martensite. The failure in the tool resulted from the high induced shear stresses which required a redesign of the tool.[37,38]

Spizzuoco, Calabrese, and Serino investigated the lateral and vertical behavior of unbounded square carbon recycled rubber–fiber reinforced bearings (RR–FRBs) when subjected to seismic loading. Currently, FRBs are manufactured with NR. The authors explored the possibility of completely utilizing RR in FRBs. For the analysis, an isolator was designed without any end plates and this was subjected to static and dynamic experimental analyses. The obtained results were verified by FEA analysis. In unbounded FRBs, the compressive loads create a moment that is balanced by the moment induced by the applied shear. The FRBs made from recycled rubber were light weight and cheaper. The compression test conducted on the specimen did not affect the vertical stiffness significantly. Debonding was also not observed after the test. Enhanced seismic performance was noted from the cyclic lateral tests. Larger damping ratios were also reported for the FRBs. The results were verified with finite element analysis (FEA) and were found to be similar.[39,40]

Thompson and Yeung conducted a multiple-pass study on layered silicate–thermoplastic olefin elastomer (TPO) to determine the impact of processing on the properties of the material. TPOs are mainly composed of polypropylene and an elastomeric resin such EPDM, ethylene-propylene rubber (EPR), or an ethylene–α–olefin metallocenecopolymer. Organoclay is used as a filler for the development of polymer layered silicate (PLS). The greatest limitation in recycling PLS is found to be enhanced intercalation/exfoliation. In a single pass of processing, PLS nanocomposites did not undergo immediate exfoliation of organoclay. A percolating network of tactoids in the matrix was reported in the subsequent steps of processing. It was concluded that the best recycling strategy for this class of material is the introduction of both additional antioxidant and compatibilizer in subsequent passes of processing through appropriate machinery.[41]

Ausias *et al.* developed a micro-mechanical model for TPEs made from polypropylene and waste rubber particles. As the cross-linked elastomers cannot be subjected to remolding, the only alternative available is recycling. The most established method of recycling rubber is to grind it by various milling processes. The output of milling processes is a fine rubber powder. The fine powder can be applied as a filler for preparing TPEs. In order to obtain better mechanical properties of TPEs, it is necessary to obtain perfect interaction between the two phases. The radical reaction in the melt-mixing process was found to improve the interaction behavior of WRP with PP. The properties were comparable to that of commercial grade TPEs. A micro-mechanical model was successfully developed to predict the behavior of TPEs through FEA simulation.[42]

Michel Biron and Odile Marichal summarized the future prospects for thermoplastic composites. It has been reported that recycling of thermoplastic composites is a tedious process due to technical, economic and environmental constraints. The major recycling techniques are mechanical recycling, chemolysis, solvent extraction, thermolysis and co-combustion. Mechanical recycling of composites involves shredding and grinding of polymer waste into a partially reusable compound. However, the performance level of the recycled polymer is not appreciable. In chemolysis, polymers such as polyurethane can be depolymerized. Soluble thermoplastics may be subjected to the solvent extraction technique. The recyclates are found to possess only moderate properties owing to the presence of contaminants/pollutants. It is suggested the mechanical properties of the recyclants can be improved by adding suitable additives/compatibilizers during processing.[43]

Fukumori *et al.* reported the development of a new devulcanization technique to recycle automotive rubber. In this technique, the odor of the recycled rubber can be successfully eliminated. The properties of rubber recycled by the new technique were found to be comparable to virgin rubber. Material recycling is still one of the most preferred methods to recycle rubber on the basis of energy stability. However, the presence of 3D cross-linked structure in rubber makes its recycling tedious. Conventional recycling processes such as the mechanical pan process, microwave devulcanization and ultrasonic devulcanization yielded inferior rubber unsuitable for practical applications. A new deodorization method was introduced in the continuous recycling technique (devulcanization), in which water at high pressure was injected into a barrel to remove the odor. It has been reported that incorporation of recycled rubber tire tread produced a wear performance similar to that of a new tire.[44]

Zaharescu *et al.* investigated the effect of accelerated electron beams or γ-rays on the behavior of a butyl rubber specimen. The study was carried out to analyze the availability of butyl rubber to radiation processing. The modifications in properties such as the average weight, molecular weight, gel content and so forth were evaluated to understand the impact of the irradiation dosage. The renewed interest in using radiation for recycling/reprocessing of polymers is attributed to advantages such as the low cost, high productivity, energy conservation and low level of contaminants. It was found that butyl rubber subjected to irradiation can be reclaimed. The butyl rubber exposed to a radiation dose of 100–150 kGy resulted in a material with medium plasticity. On the other hand, at a higher dose ($\geq$ 400 kGy) the material was transformed into an adhesive.[45]

Wu and Zhou suggested a novel way to dispose of waste tires. The waste tire rubber (WTR) is converted into a useful substance *viz.* an oil absorptive material and could be a way to solve the oil spills/pollution problem. The recycled rubber was found to possess an inherent ability to absorb hydrocarbons. However, the absorption capacity of rubber was limited to four to six times its weight. An attempt was made to make an oil absorptive material from WTR and 4-*tert*-butylstyrene (tBS) by adopting a graft copolymerization-blending technique. Divinylbenzene (DVB) and benzoyl peroxide (BPO) were used as the crosslinker and initiator, respectively. The oil absorbency of the RR depended on the DVB content. The maximum absorption value of 24.0 g g^{-1} was reported at 1 wt% of DVB. However, the absorbency started to become reduced upon further loading of DVB. It was concluded that a huge potential exists in the field of oil absorption for recycled rubber.[46,47]

Sadaka *et al.* proposed a novel method to recycle waste tires. The metathesis technique transforms waste rubber into telechelic polymers. The novel material (high value liquid rubber) obtained from this technique can be applied in many practical applications. The waste rubber was suspended in dicholoromethane or toluene and subjected to magnetic stirring. The catalyst solution was then added into the main reaction at room temperature. Finally, a large quantity of methanol was added to quench the content. In this process, the tire rubber is degraded into powders and granules. It was reported that a precise selection of catalyst concentration and solvent influenced the degradation of waste tire rubber.[48]

Sadaka *et al.* described the use of periodic acid for the degradation of ground WTR. A study has estimated that 6 kg of waste rubber is produced per individual every year in developed countries. One major hindrance in tire recycling is the resistance of tire against physical, mechanical and biological degradation. A different method for polymer degradation is proposed to overcome this difficulty. It was found that the reaction time and periodic acid quantity influenced the extent of degradation. The extent of degradation was estimated by conducting the average molecular weight analysis. The degradation of waste tire rubber was also carried out by using periodic acid through oxidative cleavage to obtain carbonyl telechelic oligomers. The emphasis of this method is on severing the carbon–carbon double bonds followed by cleavage of the oxirane units. As periodic acid is cheap and easily available, it is used for selective oxidative degradation of polyisoprene present in waste rubber.[49]

Yasin *et al.* proposed a novel method to recycle WTR using gamma rays and SBR. The influence of polyfunctional monomers (PFMs) and absorbed dose on the final characteristics of SBR/WTR blend has been discussed. The blends were prepared by varying the contents of PFMs and WTR. The blends were then exposed to gamma radiation so that a crosslink is established between the PFMs and the polymer. The PFMs were introduced in the blends to lower the effect of radiation and enhance the properties with minimal degradation. Two types of PFMs were used for this purpose namely, trimethylol propane trimethacrylate (TMPTMA) and trimethylol propane triacrylate (TMPTA). It was concluded that blends crosslinked in the presence of TMPTA exhibited better physical and mechanical properties over the blends cross-linked in the presence of TMPTMA.[50,51]

Kim *et al.* discussed the properties of TPEs using waste rubber and two polyolefins. The polymer blends were processed with the aid of a twin-screw extruder. It has been pointed out that recycling is the only option available to use rubber effectively. One proven method is the use of ultrasonic treatment for surface modification. It has been indicated that the ultrasonic waves at certain pressure and heat, effectively break the 3D cross linked structure. Blending the waste rubber with some other material such as virgin material is gaining attention over recycling. The effect of SEBS-*g*-MA compatibilizer on LDPE/waste tire powder and PP-*g*-MA/waste tire powder blends was investigated. The properties of these blends were then compared with blends exposed to ultrasonic waves. It has been reported that the blends exposed to ultrasonic waves suffered degradation and the properties were poor. The properties of untreated blends in the presence of compatibilizer exhibited superior properties.[52,53]

Kaminsky, Mennerich and Zhang proposed an indirect heated fluidized bed process for effectively recycling (pyrolysis) of natural and synthetic rubber. It has been indicated that pyrolysis of natural and synthetic rubber yields at 600–650 °C gives high amounts of isobutene and isoprene. The pyrolysis of tire rubber was carried out in a reactor at 600, 650 and 700 °C at residence times in the reactor of 1.9, 2.6 and 3.3 s with N_2 and steam as the fluidizing gas. It was observed that a lower amount of oil and more gas and CB were formed when the temperature was raised. At high temperature (700 °C), the ethane and propene contents were high. It has been reported that the pyrolysis temperature greatly influences CB production from NR. At low temperature pyrolysis (600 °C) high quality CB was obtained. At elevated temperatures, the quality of CB was not high.[54]

Llompart *et al.* investigated the presence of harmful organic chemicals in surfaces made from recycled rubber tires. Solvent extraction and GC–MS methods were used to analyze the existence of organic chemicals. The samples were collected from various playgrounds made from recycled tire rubber. A large number of volatile and semi-volatile compounds, particularly polycyclic aromatic hydrocarbons (PAHs) were found in the samples. The content of PAHs varied between 1 $\mu g\ g^{-1}$ and 200 $\mu g\ g^{-1}$ in playgrounds. However, alarming levels of PAHs were reported in the samples collected from road pavements (400–17 000 $\mu g\ g^{-1}$). The results indicated the presence of harmful organic chemicals in recycled rubber and it was concluded that extreme caution should be taken during recycling of tire rubbers.[55,56]

Feriha *et al.* investigated the optimum proportion of reclaimed rubber and crumb rubber (a granulated material derived from scrap tires) to make the various parts of a tire. The master batches were prepared by varying the proportion of reclaimed rubber and crumb rubber along with actual ingredients. The master batches were then compared with a control, *viz.* a sample prepared wholly with actual ingredients. The results indicated that the content of reclaimed rubber is limited to 15 phr in radial tread and 25 phr in radial side walls. For crumb rubber, the proportions are 10 and 20 phr for the radial tread and side wall respectively. However, addition of a higher quantity of reclaimed and crumb rubber reduced the vulcanized tension property. The incorporation of reclaimed and crumb rubber reduced the SBR content by 7.7% in the radial tread, and the NR content by 12.8% in the radial side wall respectively. It was concluded that energy conservation and cost reduction can also be achieved.[57,58]

Ferreira *et al.* investigated the possibility of forming composites with waste-polyamide (W-PA)/waste-polyethylene terephthalate (W-PET) blends and glass fiber. The physical and mechanical properties of the composites were studied. The incorporation of 5% glass fiber improved the impact strength of composites by 100% compared to pure W-PA. Impact resistance up to 105 $J.m^{-2}$ was achieved when the glass fiber content was increased up to 30%. When 5% glass fiber was added to the blend, the tensile strength increased to 24 MPa. However, no significant raise was noticed with further addition of glass fiber. Even though the mechanical performance is a function of reinforcement, optimal properties were reported at 15% loading of glass fiber. At higher loading the properties tend to follow a negative trend due to poor dispersion and high shear load. The thermal stability was found to be constant after glass fiber incorporation. Maximum impact resistance was observed for blends with 75% PET and 25% PA. Tensile and other properties exhibited negligible improvements.[59,60]

Cao *et al.* prepared carboxylated (XSBR)/cellulose nanocrystals (CNs) latex composites using a two-roll mill. The physical, thermal and mechanical performances of the composites were analyzed in detail by adding varied quantities of CNs (0 phr, 3 phr, 5 phr, 10 phr and 15 phr). A strong filler–filler network was established which enhanced the homogeneous dispersion of CNs in the XSBR matrix. An increase in the maximal torque value (MH) was reported upon addition of CNs in the matrix. The maximum value was achieved at 15 phr loading of CNs. The storage modulus (G′) of the composites exhibited the same trend *viz.* increased with filler loading. The storage modulus (E′) increased with incorporation of filler in the matrix. The glass transition temperature (T_g) increased from 48.45–50.64 °C at 3 phr filler loading. However, at further loading of filler Tg showed the reverse trend. The tensile strength, tear strength and shore A hardness increased with addition of CNs in the XSBR matrix. The augmentation in properties of composites was attributed to good interfacial adhesion between the CNs and XSBR.[61,62]

Song *et al.* prepared the montmorillonite (OMt)/*cis*-1,4-polybutadiene rubber nanocomposites using the solution intercalation method and evaluated the properties and structure of the samples. The influence of recovered solvent on the change of butadiene polymerization was investigated. The OMt particles exhibited a homogeneous distribution in the BR matrix, as was evident in the transition electron microscopy (TEM) images. At 6 wt% filler loading, the tensile strength of the OMt/BR nanocomposites improved considerably from 2.2–6.2 MPa and the elongation at break reflected the same trend. The glass transition temperature of the nanocomposites was 3.5 °C higher than pure BR. On the other hand, the tan δ value was lower. The improvement in mechanical properties is due to the homogeneous distribution of filler in the matrix. The recovered solvent was found to have no influence on the polymerization of butadiene and reuse of this was recommended.[63,64]

Chen *et al.* investigated the feasibility of using methacrylic acid (MAA) as a novel *in situ* interfacial modifier for SBR/boehmite composites. The samples were prepared by the direct blending method. It was found that MAA reaction was confined to the coordination reaction and bridging bidentate structure. It was evident that MAA can organically modify boehmite (BM). When 40 phr BM was added to SBR, the scorch time and vulcanization time remained unchanged, but the maximum and minimum torque (MH and ML) values showed variations. Higher loading of MAA tended to decrease the value of MH. The mechanical properties of the composites improved significantly through *in situ* modification of BM by MAA. The tensile strength increased up to 24 phr loading of BM and started decreasing at higher loading. The tear strength and hardness exhibit similar trends. The improvements in mechanical performance were attributed to better dispersion of BM and increased interfacial bonding introduced by *in situ* modification.[65]

Zanchet *et al.* prepared SBR composites filled with SBR industrial scraps (SBR-r) at 80 phr loading. The rheometric and physical–mechanical properties of the revulcanized composites were evaluated. The samples were subjected to two kinds of aging processes, namely thermo-oxidation and photo-oxidation. The SBR-r was devulcanized before incorporation in SBR. The treatment time of SBR-r affected the devulcanization process. The SBR-r which was devulcanized for 2 mins in a microwave oven was found to considerably improve the properties of the composites. However addition of SBR-r had no effect on the processing characteristics of SBR composites. With addition of SBR-r, the hardness remained unaffected, but the tear strength was reduced. The tensile strength and elongation at break improved significantly with addition of SBR-r.[66]

Al-Aqeeli investigated the possibility of using crumb rubber (CR)-modified coatings on metal substrates. The coatings, reportedly possessing anticorrosive properties, were prepared by adding varied amounts of CR, liquid epoxy resin and hardener. The effects of CR particle size and concentration on the mechanical properties of coatings were evaluated. At the CR concentration of 24–50%, better adhesion was observed. However, when the concentration was less than 24% or more than 50%, non-uniformity and voids were reported. An inverse relationship existed between hardness and CR concentration. Smooth particles produced fine surfaces but coarse particles were also included to improve hardness. Optimum hardness was found in the coatings with both fine and coarse particles.[67]

Zaharia *et al.* evaluated the effect of adding WTR on the incineration of WT/metallurgical coke (MC) blends in an electric arc furnace. The blends were prepared by varying the contents (10, 20 and 30%) of WT in the blends and they were subjected to pyrolysis and combustion at 1473 K. The weight loss of 30% in the rubber blends was swifter than the other blends and 100% MC. The degradation time of MC also decreased with increasing rubber content. The weight loss was also very high in the case of 30% rubber mixture. Practical difficulties prevented further addition of rubber in the MC.[68]

Amari *et al.* calculated the life cycle energy budget based on the energy consumed for producing tires and the energy recovered while burning the tires. The tires can be used as fuel directly or tire derived fuel (TDF). It was evident that rubber had a smaller moisture content, greater combustible matter and less carbon when compared with coal. The TDF samples produced 9–23% of ash when compared to 11% of ash in the case of coal. It was reported that the disposal of ash produced by TDF combustion was easier owing to the absence of heavy metals. The incineration of TDF also reduced the emission of NO_x gases. Retreading was suggested as a viable means of effectively recycling rubber. It was concluded that only 40% energy recovery was possible for tyres.[69]

Kaminsky and Mennerich investigated the possibility of conducting pyrolysis of rubber at lower temperatures, preferably lower than 700 °C. The SBR tire was subjected to pyrolysis at a temperature range of 500–600 °C with steam as the fluidizing gas. The effect of temperature and the fluidizing gas on the quality of CB was investigated. The production of gas was reduced from 9 to 5% at 600 and 500 °C respectively. Maximum gas production was reported at 550 °C. The yield of CB was higher at 600 °C (40%) when compared to the yield at 500 °C (30%). The steam as a fluidizing gas did not influence the production of CB in any way. It was concluded that the best quality of CB was obtained at 600 °C with inert N_2 as the fluidizing gas.[70]

Merchant and Petrich discussed the conversion of scrap tires into activated carbon. Activated carbon with a surface area greater than 500 m^2g^{-1} can be obtained from scrap or used tires. The utilization of reclaimed rubber in virgin rubber is inhibited by the incompatible properties of scrap rubber. The process of using reclaimed rubber with virgin rubber is also costly. In a typical tire, the composition of SBR and CB are 62 and 31% respectively, along with other constituents. Therefore, when tires are subjected to controlled pyrolysis at about 850 °C, activated carbon can be obtained.[71]

Naskar *et al.* proposed a novel method for value addition of waste tire rubber derived materials. Waste tire rubbers can be crushed into rubber particles with a size in the micron range through cryogenic techniques. The rubber in the micronized structure was dissolved in a hot oleum bath to yield a sulfonated rubber slurry. The slurry, when filtered, washed and compressed, yielded a solid cake. Thus, carbon was obtained from the modified rubber cake through pyrolysis in a nitrogen atmosphere. Although many methods are available for recycling/reclaiming tire rubber, their applications are limited. The CB obtained is not fit for reinforcing, but can be used as a cheap additive for certain electrode applications.[72]

Wang *et al.* proposed a cost effective method to make electrode materials for microbial fuel cells (MFCs) with graphite paint coated crumb rubber. The crumb is derived from scrap/recycled tires. The scrap tires which are stockpiled are prone to fire accidents and occupy a large area, in addition to the health hazards. A filtration system prepared from crumb rubber is effectively used to treat waste water and ballast water from ships. The rubber particles derived from tires exhibit better bonding when used in bioreactors for biofilm attachment. Crumb rubber with 2–4 layers of graphite coating offers a huge potential as anodes in MFCs. The use of CR in MFCs will have greater economic and environmental impacts.[73]

Karabork *et al.* studied the cure characteristics and tensile properties of SBR/GTR composites with different contents of GTR. The GTR was devulcanized with the help of a microwave technique with different exposure times to yield devulcanized rubber (DV-R). The main hurdle in rubber recycling is the presence of vulcanization induced 3D cross-links. Hence, the cross-links should be broken down before using the scrap rubber. The recycling of GTR is also gaining importance due to heightened environmental awareness. In vulcanization induced by microwave, only the carbon–sulfur and sulfur–sulfur bonds are broken and the carbon–carbon bond is left untouched. This ensures that the GTR is reclaimed without depolymerization and the resulting material will be identical to virgin rubber.[74]

Xin *et al.* studied the influence of WGRT and molding conditions on the properties of PP/WGTR composites. The polymer industry is exploring the use of WGTR as a means of reducing material consumption and improving impact resistance. Thermoplastic elastomers can be obtained by incorporating scrap rubber in a polymer for foaming applications. PP/WGRT blends offer distinct advantages such as light weight, low cost and fuel economy. The foamed polymers are prepared by the injection molding process. Hence, it has been rightly indicated that value-added products obtained from PP/WGRT can reduce waste handling to some extent.[75]

11.5 Application of Rubber in Construction

Meddah *et al.* experimentally investigated the possibility of incorporating shredded rubber tire in roller compacted concrete pavement (RCCP). The rubber particle was added in some parts of natural crushed particle. Thermal and mechanical properties were evaluated and compared according to the

Table 11.5 Tensile strengths. Reproduced from ref. 76 with permission from Elsevier, Copyright 2015.

	Splitting tensile strength MPa			Bending strength MPA		
Mix	7d	14d	28d	7d	14d	28d
RCCP0	2.53	2.75	2.79	4.9	5.03	5.82
RCCP5	2.54	2.73	2.76	4.7	4.68	4.42
RCCP10	1.87	2.02	2.47	4.25	4.7	5.3
RCCP15	1.74	1.77	2.24	4.12	4.5	4.7
RCCP20	1.69	1.96	2.17	4.15	4.34	4.49
RCCP25	1.14	1.89	2.15	2.89	3.19	4.25
RCCP30	1.63	1.81	2.13	3.23	3.57	4.0

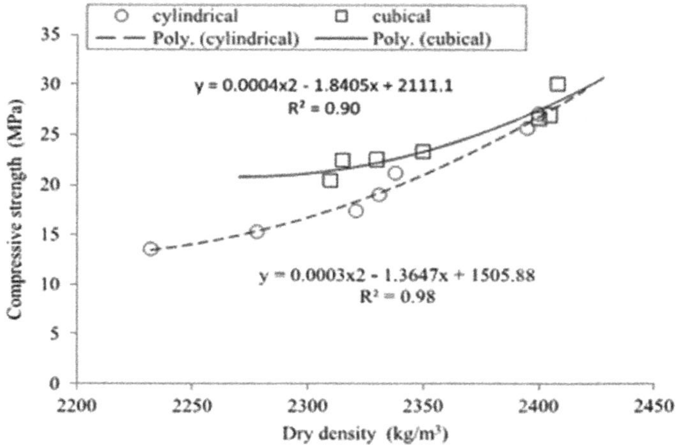

Figure 11.13 Relationship between compressive strength and density. Reproduced from ref. 76 with permission from Elsevier, Copyright 2015.

proportion added. Addition of rubber particle in RCCP changed their characteristics in the fresh and hardened state. The tensile strengths of RCCP mixes with different rubber contents are shown in Table 11.5. Figure 11.14 shows the type of specimen failure under flexural load. An increase in rubber content tends to decrease the mechanical properties, with simultaneous enhancement in porosity, ductility and crack resistance due to the modified roughness of rubber particle surfaces. A better correlation between compressive strength and density is observed in Figure 11.13. Finally, some parts of the waste can be removed and the environment is protected.[76]

Pettinari *et al.* investigated the effects of cryogenic crumb rubber (CCR) on the cold mixed reclaimed asphalt pavement (RAP). The RAP included bitumen emulsion and cement. The CR was produced cryogenically at the granular level. It was found that RAP can be effectively replaced by CCR with better results such as self-compaction and controlled volumetric change. Addition of CCR influenced the indirect tensile strength (ITS) and the ITS was reduced with a higher presence of CCR. The impact of cement on ITS was limited to a minimal level when 5% bitumen emulsion was used. It was found that rubber did not modify the curing characteristics or the thermally dependent nature of the mix. It was concluded that incorporation of CCR in a controlled manner greatly improves crack initiation resistance.[77]

Dondi *et al.* studied the feasibility of using CR inside a blend of 100% RAP, bitumen emulsion and cement. Two types of crumb rubber, traditional crumb rubber (TCR) and CCR, were used in preparation of the cold recycled mixtures and the resulting mechanical, physical properties were evaluated for both the conditions. Upon experimental evaluation, it was found that incorporation of CR instead of RAP significantly improved the self-compaction of the mixtures. About 1.1% and 2.1% improvements in compaction

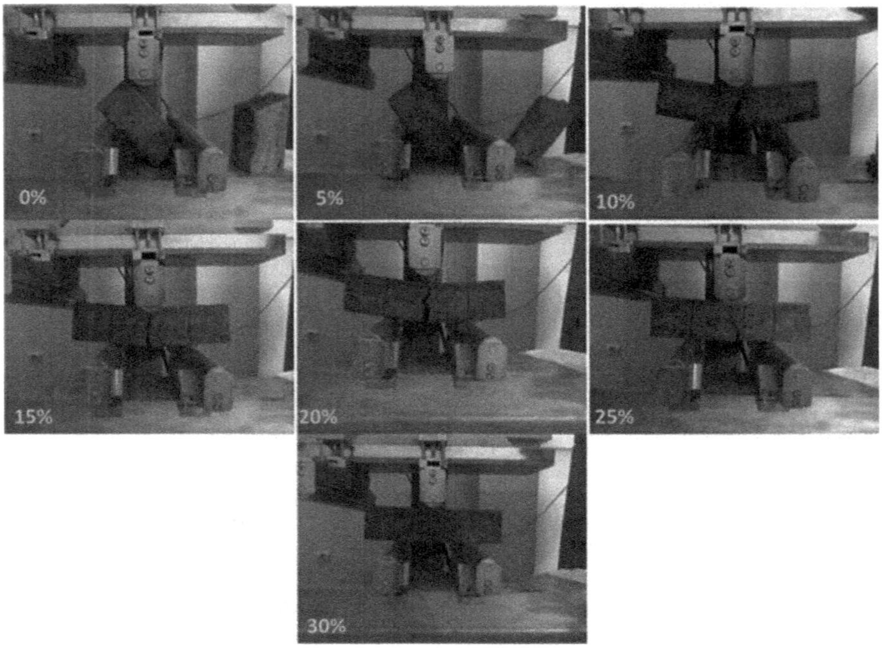

Figure 11.14 Specimens at failure. Reproduced from ref. 76 with permission from Elsevier, Copyright 2015.

resulted upon addition of TCR and CCR respectively. This improvement was attributed to the difference in the gradation. The volumetric characteristics also exhibited significant improvements upon addition of CR. On the other hand, addition of CR resulted in reduced mechanical characteristics.[78]

Shu and Huang reviewed the current advances for using WTR in asphalt and Portland cement. The maximum quantity of CR was employed in asphalt paving mixtures. The size of CR was limited between 0.075 mm and 4.75 mm. The addition of CR to asphalt paving mixtures was found to improve the resistance against failures such as distress, rutting, and fatigue cracking. The biggest challenge encountered in making rubberized asphalt was reported storage stability. However, the usage of rubber in Portland cement is not widespread owing to limitations such as stiffness and chemical composition.[79,80]

Bignozzi and Sandrolini explored the possibility of making concrete tough with the addition of different amounts of untreated tire waste. The mechanical and microstructural behavior of such SCRC was investigated and compared with that of SCC. A slightly higher amount of superplasticizer than SCC was required to induce self-compacting properties in SCRC. It was also found that addition of rubber phases in SCC reduced the compressive strength and stiffness. However, the capability to withstand post-failure loads with deformation was exhibited by SCRC.[81,82]

Karacasu, Okur and Er studied the rheological properties of rubber-modified asphalt (RMA) concrete subjected to static and dynamic loads. The samples were prepared by varying the rubber content and the size of rubber particles. The creep tests were conducted on the RMA samples under static and dynamic loading conditions. The shear modulus and damping values were determined through resonant column tests. The static creep stiffness was found to be unaffected by loading time. The strength of the samples was altered by the addition of rubber. The change was attributed to the change in the homogeneous nature of asphalt after addition of rubber. As asphalt is composed of aggregates and bitumen, the addition of rubber increases the binding and reinforcement effects. Under static loading, the presence of rubber induced the slip-down movement easily. The dynamic creep stiffness decreased with an increase in the number of cycles. It was concluded that creep stiffness is a function of the type of loading and shear modulus and damping depended on the particle size.[83,84]

Liu *et al.* investigated the rheological properties of crumb rubber modified rubber (CRMA) with the addition of *trans*-polyoctenamer rubber (TOR) additives. The mechanism of modification between CRMA and TOR was also evaluated. The samples were prepared by incorporating CR in asphalt at different weight proportions (15, 20 and 25%). TOR was added at two levels, 0 and 45%. The penetrations of CRMA were found to decrease with the addition of CR but simultaneously, the softening point increased. The elastic recovery of CRMA showed a positive trend with CR and TOR addition. TOR was found to decrease the viscosity and improve the workability of CRMA. Homogeneous distribution of CR particles in asphalt was enhanced with the addition of TOR. In conclusion, it was reported that TOR and CR improved both the physical and mechanical properties of the mixture.[85]

Murray *et al.* investigated the hydraulic conductivity, mechanical properties, and the pollutant absorbing capacity of permeable pavement materials based on recycled/crumb rubber and chitosan. Chitosan, a waste product from the seafood industry, is a biopolymer derived from chitin. A number of samples were prepared by mixing weakly acidic aqueous chitosan solution. The permeable pavements were prepared for filtering storm water effectively. When 12–13 ml of chitosan solution is used, optimum void spaces are found in the rubber. When the quantity of solution exceeds 14 ml, the sample was found to be highly porous and brittle. On the other hand, addition of a high concentration polymer solution resulted in a stronger filtration capacity. The chitosan/crumb rubber composite also exhibited a metal removal capacity.[86]

Xiang *et al.* prepared a CR/SBS composite modified asphalt and evaluated the thermal aging mechanism of the prepared samples. The following samples were also prepared: CRMA and SBS modified asphalt (SBSMA). A novel method was proposed to reduce the cost of the asphalt with an increase in performance by using WR. The waste utilization can also be increased in this way. Better effects were noticed in the CR/SBSMA in the distribution, colloid and the macroscopic properties.[87,88]

Yu *et al.* prepared plastic–rubber asphalt (PRA) by powdering and mixing rubber and PP in asphalt. The prepared PRA mixture was compared with an SBS asphalt mixture and the engineering properties and environmental concerns were investigated. The elastic recovery and softening point of PRA were inferior to that of SBS asphalt. However, life-cycle assessment proved that PRA is more bio-friendly. PRA also exhibited better performance at both low and high temperature conditions. Thus, it was concluded that PRA can be effectively used as pavement materials.[89,90]

Mohammadi *et al.* proposed an innovative method to improve the performance of crumb rubber concrete (CRC) by introducing crumb rubber in wet form. Even though rubber cannot be used for structural applications, it can be effectively used in concrete pavements. However, the utilization of rubber was limited due to want of a better method of incorporation in concrete. The low density, high toughness, better insulation and fire resistance properties of CR make it suitable for pavement applications. Around 40% rubber, along with water-cement (WC) resulted in better concrete. Also, the compressive strength improved by 22% with rubber incorporation. The fatigue strength also increased considerably. Thus, CR is suggested as a better reinforcement for concrete.[91,92]

Holmes *et al.* investigated the acoustic performance of CRC panels at low and high frequencies. It was found that CR, when incorporated in concrete, is effective in sound absorption and exhibited better insulation than concrete in plain form. The insulation properties of CRC were not affected much by heating or freezing. At 15% rubber addition in concrete, a high absorption coefficient was achieved. ALthough the flowability of concrete decreased with the presence of rubber, the density was low with minimum air content. Thus, crumb rubber has a huge potential as the right material for cladding against noise pollution.[93]

Liang *et al.* investigated the relationship between microstructure and rheological properties and storage stability of tire scrap rubber-modified asphalt (TSRMA). TSRMA is gaining greater attention as a potential material for road construction. However, the applicability is minimized by poor storage stability, which in turn disturbs the performance of pavements. The rutting or permanent deformation resistance improved considerably when tire scrap rubber was incorporated in asphalt. A non-Newtonian type of behavior was induced by large size rubber particles, as was evident by an increase in viscosity. Therefore, addition of rubber in asphalt tends to reduce the maintenance costs and stresses induced in the pavements.[94]

Shafabakhsh *et al.* discussed the effect of WRP on the rutting performance of asphalt mixture. The inherent problems of road pavements such as low resistance against dynamic loads and short service life can be successfully overcome by adding some waste materials such as tire rubber/CR. The incorporation of CR improves the service life of road pavements and also reduces overall costs. The rut depth was greatly reduced when WRP was incorporated in hot mix asphalt (HMA). The properties of HMA with around 10% rubber exhibited superior properties compared to traditional asphalt mixture with bitumen. Thus, the addition of WRP to HMA resulted in better properties and the recycling/reuse of rubber was also effectively carried out.[95]

Meyer summarized the various aspects of the concrete industry and the ways to effectively reduce its effect on theenvironment. It has been reported that the impact of the concrete industry is enormous owing to the constituent materials involved. One practical way of recycling tire rubber is to reuse it properly after retreading. Waste tires are still used in cement kilns to extract their thermal energy. Another promising route for tire reuse is in HMA or asphalt pavements in the form of CR or WRP. Either shredded rubber (450 mm) or the powder form (75 μm) can be used to significantly improve the asphalt properties such as strain capacity, ductility and crack propagation resistance. The rubber also enhances the sound absorption and thermal behavior of the mixture.[96]

Sridhar *et al.* prepared novel thermoplastic composites (TPC) from fly ash and WTP and investigated the effect of fly ash content on the physical and mechanical properties of the composites. Blending waste rubber with another polymer is a promising way of recycling/reusing waste rubber economically and effectively. Waste ground rubber tire can be employed to toughen plastics which are brittle in nature. For effective thermoplastic composites, the constituents must be different thermodynamically in a manner to induce proper mixing with distinct phase separation. Fly ash was effectively used to reinforce the thermoplastic, although the fly ash exhibited better bonding with rubber than plastic. The thermal property of TPC increased with filler addition and started to decrease after a certain level. The thermal property also improved considerably.[97]

Yilmaz and Degirmenci evaluated the possibility of utilizing fly ash and waste rubber with Portland cement as composites in masonry applications. Normally tire rubber particles are used as fine or coarse aggregates in concrete. Rubbers incorporated in concrete are found to decrease the density with subsequent increase in toughness, ductility and sound insulation. The workability of concrete also improved with the addition of rubber. The waste rubber can also absorb organic wastes by acting as an effective absorbent. The usage of waste rubber in masonry applications reduces waste accumulation and preserves valuable resources. Compressive strength of the mixture increased with fly ash addition, but decreased with the presence of rubber. The flexural strength was high, up to 20% of rubber and was then reduced. Thus, it was concluded that utilization of waste rubber and fly ash in cement mixture is an effective way of recycling tyres.[98]

Segre and Joekes evaluated the effect of surface modification on the adhesion behavior of rubber with cement. Material and energy recovery are proposed as feasible alternatives for recycling WTR. Another practical way is to incorporate waste rubber as an aggregate in concrete. However, addition of rubber as such in concrete tends to decrease the strength of the concrete irrespective of size (2–16 mm) or nature (fine or coarse). Rubber particles, after undergoing surface modification, display better adhesion with concrete. The mechanical and flexural properties were improved considerably with the addition of rubber as particles, rather than as aggregates. Thus, WTR can be used as a potential material for concrete in low strength applications.[99]

Cao studied the properties of recycled tire rubber modified asphalt using the dry process. Recycled tire rubber can be used in pavements and reduces the environmental issues associated with waste tire disposal. The addition of rubber to concrete dates back to the 1940's when devulcanized recycled rubber was used in an asphalt paving mixture. Crumb rubber can be incorporated into asphalt in two ways, namely the wet process and the dry process. The incorporation of recycled tire rubber in asphalt mixtures using the dry process was found to enhance the engineering properties of asphalt mixtures. At elevated temperature, the resistance to permanent deformation increased with rubber content and cracking at low temperature decreased with the addition of rubber.[100,101]

Gugliemotti *et al.* proposed a novel method for tire recycling, namely direct powder molding. In this method, large rubber parts can be made directly from scrap tires without the addition of virgin rubber or additives. The rubber powder was first obtained by grinding and the ground rubber powder was used to make rubber pads by the compression molding process. Although tire remanufacturing or retreading offers great potential for scrap tires, the difficulties involved in the process, combined with the material loss, inhibits its widespread use. In grinding, the tire is cut into small pieces and ground to the desired shape. Then, the steel and other organic materials are removed with the help of cyclone and mechanical separators. The CR when subjected to pyrolysis yields solid, liquid and gaseous products. Combustion is another process to recycle rubber as TDF. Recycled rubber can also be used along with virgin rubber. However, the scrap rubber can be made into useful products for damping, flooring and buffer components.[102]

Siddique and Naik reviewed the various methods in which scrap tires can be effectively used in Portland cement concrete. The scrap tires can be used in asphalt–concrete mixtures, as TDF in cement kilns, and for CB derivation. The above methods have serious limitations in economical and practical planes. The concrete workability was significantly enhanced when rubber was incorporated into the concrete. Fine CR was found to improve workability compared to coarse type. Addition of rubber induced higher air content in the rubberized concrete mixture. The unit weight was reported to be less for concrete with rubber owing to the low specific gravity of rubber. Thus, it has been concluded that utilization of tires in construction applications can drastically reduce the disposal problem of used tires. Scrap tire can also be employed in flowable fills.[103]

11.6 Conclusion

The need for recycling of rubber nanocomposites, and used tires in particular, is very high. With intense utilization of automobiles and truck transportation, particularly in developing countries with large populations, the environmental problems associated with waste/scrap disposal will be harsh. In addition, disposal of waste tires by land-filling is not viable, as it not only occupies large areas of valuable land, but will also waste considerable amounts of organic structural materials that are expensive and have restricted availability. Therefore, recycling of waste rubber from used tires is becoming crucial, and the speed at which the necessary technologies are developing may not be adequate. As newer nanocomposites are used as industrial products, a faster pace is needed for research in this context.[25]

Recycling of waste tire rubber can be accomplished by reclamation/retreading, in which the waste rubber is used as the tread part. Another way of recycling is to blend it with thermoplastic and thermoset materials in a cost-effective and efficient manner. With increasing use of polymers and rubber in particular in applications from packaging to civil applications, TPEs have a huge potential as alternate materials. The current trend shows great progress in tire-rubber recycling, particularly in construction activities such as asphalt, concrete and so forth. Therefore, it is hoped that waste-tire recycling will soon be an accelerated field which will remove environmental impacts and also contribute towards sustainable manufacturing.[7,8,11]

References

1. R. Ansari and A. H. Alikhani, *J. Coat. Technol. Res.*, 2009, **6**, 221.
2. D. A. Baeta, J. A. Zattera, M. G. Oliveira and P. J. Oliveira, *Braz. J. Chem. Eng.*, 2009, **26**, 23.
3. S. Sadhu and A. K. Bhowmick, *J. Polym. Sci., Part B: Polym. Phys.*, 2004, **42**, 1573.
4. S. Thomas and R. Stephen, *Rubber Nanocomposites: Preparation, Properties and Applications*, John Wiley & Sons, 2010.
5. S. Varghese and J. Karger-Kocsis, *Polymer*, 2003, **44**, 4921.
6. S. Joly, G. Garnaud, R. Ollitrault and L. Bokobza, *Chem. Mater.*, 2002, **14**, 4202.
7. K. Hamad, M. Kaseem, Y. Gun Ko and F. Deri, *Polym. Sci., Ser. A*, 2014, **56**, 812.
8. M. T. García, I. Gracia, G. Duque, A. Lucas and J. Francisco Rodríguez, *Waste Manage.*, 2009, **29**, 1814.
9. M. Sadat-Shojai and G. R. Bakhshandeh, *Polym. Degrad. Stab.*, 2011, **96**, 404.
10. A. I. Isayev, in *The Science and Technology of Rubber*, ed. J. E. Mark, B. Erman and C. M. Roland, Academic Press, Boston, 4th edn, 2013, p. 697.
11. A. Isayev, *Chem., Manuf. Appl. Nat. Rubber*, ed. S. Kohjia and Y. Ikeda, Woodhead Publishing, 2014, p. 395.
12. I. Khais, A. Reali, M. Reali and I. Reznik, US Pat US5634599, 1997.
13. K. Hamad, M. Kaseem and F. Deri, *Polym. Degrad. Stab.*, 2013, **98**, 2801.
14. A. Soroudi and I. Jakubowicz, *Eur. Polym. J.*, 2013, **49**, 2839.

15. E. Abraham, P. A. Elbi, L. A. Pothen, S. Thomas and B. M. Cherian, *Recent Dev. Polym. Recycl.*, 2011, **47**, 100.
16. Q. Zhang, V. N. Mochalin, I. Neitzel, I. Y. Knoke, J. Han, C. A. Klug, J. G. Zhou, P. I. Lelkes and Y. Gogotsi, *Biomaterials*, 2011, **32**, 87.
17. Y. Fang, M. Zhan and Y. Wang, *Mater. Des.*, 2001, **22**, 123.
18. J. W. Jang, T. S. Yoo, J. H. Ohb and I. Iwasakic, *Resour., Conserv. Recycl.*, 1998, **22**, 1.
19. S. L. Zhang, Z. X. Zhang, K. Pal, Z. X. Xin, J. Suh and J. K. Kim, *Mater. Des.*, 2010, **31**, 3624.
20. E. Lievana and J. Karger-Kocsis, *Prog. Rubber, Plast. Recycl. Technol.*, 2004, **20**, 1.
21. Y. Yang, R. Boom, D. J. Heerden, P. Kuiper, H. Wit and B. Irion, *Chem. Eng. Process.*, 2012, **51**, 53.
22. Y. Bai, Z. Wang and L. Feng, *Mater. Des.*, 2010, **31**, 999.
23. L. Reijnders, *Recent Adv. Elastomeric Nanocomposites*, ed. V. Mittal, J. K. Kim and K. Pal, Springer, 2011, p. 179.
24. B. Adhikari, D. De and S. Maiti, *Prog. Polym. Sci.*, 2000, **25**, 909.
25. S. K. Dey, A. Isayev and K. Khait, in *Rubber Recycling*, ed. Taylor & Francis Inc, CRC Press Inc, Bosa Roca United States, 2005, p. 528.
26. G. Marković, O. Veljković, M. Marinović-Cincović, V. Jovanović and J. Budinski-Simendic, *Composites, Part B*, 2013, **45**, 178.
27. C. Vladuta, M. Voinea, E. Purghel and A. Duta, *Mater. Sci. Eng., B*, 2009, **165**, 221.
28. N. Noriman, H. Ismail and A. Rashid, *Polym. Test.*, 2010, **29**, 200.
29. L. Mészáros, T. Bárány and T. Czvikovszky, *Radiat. Phys. Chem.*, 2012, **81**, 1357.
30. S. O. Mohaved, A. Ansarifar, S. K. Nezhad and S. Atharyfar, *Polym. Degrad. Stab.*, 2015, **111**, 114.
31. R. O. Aly, *Arabian J. Chem.*, 2012, **9**, 1547.
32. T. Saleesung, P. Saeoui and C. Sirisinha, *Polym. Test.*, 2010, **29**, 977.
33. H. Han, X. Wang and D. Wu, *Composites, Part A*, 2012, **43**, 1947.
34. C. Teng, C. M. Maa, B. D. Cheng, Y. Shih, J. Chen and Y. K. Hsiao, *Composites, Part A*, 2011, **42**, 928.
35. C. Bendjaouahdou and S. Bensaad, *Energy Procedia*, 2013, **36**, 574.
36. A. Shakun, J. Vuorinen, M. Hoikkanen, M. Poikelispää and A. Das, *Composites, Part A*, 2014, **64**, 49.
37. A. Maragiannis, D. Papageorgiou, C. Medrea and I. Chicinas, *Eng. Failure Anal.*, 2013, **35**, 247.
38. D. Statharas, D. Papageorgiou, J. Sideris and C. Medrea, *Int. J. Mater. Form.*, 2008, **1**, 431.
39. M. Spizzuoco, A. Calabrese and G. Serino, *Eng. Struct.*, 2014, **76**, 99.
40. A. Karimzadeh Naghshineh, U. Akyüz and A. Caner, *Earthquake Eng. Struct. Dynam.*, 2014, **43**, 301.
41. M. Thompson and K. Yeung, *Polym. Degrad. Stab.*, 2006, **91**, 2396.
42. G. Ausias, S. Thuillier, B. Omnès, S. Wiessner and P. Pilvin, *Polymer*, 2007, **48**, 3367.

43. M. Biron, in *Thermoplastics and Thermoplastic Composites*, ed. M. Biron, William Andrew, 2012.
44. K. Fukumori, M. Matsushita, H. Okamoto, N. Sato, Y. Suzuki and K. Takeuchi, *JSAE Rev.*, 2002, **23**, 259.
45. T. Zaharescu, C. Cazac, S. Jipa and R. Setnescu, *Nucl. Instrum. Methods Phys. Res., Sect. B*, 2001, **185**, 360.
46. B. Wu and M. Zhou, *Waste Manage.*, 2009, **29**, 355.
47. J. Cai and R. Salovey, *J. Mater. Sci.*, 2001, **36**, 3947.
48. F. Sadaka, I. Campistron, A. Laguerre and J. Pilard, *Polym. Degrad. Stab.*, 2013, **98**, 736.
49. F. Sadaka, I. Campistron, A. Laguerre and J. Pilard, *Polym. Degrad. Stab.*, 2012, **97**, 816.
50. T. Yasin, S. Khan, M. Shafiq and R. Gill, *Radiat. Phys. Chem.*, 2015, **106**, 343.
51. M. Awang, H. Ismail and M. Hazizan, *Polym. Test.*, 2008, **27**, 93.
52. J. K. Kim, S. H. Lee and M. Balasubramanian, *Polimeros*, 2006, **16**, 263.
53. H. S. Liu, J. L. Mead and R. G. Stacer, *32nd Int. SAMPE Tech. Conf.*, 2000.
54. W. Kaminsky, C. Mennerich and Z. Zhang, *J. Anal. Appl. Pyrolysis*, 2009, **85**, 334.
55. M. Llompart, L. Sanchez-Prado, J. Pablo Lamas, C. Garcia-Jares, E. Roca and T. Dagnac, *Chemosphere*, 2013, **90**, 423.
56. G. S. Kumaran, N. Mushule and M. Lakshmipathy, *Am. J. Eng. Appl. Sci.*, 2008, **1**, 40.
57. K. M. Feriha, R. A. Hussein, G. A. Ismail, H. M. El-Naggar and O. D. El-Sebaie, *J. Environ. Eng. Ecol. Sci.*, 2014, **3**, 5.
58. N. Sunthonpagasit and M. R. Duffey, *Resour., Conserv. Recycl.*, 2004, **40**, 281.
59. C. T. Ferreira, C. A. B. Perez, D. Hirayama and C. Saron, *J. Environ. Chem. Eng.*, 2013, **1**, 762.
60. M. Zaharia, V. Sahajwalla, B. C Kim, R. Khanna, N. S. Chaudhury, P. O'Kane, J. Dicker, C. Skidmore and D. Knights, *Energy Fuels*, 2009, **23**, 2467.
61. X. Cao, C. Xu, Y. Liu and Y. Chen, *Carbohydr. Polym.*, 2013, **92**, 69.
62. M. Du, B. Guo, Y. Lei, M. Liu and D. Jia, *Polymer*, 2008, **49**, 4871.
63. G. Song, Z. Gu, P. Li, L. Wang and L. Gao, *Appl. Clay Sci.*, 2012, **65**, 158.
64. Ľ. Jankovič, J. Madejová, P. Komadel, D. Jochec-Mošková and I. Chodák, *Appl. Clay Sci.*, 2011, **51**, 438.
65. W. Chen, S. Wu, Y. Lei, Z. Liao, B. Guo, X. Liang and D. Jia, *Polymer*, 2011, **52**, 4387.
66. A. Zanchet, L. N. Carli, M. Giovanela, R. N. Brandalise and J. S. Crespo, *Mater. Des.*, 2012, **39**, 437.
67. N. Al-Aqeeli, *Materials*, 2015, **8**, 181.
68. A. Khalilitabas, C. Dehghanian and M. Aliofkhazraei, *Anti-Corros. Methods Mater.*, 2011, **58**, 76.
69. T. Amari, N. J. Themelis and I. K. Wernick, *Resour. Policy*, 1999, **25**, 179.
70. W. Kaminsky and C. Mennerich, *J. Anal. Appl. Pyrolysis*, 2001, **58**, 803.

71. A. A. Merchant and M. A. Petrich, *AIChE J.*, 1993, **39**, 1370.
72. A. K. Naskar, Z. Bi, Y. Li, K. Sam, K. Akato, D. Saha, M. Chi, C. A. Bridges and M. P. Paranthaman, *RSC Adv.*, 2014, **4**, 38213.
73. H. Wang, M. Davidson, Y. Zuo and Z. Ren, *J. Power Sources*, 2011, **196**, 5863.
74. F. Karabork, E. Pehlivan and A. Akdemir, *J. Polym. Eng.*, 2014, **34**, 543.
75. Z. X. Xin, Z. X. Zhang, K. Pal, J. U. Byeon, S. H. Lee and J. K. Kim, *Mater. Des.*, 2010, **31**, 589.
76. A. Meddah, M. Beddar and A. Bali, *J. Cleaner Prod.*, 2014, **72**, 187.
77. M. Pettinari, G. Dondi, C. Sangiorgi and O. Hededal, *Constr. Build. Mater.*, 2014, **63**, 249.
78. G. Dondi, P. Tataranni, M. Pettinari, C. Sangiorgi, A. Simone and V. Vignali, *Constr. Build. Mater.*, 2014, **68**, 370.
79. X. Shu and B. Huang, *Constr. Build. Mater.*, 2014, **67**, 217.
80. B. Huang, L. Mohammad, P. Graves and C. Abadie, *Transp. Res. Rec.*, 2002, **1789**, 1.
81. M. Bignozzi and F. Sandrolini, *Cem. Concr. Res.*, 2006, **36**, 735.
82. L. Guoqiang, M. A. Stubblefield, G. Garrick, J. Eggers, C. Abadie and B. Huang, *Cem. Concr. Res.*, 2004, **34**, 2283.
83. M. Karacasu, V. Okur and A. Er, *Sci. World J.*, 2015, **2015**, 258586.
84. J. Borah, G. Lin and C. S. Wang, *Adv. Mater. Res.*, 2010, **87–88**, 271.
85. H. Liu, Z. Chen, W. Wang, H. Wang and P. Hao, *Constr. Build. Mater.*, 2014, **67**, 225.
86. C. A. Murray, K. S. Snyder and B. A. Marion, *Constr. Build. Mater.*, 2014, **69**, 221.
87. L. Xiang, J. Cheng and S. Kang, *Constr. Build. Mater.*, 2015, **75**, 169.
88. X. Li, L. G. Tabil, S. Panigrahi and W. J. Crerar, *Can. Biosyst. Eng.*, 2009, **8**, 1.
89. B. Yu, L. Jiao, F. Ni and J. Yang, *Constr. Build. Mater.*, 2014, **71**, 416.
90. T. B. Moghaddam, M. R. Karim and T. Syammaun, *Constr. Build. Mater.*, 2012, **34**, 236.
91. I. Mohammadi, H. Khabbaz and K. Vessalas, *Constr. Build. Mater.*, 2014, **71**, 456.
92. A. R. Khaloo, M. Dehestani and P. Rahmatabadi, *Waste Manage.*, 2008, **28**, 2472.
93. N. Holmes, A. Browne and C. Montague, *Constr. Build. Mater.*, 2014, **73**, 195.
94. M. Liang, S. Liu, W. Fan, X. Xin, H. Luo and B. Xing, *Constr. Build. Mater.*, 2015, **74**, 124.
95. G. Shafabakhsh, M. Sadeghnejad and Y. Sajed, *Case Stud. Constr. Mater.*, 2014, **1**, 69.
96. C. Meyer, *Cem. Concr. Compos.*, 2009, **31**, 601.
97. V. Sridhar, Z. Z. Xiu, D. Xu, S. H. Lee, J. K. Kim, D. J. Kang and D. S. Bang, *Waste Manage.*, 2009, **29**, 1058.
98. A. Yilmaz and N. Degirmenci, *Waste Manage.*, 2009, **29**, 1541.

99. N. Segre and I. Joekes, *Cem. Concr. Res.*, 2000, **30**, 1421.
100. W. Cao, *Constr. Build. Mater.*, 2007, **21**, 1011.
101. L. Xiang, J. Cheng and G. Que, *Constr. Build. Mater.*, 2009, **23**, 3586.
102. A. Gugliemotti, C. Lucignano and F. Quadrini, *Plast., Rubber Compos.*, 2012, **41**, 40.
103. R. Siddique and T. R. Naik, *Waste Manage.*, 2004, **24**, 563.

CHAPTER 12

Hybrid Nano-filler for Value Added Rubber Compounds for Recycling

KISHOR KUMAR SADASIVUNI*[a], SUNITA RATTAN[b], KALIM DESHMUKH[c], AQIB MUZAFFAR[c], M. BASHEER AHAMED[c], S. K. KHADHEER PASHA[d], PAYAL MAZUMDAR[b], SADIYA WASEEM[e], YVES GROHENS[f] AND BIJENDRA KUMAR[g]

[a]Center for Advanced Materials, Qatar University, P.O. Box 2713, Doha, Qatar; [b]Amity Institute of Applied Sciences, Amity University, Sec-125, Uttar Pradesh, Noida, India; [c]Department of Physics, B.S. Abdur Rahman Crescent Institute of Science and Technology, Chennai, 600048, TN, India; [d]Department of Physics, VIT-AP University, Amaravati, Guntur, 522501, Andhra Pradesh, India; [e]Advance Carbon Products, CSIR-NCL, New Delhi, 686574, India; [f]IRDL, FRE-CNRS 3744, Laboratory, Université de Bretagne Sud, Rue St Maudé, 56100, Lorient, France; [g]Department of Technology, Elizabeth City State University, Elizabeth City, NC, 27909, USA
*E-mail: kishor_kumars@yahoo.com

12.1 Introduction

In a modern technological era, recycling of materials presents itself as a process to improve environmental conditions and generate new composite materials. The recycling of materials guides the economy towards price

Green Chemistry Series No. 59
Rubber Recycling: Challenges and Developments
Edited by Jin Kuk Kim, Prosenjit Saha, Sabu Thomas, Józef T. Haponiuk and M. K. Aswathi

Published by the Royal Society of Chemistry, www.rsc.org

control and proper disposal of materials. Especially with regard to those materials having longer decomposition times, recycling could be the remedy.[1] The recycling process avoids the potential threat of waste materials to both the human race and the environment and plays a pivotal role in waste management.[2] The process can be divided into three steps, beginning with the collection of waste materials, followed by a breakdown process and finally re-manufacturing the material.

Rubber is considered to be one of the most prominent products with widespread applications and huge industrial value.[3] As a result of this, rubber waste is an important aspect to be dealt with. Rubber as a material has idiosyncratic physical and chemical properties in addition to hyperelasticity. Natural rubber occurs in the form of latex from a rubber plant. This latex extraction is a consequence of incisions made on the stem of the rubber tree. The rubber making process starts from latex collection and is followed by latex straining. The strained latex is then diluted with water, followed by acid treatment to form a coagulate of rubber particles. The rubber coagulate is then allowed to pass through a series of rollers in which the rubber is pressed and drawn into the form of thin sheets followed by drying. After drying the rubber is ready to be used for different applications.

The enhancement in the mechanical properties of rubber is achieved by a process known as the vulcanization. Vulcanization can be defined as an additive chemical process in which natural rubber is treated with different chemicals or polymers to form a mechanically strong rubber material. The main mechanical property transformation of rubber material consists of polymer cross-linked chains adhered to the rubber substrate.[4] The incorporation of other materials to rubber results in changing the behavior of the rubber material and opens up the possibility for a new approach to the development of technology.[5]

Nanofillers are nanoparticles termed as high-value materials which lead to improvement of the physical and mechanical properties of polymer based composites.[6] The reason for the high value of nanofillers is their defect free occurrence. The nanofillers provide a greater area when dispersed homogeneously, resulting in changing the key physical properties such as the mobility, mechanical and thermal properties of the composites.[7] Additionally, the nanofillers show superior reinforcement in nanocomposite production as a consequence of their higher aspect ratio. The incorporation of nanofillers is considered on the basis of weight to produce better nanocomposites.[8]

Nanofillers are classified on the basis of their relative dimensions as one, two or three-dimensional particles on the nanoscale.[7] One-dimensional nanoparticles include halloysite and carbon nanotubes (CNTs) while flat nanostructures such as clay and graphene are considered as two-dimensional nanomaterials.[7] Rubber nanoparticles and silica are examples of three-dimensional nanoparticles.[7] The impact of the shape of nanofillers plays a vital role in the interactions between the filler and matrix on the overall interface area, whereas the size of nanofillers impacts on the volume fraction in the interphase region.[9] The interphase region consists of a polymer matrix with different properties compared to the bulk polymer. The reason for different properties lies in the presence of the interphase layer surrounding the particles containing polymer chains with different conformations to those on the bulk polymer.[9]

The amalgamation of hybrid nanofillers with rubber material provides new theories for research, especially in the recycling process. There are various reports regarding the addition of hybrid nanofillers to rubber while recycling. The commonly used nanofillers include titanium dioxide (TiO_2), calcium carbonate ($CaCO_3$) and carbon black (CB).[8] Other fillers added to rubber material in nanocomposite formation include silicon dioxide (SiO_2)/polyhedral oligomericsilsesquioxanes (POSS),[10] CNT/epoxy,[11] and multi-wall carbon nanotubes/nanoclay hybrids.[12] In addition to those there are biological fillers including coir nanofiller, cellulosic nanofillers, sugarcane bagasse ash, rice hull ash (residue production), oilshare residue, defatted soy flour and organoclay as a substitution for silica in rubber composites.[13–17] The most commonly used nanofiller in the rubber industry is silica, which is added to rubber to enhance abrasion resistance, wet grip and to reduce the rolling resistance.[18]

The rubber that is generally considered for recycling comes from the tyres of vehicles. After their use, the waste tyres can lead to hazardous effects to both public health and the environment due to their inappropriate disposal or storage. The rubber used for tyres takes longer time periods to decompose naturally, thereby increasing landfill pollution. Additionally, if rubber is burnt, it produces toxic and carcinogenic gases causing air pollution and risk to the health of living beings. Thus, recycling of rubber provides an alternative to avoid the above mentioned negative impacts.

The hybridization method is one of the common methods applied to rubber recycling with nanofiller addition. The most common nanofiller for such method is CB due to its profusion, cheap price and better reinforcement effects.[19] CB nanofillers are used in higher amounts for hybridization in order to achieve a conductive network in the polymer matrix.[20] The consequences of a large amount of CB nanofillers are deprived flexibility in the processing of nanocomposites and degeneration of mechanical properties.[20] On the other hand, CNTs work well as nanofillers in the formation of nanocomposites due to their one-dimensional structure and higher aspect ratio.[21] There are reports of the use of CB as well as CNT as dual hybrid fillers in rubber-based nanocomposites. The addition of both CB and CNT to rubber during recycling leads to an enhancement of the mechanical, thermal and electrical properties of the nanocomposite. The interaction of CNTs takes place along the surface of the CB particles and forms a conducting path in the resulting CB/CNTs/rubber nanocomposite.[22] Prior to that, the rubber is treated with styrene-butadiene to form styrene-butadiene rubber. This nanocomposite revealed a lower percolation electrical conductivity.[22] Generally, nanoparticles exhibiting an anisotropic structure have the ability to self- coordinate in the nanocomposite inside the rubber matrix.[23] These anisotropic nanoparticles improve the mobility of rubber particles at the nanofiller and rubber interface resulting in enhancement of the mechanical properties.[24] Therefore, the nanofiller addition perks up the filler networking and rubber–filler interaction, resulting in improvement of the functionality and enhancement of the properties.[25–27]

The hybrid nanofiller comprising of POSS and SiO_2 is another promising candidate for hybridization during rubber recycling.[28] POSS has a high tendency to interact with the polymer matrix due to organic functionalities consisting of inorganic silica and a zero-dimensional molecular material.[29] The core is formed by silica which is covalently bonded to eight organic groups. The presence of organic groups favours the polymer matrix formation. As a result of that, the polymer–POSS interface provides enhanced physical and chemical interactions.[30,31] The nanocomposite formed by SiO_2/POSS/rubber has enhanced rigidness, improved wear and tear, higher shear modulus, lower flammability value, insubstantial weight and improved ductile properties.[31] The SiO_2/POSS/rubber is synthesized using silanization of SiO_2 in the presence of methacrylsilane with the incorporation of a small amount of grafted POSS. The reaction is aided by dicumylperoxide at the POSS/SiO_2 interface.[31] In the nanocomposite structure, POSS is bonded to the polymer by means of grafting or *via* polymerization.[32] A POSS-based rubber epoxy hybrid composite comprising of di-epoxy-POSS monomer as a polymerizing agent has also been reported.[33]

Layered clay platelets such as montmorillonite, bentonite, and hectorite, along with rubber nanoparticles, have been used as reinforcing nanofillers.[34] In layered nanoclay, the overall properties of the nanocomposite are based on the volume ratio. The enhancement in electrical, barrier and thermal properties arise due to the finely dispersed interfacial area among the rubber and nanoclay fillers. However, the macroscopic properties exist due to an increasing interfacial surface area at lower concentrations.[35] The interaction between the rubber and nanoclay filler can be improved by increasing the interlayer spacing, owing to long organic chains and modifications, thereby providing only those functional groups that have a better interaction property with the rubber matrix. The increased interlayer spacing also has implications on the surface hydrophobic property of the nanoclay and the nanocomposite. The development of a highly water repellent rubber toughened biopolymer and organoclay nanocomposite has been reported. The nanocomposite has a water-repellent property and is highly adhesive to metal surfaces due to the nano hierarchal surface of the nanoclay.[36] This effect (hydrophobicity) produces a self-cleaning quality in the hybrid rubber/nanofiller nanocomposite (mainly synthetic and fluoroacrylic rubbers).[37] The rubber can also be treated with bentonite clay as a nanofiller to improve the tensile strength, thermal stability, barrier property, decrease in flammability and solvent permeability of the rubber matrix.[38] Before treatment with rubber, bentonite clay needs organic modification to enable it to be compatible with organic rubber. In hybrid nanofillers, the most attention has been focused on reduced graphene oxide (RGO) and rubber nanocomposites. The addition of RGO to the rubber matrix enhances the mechanical, electrical and thermal properties, especially when added during recycling.[39,40] The process involved in nanocomposite formation comprises of mixing, coagulation, treatment and processing steps.

This chapter describes the different recycling methods of rubber with the addition of hybrid nanofillers, the fabrication of the recycled nanocomposites, their advantages, disadvantages, challenges and the future of such nanocomposites.

12.2 Fabrication of Hybrid Nanofillers/Rubber Nanocomposites

Rubber as a single entity cannot form nanocomposites. Therefore, it is necessary to add nanofillers to shed the low modulus and potency of rubber to form nanocomposites. The incorporation of nanofillers provides strength and improves the applicability of the overall nanocomposite. The reinforcement of rubber is one of the prime effects caused by the nanofillers. Reinforcement of rubber can be obtained by common nanofillers such as CB and silica.

Nanocomposites consisting of a rubber matrix and nanofillers are among the most popular research fields due to their multi-functionality. The multi-functionality of such nanocomposites lies in their dispersion. In the case of inorganic nanofillers, the dispersion with the rubber matrix is achieved by melting the rubber with nanofillers. The direct melting of nanofillers in a single polymer matrix produces micron-sized filler bunches due to the higher surface reactivity of nanofillers.[41] This creates a bigger issue concerned with the fabrication of rubber-nanofiller nanocomposites and causes a non-uniform dispersion of nanofillers. However, the problem can be solved to some extent by uniformly dispersing the nanofiller in the rubber matrix using an organic compound for modification of the nanofiller.[42] The nanofillers are subjected to surface or interlayer modification using polymerization reactions or the sol–gel method.[43] The surface modification avoids the micron sized filler-bunch formation. On the other hand, these surface modifications form a set of complex reactions as a limiting factor for large-scale production.[44] The alternative, to avoid complex chemical reactions, comprises of uniform inorganic nanofiller dispersion into various polymers by a direct melting procedure without influencing any surface or interlayer modification.[45]

The methods for fabrication of nanofiller and rubber based nanocomposites are categorized into four types.[46] These methods include the intercalation method, *in situ* polymerization, (*in situ* nanofiller formation followed by *in situ* polymerization) direct mechanical mixing and the sol–gel method.

12.2.1 Intercalation Method

In the intercalation method, the incorporation of exfoliated inorganic nanofillers is achieved by insertion of an organic compound.[47] The insertion occurs in the interlayers of the nanofillers leading to the formation of the uniformly dispersed plate like nanofiller.[48] The nanofillers to be incorporated into the rubber matrix require modification attained by addition of surfactants such as alkyl ammonium, imidazolium, phosphonium salts and amino acids.[49–51] The surfactant material provides required increments in hydrophobicity, resulting in their proper dispersion with nanofiller materials. The addition of the surfactant is only applicable with the hydrophilic nanofillers. The surfactant-nanofillers are then finally intercalated with the rubber matrix using chemical or mechanical techniques. The chemical technique involves polymerization of the monomer within surfactant-nanofiller layers.[48,52] The chemical technique can be utilized to form carbon (*e.g.*, graphene, CB and CNTs) based rubber nanocomposites. On the other hand, the mechanical technique for intercalation involves either direct insertion of rubber to a layered nanofiller in an appropriate solvent or melting rubber with nanofillers under high cut off conditions without using any solvent.[52] This technique is best suited for the fabrication of those rubber nanofiller nanocomposites which cannot be fabricated with the *in situ* polymerization technique.

12.2.2 *In situ* Polymerization

The second method of fabrication of nanocomposites involves the polymerization reaction in nanocomposite formation and is termed as the *in situ* polymerization method. The fabrication starts with a dispersion of inorganic nanofillers in a monomer solution followed by standard polymerization of the resulting hybridised nanocomposite.[53] Using this method, metal-based nanocomposites have been successfully synthesised. The nanocomposite formation is a consequence of synchronised metal particle formation from metallic precursors and the polymer matrix.[54] This method of nanocomposite fabrication requires a protective polymer in order to limit the size of the metallic particles. However, the most important parameter for the *in situ* polymerization method is the proper dispersion of nanofillers in the monomer requiring surface modifications of the particles or the metallic precursors.

12.2.3 Mechanical Mixing Method

The third fabrication method is the direct mechanical mixing of rubber and nanofillers to form nanocomposites. This method is based on the mechanical breakdown of the amassed nanofillers while mixing. The direct mixing method is mainly employed for nanocomposites containing slightly higher dimensions of fillers. The higher dimensions could result in improper dispersion and hinder nanocomposite formation. Therefore, the mechanical treatment is applied to bring the fillers into a suitable size for better dispersion to form hybrid materials.[46] In mechanical mixing the nanofiller and rubber are either mixed in the presence or absence of a solvent.[55]

When the mixing takes place in the presence of a solvent, the nanofillers undergo dispersion in a polymer solution. The nanofiller–polymer solution is then evaporated to remove the solvent. The solvent acts as a dispersion medium in which nanofillers finely disperse with the polymer matrix.[56] The uniform dispersion is as a consequence of the larger shear force induced in the nanofiller and polymer solution. A better dispersion of nanofillers in the solution can also be achieved by application of ultrasonic waves or filler surface modification.[56] Using this method, surface modification of the nanofiller is attained without drying.

In the absence of solvent, the hydrodynamic force is used to breakdown the fillers up to nanoscale dimensions accompanied by melt-compounding.[57] The polymer is first melted and then shear stress force induces the filler breakdown in the polymer melt. The dispersion in the melt-compounding is dependent on the internal shear stresses developed due to viscous drag on the nanofillers. The surface modification of nanofillers can be negotiated if a melt compounding method combines an induced shear mixing with ultrasonic mixing.[57]

12.2.4 Sol–Gel Method

The methods mentioned above followed the top-down sequence of nanofiller polymer fabrication, whereas the sol–gel method follows the bottom up pattern of fabrication of nanocomposites. In the sol–gel fabrication method, the combination of *in situ* nanofiller formations and polymerization occurs.[58] This method of nanocomposite fabrication enables dispersion of inorganic fillers having dimensions less than the chain length of the polymer matrix molecules.[59] The sol–gel reactions produce some dramatic effects on the structure of the inorganic network formed. These effects can lead to improper size and alignment of the inorganic fillers in hybrid nanocomposite structures.[60] To overcome these effects the side selective hybrid method can be deployed, which uses pre-modified polymers.[61] In the side selective hybrid method there is no interaction between the metallic particles and polymer matrix and no control over the size of the nanocomposite as per the desired applications.

The rubber nanocomposites can be fabricated by using pre-modified rubber along with hybrid nanofillers such as halloysite nanotubes. In rubber halloysite nanocomposites, halloysite acts as a perfect component for fabrication in comparison to CNTs.[62] Halloysite is affordable and easily available. The rubber matrix and halloysite are strongly bonded through interfacial bonds in the presence of sorbic acid to enhance the performance of the nanocomposite.[62] In the fabrication of rubber based nanocomposites, the other chemicals used are methacrylic acid, calcium carbonate, crotonic acid, and so forth.[62] These chemicals, when added to styrene-butadiene rubber, improve the shear modulus and tensile strength thereby enhancing the interfacial reactions between the rubber and halloysite.

To be precise, for the fabrication of rubber nanocomposites, two main methods are used *viz.* melt compounding and solution blending. The melt compounding forms a complex method when concerned with rubber nanoclay based materials as a consequence of the higher viscosity due to the higher molecular weight of the polymer.[63] Alternatively, the solution method provides an easy way for rubber nanoclay formation, accompanied by polymer and clay mixing pursued by solvent evaporation. However, there arises complexity in the intercalation of polymer chains and the removal of solvent in solution blending. In addition to this, one of the recently developed methods regarding fabrication of rubber nanocomposites, named as the "latex method" was introduced to overcome the limitations of the two main methods.[63] This method is a two-step process in which the first step comprises of the nanoclay dispersion in deionized water and the second step comprises of the mixing of the dispersed clay with rubber latex. The mixing is accompanied by fluffy co-coagulation yielding exfoliation of a few silicate layers. The overwhelming feature of this method is the non-involvement of solvents and the coincidence of intercalation and exfoliation.

12.2.5 Melt Compounding Method

The prime requirement for the rubber based nanocomposite fabrication is the vulcanization prior to compounding at the nanoscale of rubber and fillers.[63] The consequences of vulcanization tends to alter the chain flexibility and cross-linking rate among the internal and external regions of the fillers. Melt compounding offers a rubber processing ability of vast practical importance, in addition to being the least costly commercially and the eco-friendly nature. In melt compounding, the mixing can take place excluding the use of an inorganic solvent.[64] A nanocomposite composed of attapulgite and rubber latex fabricated by melt compounding displayed outstanding characteristics along with a lower fabrication cost.[65] In the melt compounding method the stiffness is directly related to the concentration of nanofillers. The introduction of compatibilizer enhances the mechanical properties while compounding the nanofiller into the rubber matrix, this is accredited to homogenous dispersion and filler surface modification.

In the melt compounding method it is essential to attain better dispersion of nanofillers in a rubber matrix. The types of nanocomposites obtained using melt compounding include conventional, intercalated and exfoliated nanocomposites, out of which the exfoliated ones are considered as the nanocomposites with better reinforcing characteristics. The completely exfoliated nanofillers act as enhancers of barrier characteristics and exfoliation resistance, but hard to prepare.

The morphology of the rubber nanocomposites fabricated by melt compounding is dependent on the nature of intercalates, curatives temperature, pressure, shear conditions, cross-linking reactions and rubber matrix polarity.[64] However, the final morphology of the nanocomposites is related to the vulcanization. The overall process occurs in a two roll mill, with the maintenance of least separation between the rolls when rubber is in the fluidic form. The modified nanofillers are inserted in a stepwise manner to achieve better dispersion followed by addition of a vulcanizer and accelerator to obtain the final compound.[61]

12.2.6 Solution Blending Method

In solution blending, fabrication of nanocomposites requires an inorganic solvent. This method is accompanied by the dispersal of dry rubber along with the modified nanofillers under vigorous stirring. To obtain rubber nanocomposites the inorganic solvent is evaporated. For example, fabrication of styrene–butadiene rubber (SBR) containing organoclay as nanofillers and toluene as the solvent.[65] The nanofiller dispersion is excellent in the case of solution blending. The rubber nanocomposites fabricated by solution blending exhibit enhanced tensile strength, shear modulus and tear strength.[61] Like melt compounding, the morphology in solution blending is dependent on the concentration, temperature, pressure and so forth.[66,67] The mechanical properties of nanocomposites are linked to the nature of the solvent used and concentration of rubber and nanofillers.[66] In solution blending, the exfoliated nanocomposites are fabricated accompanied by treatment of solution followed by polymer blending.[68]

12.2.7 Latex Compounding Method

Usually rubber exists in the latex form and the co-aggregation of rubber latex and nanofillers forms the latex compounding method. The method involves dispersion of rubber latex and nanofillers in water under vigorous stirring.[69] The rubber nanocomposites fabricated by this method display brilliant mechanical characteristics and gas barrier properties.[70] The procedure of latex compounding follows the same two roll mill method as followed in melt compounding. However, in latex compounding, a solvent (organic or inorganic) acting as a flocculating agent is used. The nanofillers are directly dispersed into the rubber latex along with other components (vulcanizing constituent and modifiers) during the rolling process. This method requires the least modifications and is easy and viable, producing high-performance rubber nanocomposites with better interactions between the nanofiller and rubber latex interfaces. The final compound is vulcanized using a standard casting mould maintained at high pressure.

12.3 Methods of Recycling

Rubber can be recycled by using various processes as described in detail elsewhere.[71] Initially, all the scrap rubber needs to be shredded and ground into crumb as shown in Figure 12.1 followed by chemicals and swelling agents to react adequately with the vulcanized structure to promote good heat transfer

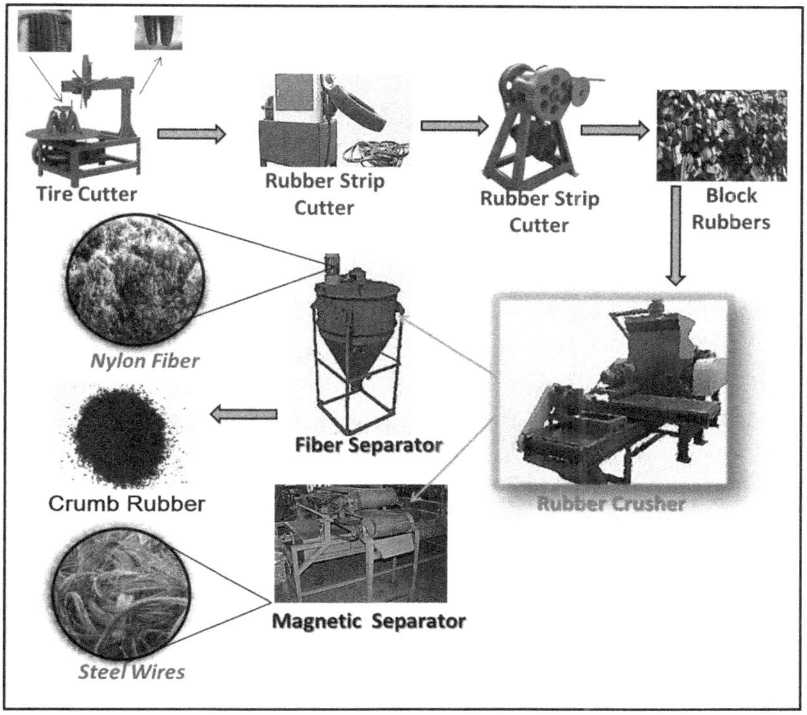

Figure 12.1 Separation technique of steel wires and nylon fibres from crumb rubber.

and to remove the fibres by mechanical or chemical action. Final processing follows each of the processes described below.

12.3.1 Biological Method

The biological method involves the breaking of sulfur bonds or devulcanization with the aid of microorganisms as shown in Figure 12.2a. There are many types of microorganisms (mesophilic bacteria) include *T. thioparus, T. thiooxidans, Leptospirillum ferrooxidans, thiobacillus ferrooxidans, T. organoparus* and so on used in the rubber devulcanization process. The energy involved in the degradation process is obtained from the oxidation of the substrates and their carbon for cell material construction from carbon dioxide, due to the chemolithoautotrophs nature.

12.3.2 Ambient Mechanical Recycling Method

In this method, the scrap material is made to break above room temperature as shown in Figure 12.2b. The reclaiming chemicals are mixed with the scrap rubber and the mixture is fed continuously into an extruder at

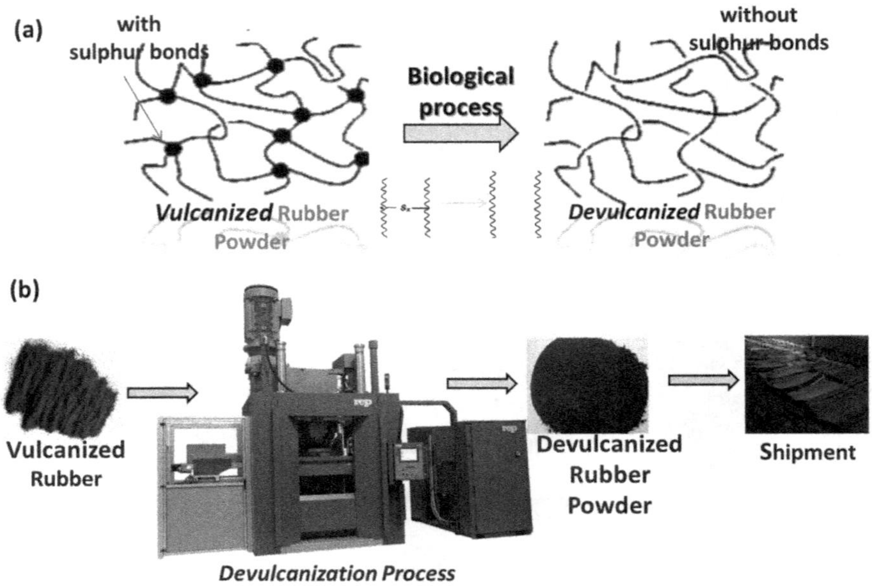

Figure 12.2 Rubber recycling (a) biological process and (b) mechanical process.

about 200 °C for a few minutes, which results in the devulcanization of rubber.

Ambient grinding is a multi-step technology and uses whole or pre-rubber composites in the form of shred or chips. An open two-roll mixing mill is used for the mechanical recycling process in which the crumb rubber is placed and milling is carried out at high temperatures. Mechanical shearing above 200 °C leads to drastic molecular weight breakdown.[45,72] The composite system, including rubber, metals, and textiles, are sequentially separated out one after the other. The rubber materials are then subjected to a shredder, which breaks the scrap rubber into chips. The scrap rubber chips are further fed into a granulator that breaks them into small pieces with the removal of metals and fibre in the process.[73] The remaining metals in the mixture are removed magnetically, and the fibres are removed through a combination of shaking screens and wind sifters. Furthermore, the grinding of rubber material in secondary granulators, followed by high-speed rotary mills results in the formation of finer rubber particles. It was found that the recycled rubber holds very high viscosity, thus indicating the low plasticity of rubber which is attributed to the crosslinking.[74]

A comparison of the curing characteristics of blends of fresh rubber with the recycled rubber indicates that there is an overall increase in the cure rate, whereas the optimum cure time decreases. The decrease is also observed in the scorch time and reversion resistance. Furthermore, as the percentage of recycled rubber in the blended rubber increases, the properties such as abrasion loss, modulus, compression set and hardness show an increase, whereas a decrease in the properties such as tensile strength, elongation at break, tear strength, flex resistance and resilience is also observed. This is explained on the basis of an increase in the density of the crosslinks caused by the increase in the percentage of the recycled rubber. Due to the low tensile strength and flex properties in the blends of natural rubber/recycled rubber (25/75), it was concluded that the recycled rubber acts as a non-reinforcing filler in this study.[75]

12.3.3 Thermal Process of Recycling

In the thermal process, the scrap materials break and are reclaimed by using heat treatment followed by possible chemical processing.[6]

12.3.4 Pan Technique

The pan technique is one of the conventional techniques used in the rubber recycling industry. In this technique, the fine scrap rubber materials are first mixed with reclaiming chemicals by using mechanical cyclones. The mixture is then placed in open pans in an autoclave and heated for 4–12 h at 15 bars and 180 °C. The residual water is completely removed from the mixture during autoclaving. Furthermore, the recycled product is ready for the final processing with reclaiming agents. The chemicals used for the recycling technique are aromatic thiols, disulfides, and aromatic oils. The usage of these chemicals allows lower temperatures and shorter recycle times and produces a product with superior mechanical properties.

12.3.5 Digester Technique

The digester process for recycling rubber is one of the batch processes. In this technique the scrap rubber is mixed with the chemicals which dissolve the fibre along with plasticizing oils, water and reclaiming agents. The mixture is heated in an autoclave for about 12 h at 180 °C. The softened rubber material is separated from the aqueous phase with the formation of regenerate, which is further dried for final refining. With recycling using the digester process, significant amounts of residual may remain. A very complex mixture of solvents, hydrocarbon resins, pine tar, softening oils, zinc and calcium chloride are used as the recycling chemicals along with the reclaim catalyst.

12.3.6 Alkaline Technique

This technique utilizes the usage of an alkaline solution. The scrap rubber is digested by using a high concentration of sodium hydroxide in the mixture followed by hydroxylation. After recycling, the excess de-fiberizing agent was removed by a repeated washing procedure. The crumb is then dried and refined. Moreover, for vulcanized SBR in alkaline reclaiming processes, *N*,*N*-dialkyl aryl amine sulfides were found to be highly active recycling agents. The thickness, body and tack evaluation were used to manually assess the state of recycling.[76,77]

12.3.7 High-pressure Steam Technique

The high-pressure steam technique leads in the development of a recycling process with a shorter reaction time. In this technique fibres and scrap rubber is mixed under high-pressure steam with recycling agents and autoclaved at 280 °C for 10 min.

12.3.8 Thermo-mechanical Recycling Process

In this process the breaking of carbon-to-carbon bonds and sulfur cross-links takes place through thermal and mechanical breakdown. This technique results in the development of soluble branched structures. In recent years, the thermo-mechanical method has been preferred over other conventional recycling techniques. The common recycling chemicals used are disulfides, thiols, amines and unsaturated compounds (1 wt%) along with oils.[8] The addition of softeners lowers the thermal degradation resistance of vulcanization and also hampers the interaction between the filler and rubber chains.

12.3.9 Cryogenic Grinding Process

Cryogenic grinding takes place at temperatures near −80 °C using liquid nitrogen or commercial refrigerants. The scrap rubber in the form of chips or ambient produced granulates is used as the feed stock. The processing of the rubber at low temperature leads to embrittlement of the rubber. The process takes place in four-phases including size reduction, cooling, separation, and milling. The rubber scrap materials enter a freezing chamber where the liquid nitrogen is used for cooling from −80 to −120 °C. Under this low temperature, rubber does not behave as a flexible material and thus can be easily crushed and broken. The granulate is made to pass through a series of magnetic screens and sifting stations to remove the impurities. This process produces a crumb of rubber of much finer quality.[78,79] The finer particles can also be produced by repeating both ambient and cryogenic processing. In fact, the overall cost can be reduced, if the two processes can be combined into one system and benefits from the characteristics of each system can be taken. The outputs from either or both systems can be used as a feedstock for further processing. However, the cryogenic process is expensive, but it results in smoother and smaller crumbs.[80]

12.3.10 Pyrolysis Process

Pyrolysis is the term used for thermal decomposition of the materials in the absence of oxygen. In this process the scrap rubber is subjected to a temperature of 450–500 °C and above in the absence of oxygen. The cracking and post cracking take place progressively at such a high temperature.[31–83]

12.3.11 Microwave Recycling Technique

Microwave recycling is a technique in which the quantity of microwave electromagnetic energy is controlled at a specified frequency for breaking the sulfur–sulfur or carbon–carbon bonds in a cross-linked rubber powder.[84] The rapid increase in the temperature (260–350 °C) of the material can be acquired by using microwaves.[2] Elastomers vulcanized using sulfur, contain polar groups which can be devulcanized through microwaves.[85] The microwave method generates heat at a temperature in excess of 260 °C to yield a mass which is fed to an extruder which extrudes the rubber at a temperature of 90–125 °C. This method transforms the scrap rubber to refined stock which can be readily used for remixing with almost 90–95% recovery of the rubber.[86,87] There are different ways to utilize the waste released from the rubber recycling as shown in Figure 12.3.

General rubber recycling has many advantages, as shown in Figure 12.4. In the nanocomposite, the nanofiller represents the main component. In general, the advantages of nanocomposites are lighter weight, affordability and enhanced properties. The advantages of rubber based nanocomposites are

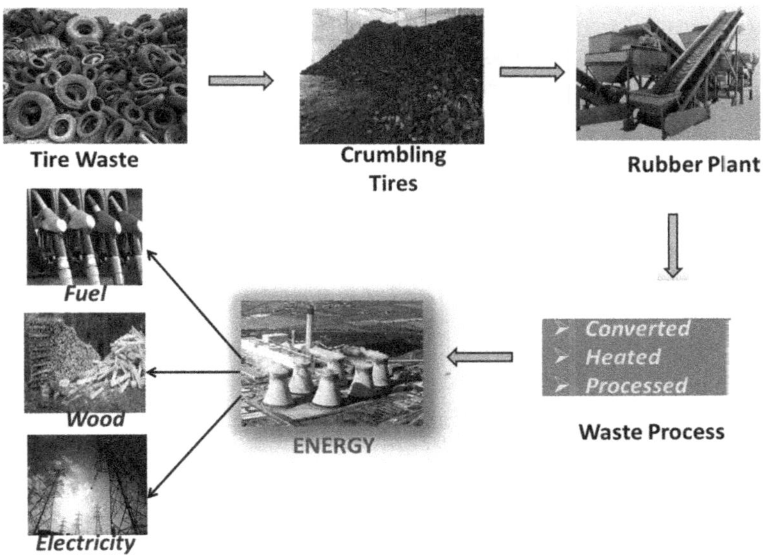

Figure 12.3 Energy utilization of the rubber recycling process.

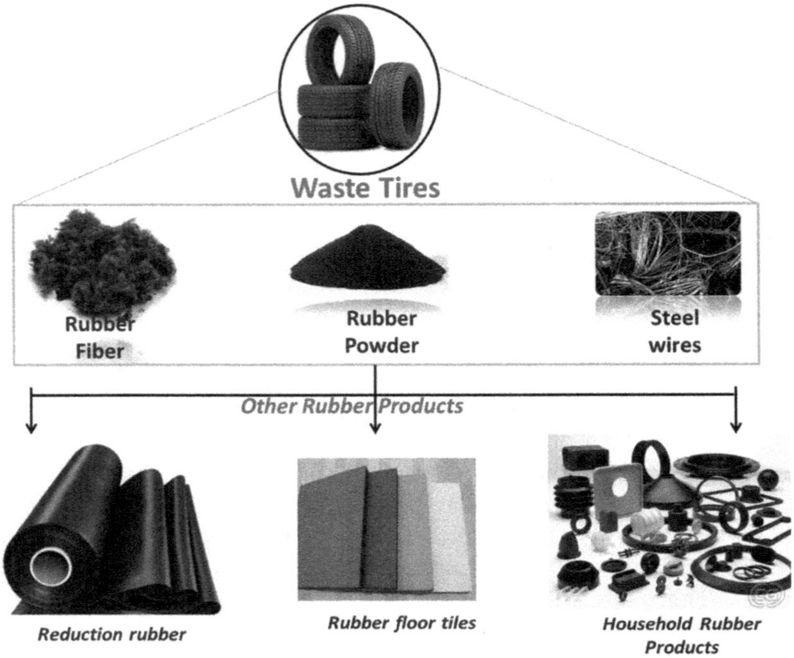

Figure 12.4 Products of the tyre recycling process.

high surface activity and smaller particle size, making them ideal for reinforcing effects, nanocomposites possess amplified surface area, improvement in chemical properties,[88] anti-blocking and promotional effect on the adhesiveness of rubber, abrasion enhancement,[89] improvement in the barrier effect (improvement in resistance),[90] better electrical conductivity and flame behaviour in vulcanized rubber,[91] excellent air permeability alignment along the nanofiller dispersion,[91] flexibility,[92] extradentary mechanical properties (improvement in stiffness, higher tensile strength and higher wear and tear strength) and preservation of clarity of the polymer matrix.[93]

12.4 Effect of Nano-fillers on Rubber Recycling

Rubber has become a prominent part of the industrial sector as far as its production is concerned. It is necessary to recycle rubber as it increases landfill pollution. Inexpensive and environmentally friendly recycling should be preferred. Thus, understanding the recycling behaviour of rubber based nanocomposites is a critical part of their manufacturing process because vast numbers of recycled rubber nanocomposites are entering the market for a wide range of potential applications. The recycling significantly affects the chemical, mechanical and rheological properties of rubber. High reinforcement of nanofillers causes less resistance to recycling in rubber nanocomposites as compared with that of pure rubber. This is because the degradation of the rubber nanocomposites utilizes a large amount of energy as compare to that of pure rubber. The recycling of rubber based nanocomposites is widely accepted for the manufacturing process and consumer scrap. However, little attention has been given to the influence of the nanofiller on the recycling process, which may possess some unique challenges such as the exposure of human beings to nanofillers during the recycling process.

The advantages of recycling rubber nanocomposites with incorporated nanofillers are:

- Reduction in landfill pollution.
- Production of new nanocomposites with above-mentioned advantages.
- Reduction in health risks.
- The proper method for waste management.
- Environmental hygiene and so on.

On the other hand, the disadvantages of rubber nanocomposites recycling are:

- Tedious remanufacturing process.
- The property enhancement is limited to a maximum range of 40 parts per hundred (pph) loading.
- Lack of rebound resilience in the nanocomposites.
- Weaker interactions of some nanofillers with the rubber matrix.
- The requirement of bigger investment to set up a recycling plant.
- The requirement of large scale waste product collection, taking a longer time duration and so on.

12.5 Conclusion

The improvement in the awareness of waste management and environment related issues has led to a substantial increase in the utilization of rubber waste. The recycling technique avoids the potential threat of waste materials to both the human race and the environment and plays a pivotal role in waste management. The collection of waste materials, followed by the breakdown process and finally re-manufacturing are the steps needed for the development of new nanocomposite materials from waste scrap. The recycling of materials back into nanocomposites by using various processing steps is more sustainable rather than finding new applications. Recycling is very much useful because it provides an economic and ecologically sound method of reusing elastomeric waste to return it to the same process and products in which it was originally generated, and it produces a similar product with equivalent physical properties. The properties of the reclaim are reported to be better than those of rubber obtained by other reclaiming methods particularly those suitable for the process. The microwave recycling process scrap rubber waste can be reclaimed without depolymerization to a material capable of being recompounded and revulcanized, having physical properties essentially equivalent to the original vulcanizate. This chapter has presented various recycling methods of rubber with hybrid nanofiller additions, the fabrication of the recycled nanocomposites, their advantages, disadvantages, challenges and the future of such nanocomposites.

References

1. D. Ponnamma, K. K. Sadasivuni, M. Strankowski, Q. Guo and S. Thomas, *Soft Matter*, 2013, **9**, 43.
2. K. K. Sadasivuni, S. Thomas and Y. Grohens, Recent Advances in Rubber Recycling, in *Recycling and Reuse of Materials and their Products*, CRC Press, 2013, p. 53.
3. D. Ponnamma, K. K. Sadasivuni and S. Thomas, *NMR Studies of Natural Rubber Composites from Macro-to Nanoscale – A review in Natural Rubber Materials*, RSC Publications, 2013, pp. 283–702.
4. K. K. Sadasivuni and Y. Grohens, Non Linear Viscosity of Two Dimensional Filler Reinforced Rubber Nanocomposites, in *Non Linear Viscoelasticity of Rubber Composites and Nanocomposites*, Springer Publications, 2014, pp. 43–57.
5. N. V. David, X. L. Gao and J. Q. Zheng, *J. Eng. Mater. Technol.*, 2011, **133**, 011001.
6. J. Njuguna, K. Pielichowski and S. Desai, *Polym. Adv. Technol.*, 2008, **19**, 947.
7. L. S. Schadler, L. C. Brinson and W. G. Sawyer, *JOM*, 2007, **59**, 53.
8. D. Thomas, K. A. Vijaylakshmi, J. J. Mathen, S. Augustine, D. Ponnamma, K. K. Sadasivuni and J. J. Cabibihan, *Polym. Bull.*, 2017, **74**, 4989.
9. D. Ponnamma, K. T. Varughese, M. A. A. Al-Maadeed and S. Thomas, *Polym. Int.*, 2017, **66**, 931.
10. M. D'Arienzo, M. Redaelli, E. Callone, L. Conzatti, B. D. Credico, S. Dirè, L. Giannini, S. Polizzi, I. Schizzi, R. Scotti and L. Tadiello, *Mater. Chem. Front.*, 2017, **1**, 1441.
11. B. T. Marouf, Y. W. Mai, R. Bagheri and R. A. Pearson, *Polym. Rev.*, 2016, **56**, 70.
12. L. Sikong, K. Kooptarnond, M. Khangkhamano and W. Sangchay, *Dig. J. Nanomater. Biostruct.*, 2015, **10**, 1067.
13. R. J. D. Santos, D. L. D. S. Agostini, F. C. Cabrera, E. A. P. D. Reis, M. R. Ruiz, E. R. Budemberg, S. R. Teixeira and A. E. Job, *Polimeros*, 2014, **24**, 646.
14. D. Mittal, *Resonance*, 1997, **2**, 64.
15. Y. M. Xu, J. Qi, J. W. Shi, D. M. He, D. M. Wang and Q. M. Zhang, *Oil Shale*, 2011, **28**, 309.
16. L. Jong, *J. Appl. Polym. Sci.*, 2007, **106**, 3444.
17. L. N. Carli, R. R. Cássio, A. Zanchet, R. S. Mauler, M. Giovanela, R. N. Brandalise and J. S. Crespo, *Appl. Clay Sci.*, 2011, **52**, 56.
18. A. K. Bhowmick, *Current Topics in Elastomers Research*, CRC Press–Taylor & Francis Group, Broken Sound Parkway, NW, 2008.

19. K. Sankaran, P. Manoharan, S. Chattopadhyay, S. Nair, U. Govindan, S. Arayambath and G. B. Nando, *RSC Adv.*, 2016, **6**, 33723.
20. Y. Biao, W. Yanwei, H. Jia, J. Wang, Z. Xu and L. Ding, *J. Polym. Eng.*, 2017, **37**, 0375.
21. D. Ponnamma, R. Ramachandran, S. Hussain, R. Rajaraman, G. Amarendra, K. T. Varughese and S. Thomas, *Comp. Part A: Appl. Sci. Manuf.*, 2015, **77**, 164.
22. L. Bokobza, M. Rahmani, C. Belin, J. L. Bruneel and N. E. Bounia, *J. Polym. Sci., Part B: Polym. Phys.*, 2008, **46**, 1939.
23. M. Galimberti, M. Coombs, P. Riccio, T. Riccò, S. Passera, S. Pandini, L. Conzatti, A. Ravasio and I. Tritto, *Macromol. Mater. Eng.*, 2013, **298**, 241.
24. K. S. Usha Devi, D. Ponnamma, V. Causin, H. J. Maria and S. Thomas, *Appl. Clay Sci.*, 2015, **114**, 568.
25. J. Xu, L. Shuangxin, Y. Li and T. Xuefeng, *J. Mater. Sci.: Mater. Electron.*, 2016, **27**, 9531.
26. D. Ponnamma, J. George, M. G. Thomas, C. H. Chan, S. Valic, M. Mozetic, U. Cvelbar and S. Thomas, *Polym. Eng. Sci.*, 2015, **55**, 1203.
27. D. Ponnamma, S. H. Sung, J. S. Hong, K. H. Ahn, K. T. Varughese and S. Thomas, *Euro. Polym. J.*, 2014, **53**, 147.
28. E. Ayandele, B. Sarkar and P. Alexandridis, *Nanomaterials*, 2012, **2**, 445.
29. K. N. Raftopoulos and K. Pielichowski, *Prog. Polym. Sci.*, 2016, **52**, 136.
30. S. W. Kuo, *ACS Cent. Sci.*, 2016, **2**, 62.
31. A. Fina, O. Monticelli and G. Camino, *J. Mater. Chem.*, 2010, **20**, 9297.
32. K. Tanaka and Y. Chujo, *J. Mater. Chem.*, 2012, **22**, 1733.
33. L. Matějka, I. A. Kroutilová, J. D. Lichtenhan and T. S. Haddad, *Eur. Polym. J.*, 2014, **52**, 117.
34. M. A. LópezManchado, B. Herrero and M. Arroyo, *Polym. Int.*, 2003, **52**, 1070.
35. J. K. Nelson, *Dielectric Polymer Nanocomposites*, Springer, New York, 2010.
36. I. S. Bayer, A. Steele, P. Martorana, E. Loth, S. J. Robinson and D. Stevenson, *Appl. Phys. Lett.*, 2009, **95**, 063702.
37. A. Bhatia, K. R. Gupta, S. N. Bhattacharya and H. J. Choi, *J. Nanomater.*, 2012, **6**, 11.
38. S. Kumar, S. Chattopadhyay, R. Padmanabhan, A. Sreejesh, N. Sujith, G. Unnikrishnan and G. B. Nando, *Mater. Res. Express*, 2015, **2**, 105007.
39. A. Okada and A. Usuki, *Mater. Sci. Eng., C*, 1995, **3**, 109.
40. J. R. Potts, O. Shankar, L. Du and S. R. Rodney, *Macromolecules*, 2012, **45**, 6045.
41. K. Kyriaki, H. Fukushima and L. T. Drzal, *Compos. Sci. Technol.*, 2007, **67**, 2045.
42. M. A. Harmer, W. E. Farneth and Q. Sun, *J. Am. Chem. Soc.*, 1996, **118**, 7708.
43. G. Philipp and H. Schmidt, *J. Non-Cryst. Solids*, 1984, **63**, 283.
44. M. Hashimoto, H. Takadama, M. Mizuno and T. Kokubo, *Mater. Res. Bull.*, 2006, **41**, 515.
45. S. Katrin and W. Kaminsky, *Inf. Macromol., Symp.*, 2008, **261**, 10.

46. M. Tanahashi, *Materials*, 2010, **3**, 1593.
47. P. M. Ajayan, L. S. Schadler and P. V. Braun, *Nanocomposite Science and Technology*, John Wiley & Sons, 2006.
48. R. S. Sinha and M. Okamoto, *Prog. Polym. Sci.*, 2003, **28**, 1539.
49. J. J. Lin, I. J. Cheng, R. Wang and R. J. Lee, *Macromolecules*, 2001, **34**, 8832.
50. F. A. Bottino, E. Fabbri, I. L. Fragalà, G. Malandrino, A. Orestano, F. Pilati and A. Pollicino, *Macromol. Rapid Commun.*, 2003, **24**, 1079.
51. A. Usuki, M. Kawasumi, Y. Kojima, A. Okada, T. Kurauchi and O. Kamigaito, *J. Mater. Res.*, 1993, **8**, 1174.
52. A. Saleem, L. Frormann and A. Soever, *Polymers*, 2010, **2**, 200.
53. E. Reynaud, T. Jouen, C. Gauthier, G. Vigier and J. Varlet, *Polymer*, 2001, **42**, 8759.
54. A. B. R. Mayer, *Mater. Sci. Eng., C*, 1998, **6**, 155.
55. A. K. Anoop, U. S. Agarwal, A. Nisal and R. Joseph, *Eur. Polym. J.*, 2007, **43**, 2279.
56. W. Pan and H. Zou, *Bull. Mater. Sci.*, 2008, **31**, 807.
57. D. W. Chae and K. C. Byoung, *Polym. Adv. Technol.*, 2005, **16**, 846.
58. Y. Suetsugu and A. Sato, *Seikei Kakou*, 2006, **18**, 807.
59. A. Kaoru, A. K. Achimuthu and Y. Chujo, *Macromolecules*, 2004, **37**, 9793.
60. T. Ogoshi and Y. Chujo, *Macromolecules*, 2005, **38**, 9110.
61. G. Giuliana, M. Tortora and V. Vittoria, *J. Polym. Sci., Part B: Polym. Phys.*, 2005, **43**, 2454.
62. T. Fukuda, T. Fujiwara, H. Fujita and H. Goda, *Seikei Kakou*, 2005, **17**, 109.
63. R. Deepak and Y. K. Agrawal, *Rev. Adv. Mater. Sci.*, 2012, **30**, 282.
64. M. Jun, L. Q. Zhang and L. Geng, *Manufacturing Techniques of Rubber Nanocomposites in Rubber Nanocomposites: Preparation, Properties and Applications*, Wiley, 2010.
65. R. A. Vaia, H. Ishii and E. P. Giannelis, *Chem. Mater.*, 1993, **5**, 1694.
66. T. Ming, C. Qu, Y. Feng and L. Zhang, *J. Mater. Sci.*, 2003, **38**, 4917.
67. Y. R. Liang, J. Ma, Y. L. Lu, Y. P. Wu, L. Q. Zhang and Y. W. Mai, *J. Polym. Sci., Part B: Polym. Phys.*, 2005, **43**, 2653.
68. G. Markus, W. Gronski, P. Reichert and R. Mulhaupt, *Rubber Chem. Technol.*, 2001, **74**, 221.
69. S. Susmita and A. K. Bhowmick, *J. Appl. Polym. Sci.*, 2004, **92**, 698.
70. Y. L. Lu, Y. R. Liang, Y. P. Wu and L. Q. Zhang, *Macromol. Mater. Eng.*, 2006, **291**, 27.
71. J. Ma, J. Xu, J. H. Ren, Z. Z. Yu and Y. W. Mai, *Polymer*, 2003, **44**, 4619.
72. L. Zhang, Y. Wang, Y. Wang, Y. Sui and D. Yu, *J. Appl. Polym. Sci.*, 2000, **78**, 1873.
73. Y. P. Wu, Q. X. Jia, D. S. Yu and L. Q. Zhang, *J. Appl. Polym. Sci.*, 2003, **89**, 3855.
74. D. M. Marquis, C. J. Carine and É. Guillaume, *Properties of Nanofillers in Polymer*, INTECH Open Access Publisher, 2011.
75. M. Biswas and S. S. Ray, Recent progress in synthesis and evaluation of polymer-montmorillonite nanocomposites, in *New Polymerization Techniques and Synthetic Methodologies*, Springer, Berlin, Heidelberg, 2001.

76. W. Fernando and K. G. Satyanarayana, *J. Colloid Interface Sci.*, 2005, **285**, 532.
77. T. S. Ellis and J. S. D'Angelo, *J. Appl. Polym. Sci.*, 2003, **90**, 1639.
78. S. Thomas and R. Stephen, *Rubber Nanocomposites: Preparation, Properties and Applications*, John Wiley & Sons, 2010.
79. A. R. Tripathy, J. E. Morin, D. E. Williams, S. J. Eyles and R. J. Farris, *Macromolecules*, 2002, **35**, 4616.
80. B. Maxwell, US Patent.146,508, 1979.
81. D. L. Presti, *Constr. Build. Mater.*, 2013, **49**, 863.
82. D. De, A. K. Ghosh, S. Maiti and B. Adhikari, *Polym. Recycl.*, 1999, **4**, 151.
83. J. M. Ball, *Reclaimed Rubber*, Rubber Reclaimers Association Inc., New York, 1947.
84. L. Bateman, *The Chemistry and Physics of Rubber like Substance*, Maclaren, London, 1963.
85. R. Singleton and T. L. Davies, *Rubber Technology and Manufacture*, Butterworth, London, 1982.
86. R. Eckart, *Cryogenics Advances Ground Rubber Technology*, Modern Tire Dealer, 1980.
87. B. Klingensmith, *Rubber World*, 1991, **203**, 16.
88. S. B. Liang and Y. C. Hao, A novel cryogenic grinding system for recycling scrap tire peels, *Adv. Powder Technol.*, 2000, **11**, 187.
89. J. M. Bouvier and M. Gelus, *Resour. Conserv.*, 1986, **12**, 77.
90. G. S. Miguel, G. D. Fowler and C. J. Sollars, *Ind. Eng. Chem. Res.*, 1998, **37**, 2430.
91. A. V. Lozhechnik and V. V. Savchin, *J. Eng. Phys. Thermophys.*, 2016, **89**, 1482.
92. D. S. Novotny, R. L. Marsh, F. C. Masters and D. N. Tally, Goodyear tire and rubber, US: 4104205, USA, 1978 Patent.
93. V. M. Makrov and V. F. Drozdovski, *Reprocessing of Tires and Rubber Wastes*, Ellis Horwood, New York, 1991.

Subject Index